D0123407

Brief contents

Contents

Part 3 CONTEMPORARY ISSUES IN ECONOMIC GEOGRAPHY

List of figures

List of tables

Preface to the second edition

Our purpose in writing the second edition of this book remains that of conveying the diversity and vitality of contemporary economic geography to students. It is intended to work as an accessible introductory text for undergraduate geography students taking courses in economic geography at the equivalent of Levels 1 and 2 in the UK.

We have been very pleased by the positive reception of the first edition and the feedback that has been received from both the students and staff using it. *An Introduction to Economic Geography* was generally well regarded as an innovative textbook that appealed to undergraduates and has become a relevant and useful teaching guide for course leaders and tutors. We hope that that new edition will be equally well received.

In the period between the writing of the first and second editions, the world economic outlook changed significantly, most notably as a result of the financial crisis of 2007–8 and the subsequent economic recession. This edition attempts to speak to this changed context through the addition of new material related to these developments. We have retained the three themes running through the book from the first edition – globalization, uneven development and place – that serve to integrate the different chapters and parts. The book also remains informed by our 'open political economy' perspective that distinguishes it from other economic geography texts.

There are several important changes from the first edition. While the first edition was divided into two main sections, the new edition has three parts: Foundations, Key Actors and Processes and Contemporary Issues in Economic Geography. The first part is largely unchanged, apart from some updating and minor revision. In the second part, we have retained and revised the four most important substantive chapters that were in the first edition, covering the state, multinational corporations, labour and development. We have added a third part with three completely new chapters: 'The Uneven Geographies of Finance' (Chapter 9), 'Commodity Chains and Global Production Networks' (Chapter 10) and 'Alternative Economic Geographies' (Chapter 12). Chapter 11, 'Knowledge, Creativity and Regional Development', has been developed from Chapter 10 of the first edition, incorporating new material on creative economies. The pedagogic features remain unchanged in terms of the boxes, reflective questions and exercises.

Once again, we are grateful to our editors at Pearson for their support and patience. Andrew Taylor was responsible for the first edition and Rufus Curnow took over for the second. We also wish to thank the three anonymous reviewers who provided excellent reflections and suggestions from the first edition and gave constructive advice on the draft chapters of the new edition.

Publisher's acknowledgements

We are grateful to the following for permission to reproduce copyright material:

Figures

Figure 1.6 from www.ideas-forum.org.uk, illustration by Jan Nimmo; Figure 2.4 Christaller, Walter; Baskin, Carlisle W.; Central places in southern Germany, 1st edition, © 1966, pp. 224-225. Reprinted by permission of Pearson Education Inc., Upper Saddle River, NJ; Figure 3.2 from *Wrecking a Region*, Hudson, R., 1989, p.4, PION Limited; Figure 4.2 from *Nature's Metropolis: Chicago and the Great West*, W.W. Norton (Cronon, W. 1991) pp. iii-iii, Nature's Metropolis: Chicago and the Great West; Figure 4.4 from *Industrial Location*, 2nd edn., Blackwell (Chapman, K. and Walker, D. 1991) p. 180, Fig. 9.2; Figure 4.10 from The political economy of Britain's North-South divide, *Transactions: Institute of British Geographers* 13, pp. 289–418, p. 392, Fig. 1 (Martin, R. 1988); Figure 4.11 adapted from the Rand McNally *World Atlas*, 1992; Figure 5.2 from UK regional policy: an evaluation in *Regional Studies*, 31, Taylor & Francis Ltd, (J. Taylor and C. Wren 1997). http://www.tandf.co.uk/journals; Figure 5.3 reprinted from Progress in Planning, 44, Tuppen, J.N. and Thompson, I.B., 'Industrial restructuring in contemporary France: spatial priorities and policies', p. 126, copyright (1994), with permission from Elsevier; Figure 5.4 from *The Condition of Postmodernity*, Blackwell (D. Harvey 1989); Figure 5.6 from Triangulating the borderless world: geographies of power in the Indonesia-Malaysia-Singapore growth triangle, *Transactions of the Institute of British Geographers*, 29, pp. 485-98, Fig. 1 (Sparke, M. Sidaway, J.D., Bunnell, T. and Grundy-Warr, C.V. 2004), Sparke, M. Sidaway, J.D., Bunnell, T. and Grundy-Warr, C.V. (2004) 'Triangulating the borderless world: geographies of power in the Indonesia-Malaysia-Singapore growth triangle', Transactions of the Institute of British Geographers NS 29, pp. 485-98, Fig. 1; Figure 7.4 from 'A cross-country study of union membership', in *Institute of Labour Discussion Paper 2016*, IZA (Blanchflower, D. 2006); Figure 7.8 from Labour Force Survey, TUC; Figure 7.10 from *The Age of Migration*, 2nd edn., Palgrave Macmillan (Castles, S. and Miller, M. 1998) Map 1.1, The Age of Migration; Figure 8.2 from *Geographies of Development*, 2nd edn., Pearson Education Ltd. (Potter, R.B., Binns, T., Elliott, J.A. and Smith, D. 2004) p. 16, Fig. 1.2; Figure 8.5 from *Geographies of Development*, 2nd edn., Pearson Education Ltd. (Potter, R.B., Binns, T., Elliott, J.A. and Smith, D. 2004) p. 111, Fig. 3.17; Figure 8.10 from NACLA Report on the Americas, *The new extraction: rewriting the political ecology of the Andes.* pp. 12-20 (Bebbington, A. 2009); Figure 8.11 from http://webnet.oecd.org/oda2009; Figure 8.12 from *World Development Report 2010*, World Bank (2009) p. 39, Source: International Bank for Reconstruction and Development / The World Bank: WDR 2010; Figure 9.3 from *Environment and Planning A*, 1996, 28, pp. 1209–32, PION Limited; Figure 9.4 from *Debt and Development*, Wiley (Corbridge, S. 1992) Figure 9.4; Figure 10.4 from *Geographical Journal*, Vol. 174, no. 2, 100 (2008); Figure 10.7 from Globalising regional development: a global production networks perspective, *Transactions of the Institute of British Geographers*, 29 ed., pp. 464-84 (Coe, N., Hess, M., Yeung, H.W., Dicken, P. and Henderson, J. 2004); Figure 10.8 from Globalising regional development: a global production networks perspective, *Transactions of the Institute of British Geographers* 29 ed., pp. 464-84 (Coe, N., Hess, M., Yeung, H.W., Dicken, P. and Henderson, J. 2004); Figure 10.9 from Globalising regional development: a global production networks perspective, *Transactions* of the Institute of British Geographers, 29 ed., p. 470, Fig. 1 (Coe, N., Hess, M., Yeung, H.W., Dicken, P. and Henderson, J. 2004); Figure 10.10 by courtesy of Keith Chapman; Figure 11.11 from Urban development and the politics of a creative class – evidence from a study of artists, Environment and Planning A, 38 ed., pp. 1921-1940 (Markusen, A. 2006); Figure 11.6 from Blu-Book 2001 (Los Angeles: Hollywood Reporter) and Producers (Los Angeles: 1film); Figure 12.2 drawn by Ken Byrne, in Gibson-Graham, J. K. 2006. A Post Capitalist Politics. Minneapolis and London: University of Minnesota Press, p. 70; Figure 12.3 from Selective spatial closure and local economic development: what do we learn from the Argentine local currency systems?, World Development Vol. 36, no. 11, pp. 2489-2511 (Gomez, G. and Helmsing, A.H.J. 2008); Figure 12.5 created using data from:, http://www.fairtrade.org.uk/what_is_fair-trade/facts_and_figures.aspx

Tables

Table 4.1 from Economics, th ed., Pearson Education (Sloman, J. 1999) pp. 659–60; Table 4.2 adapted from 'Uneven development: social change and spatial divisions of labour' in Allen, J. and Massey (eds) Uneven Re-development: Cities and Regions in Transition, Hodder and Stoughton (Massey, D 1988) pp.250-276; Table 5.1 from 'The rise of the workfare state' in Johnston, R J, Taylor, P and Watts, M (eds) Geographies of Global Change: Remapping The World 2nd ed., Blackwell (Painter, J. 2002) pp. 158-73; Table 5.3 from Capitalism since 1945, Blackwell (Armstrong, P., Glyn, A. and Harrison, J. 1991) p. 118; Table 5.5 from Table compiled from data available here: http://stats.oecd.org/Index.aspx?DataSetCode=EO86_MAIN; Table 5.6 from Environment and Planning A, 2004, 36, p. 2098, PION Limited; Table 6.2 from Environment and Planning D: Society and Space, 1985, 3, p. 37, PION Limited; Table 7.2 from 'Places of work', in Sheppard, E. and Barnes, T. (eds) A Companion to Economic Geography, Blackwell (Peck, J. 2000) p. 139; Table 7.3 from World Employment Report 2004-05, p.191, copyright © International Labour Organisation; Table 7.4 after data taken from: http://stats.oecd.org; Table 9.1 from The economic geography of money, in Martin, R.L. (ed.) Money and the Space Economy, Wiley (Martin, R.L. 1999) pp. 256, Table 11.1; Table 9.2 from The global financial system: world of monies. In Daniels, P. Bradshaw, M. Shaw, D. and Sidaway, J. (eds) An Introduction to Human Geography, 3rd ed., Pearson Education (Pollard, J. 2008) p. 368; Table 9.3 from The Global Financial Centres Index 8, Z/Yen Group Limited p. 9; Table 9.4 from World Economic Outlook: Sustaining the Recovery, International Monetary Fund (2009) Tables A1-5, pp. 169-76;

Table 9.5 from The Impact of the Recession on Northern City-Regions, IPPR North (Dolphin, T. 2009) The Impact of the Recession on Northern City-Regions. Newcastle: IPPR North; Table 10.1 adapted from Whither global production networks in economic geography? Past, present and future, Environment and Planning A, 38 ed. pp. 1193-12045 (Hess, M. and Yeung, H.W-C. 2006); Table 10.2 from Economic Geography: A Contemporary Introduction, Wiley Blackwell (Coe, N. Kelly P.F and Yeung, H.W.C. 2007) p. 102; Table 11.2 from Urban development and the politics of a creative class – evidence from a study of artists, Environment and Planning A 38 ed., pp. 1921-1940 (Markusen, A. 2006); Table 12.1 from Date taken from: http://www.ica.coop/al-ica/, Information compiled by the International Co-operative Alliance (ICA)

Photos

The publisher would like to thank the following for their kind permission to reproduce their photographs:

Figure 2.6 RCAHMS Enterprises: Resource for Urban Design Information (RUDI). Licensor www.scran.ac.uk; Figure 2.7 Getty Images: Franco Zecchin; Figure 2.8 © David Cooper/Toronto Star/Corbis; Figure 3.5 Mary Evans Picture Library; Figure 7.5 © Sherwin Crasto/Reuters/Corbis; Figure 7.9 Anthony Ince; Figure 7.11 © Danny Lehman/CORBIS; Figure 8.9 Luciney Martins http://www.mst.org.br; Figure 11.3 © Dana Hoff/Beateworks/Corbis; Figure 12.1 RCAHMS Enterprises: © Royal Burgh of Lanark Museum Trust. Licensor www.scran.ac.uk; Figure 12.4 David Simonds;

All other images © Pearson Education

In some instances we have been unable to trace the owners of copyright material, and we would appreciate any information that would enable us to do so.

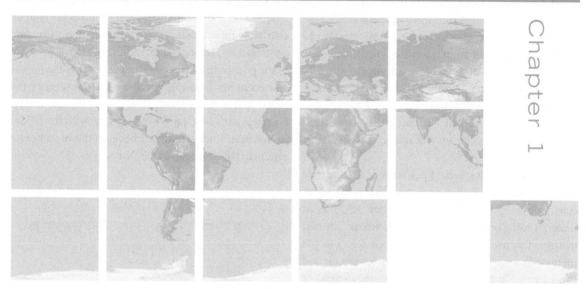

Chapter 1

Introducing economic geography

Chapter map

In the introduction to this chapter, we highlight some of the key questions addressed in this book and relate these to contemporary debates on globalization and economic development. This is followed by a discussion of each of the three main themes of the book: globalization, uneven development and place. In section 1.3, we provide a brief definition of the economy and a basic introduction to economic geography as a distinct subject area. This is followed by an outline of the political economy perspective that informs our approach in this book. Finally, section 1.5 describes the contents of the remainder of the book.

1.1 Introduction

Late in the evening of 13 September 2007, the BBC's Economic Editor, Robert Peston, broke the news that a major British mortgage bank, Northern Rock, was in serious financial trouble and had approached the Bank of England for emergency funding. Early the next morning, account-holders formed long queues outside Northern Rock branches to withdraw their money, triggering the first run on a British bank since 1866.

Northern Rock became the first high-profile casualty of the financial crisis of 2007–8 in the UK, since it could no longer access finance from the wholesale money markets to support its lending operations. These capital markets had effectively frozen in response to heavy losses derived from so-called 'subprime' loans to customers with low incomes and/or, in the US, poor credit histories. Northern Rock was nationalized by the UK government in February 2008 and has subsequently embarked on a painful process of restructuring, involving significant redundancies (Marshall *et al.*, 2010).

The story of Northern Rock serves to illustrate the main themes of this book and the value of an economic geography perspective (section 1.3.2) to an understanding of the contemporary economy. The first of these themes is **globalization**, which refers to the increased connections and linkages between people and firms located in different places, manifested in flows of goods, services, money, information and people across national and continental borders. Here, Northern Rock is indicative of the process of contagion whereby losses from subprime mortgages in the US spread rapidly throughout the financial system, resulting in a freezing of the capital markets.

Second, uneven geographical development, whereby some countries and regions are more prosperous and economically powerful than others, affected Northern Rock. After becoming a public limited company (plc) in 1997, Northern Rock was drawn into the orbit of the wider financial sector, centred upon the City of London. It pursued a particularly aggressive, high-growth business model, relying on wholesale funding rather than deposits to fuel its growth (Marshall *et al.*, 2010). In many ways, this strategy represented an effort to outgrow its northern roots and become a player in the City. This ambition was shared by other regional banks in the UK, such as Halifax/Bank of Scotland and the Royal Bank of Scotland, which also became casualties of the financial crisis and had to be rescued by the government.

Third, the Northern Rock story also highlights the theme of place, in terms of how particular areas become entangled in wider economic processes. Northern Rock is deeply embedded in the regional economy of north-east England, having been formed from the merger of two building societies in 1965. In addition to being a major employer in the region, Northern Rock sponsored the local football team, Newcastle United, and supported other local ventures through its charitable foundation. Its dynamism and growth were portrayed by regional elites and the media as symbolic of the post-industrial rebirth and renewal of the north-east economy following the industrial closures of the 1970s and 1980s (Marshall *et al.*, 2010). In reality, however, such renewal proved illusory, with the failure of Northern Rock echoing earlier episodes of corporate rationalization and job loss in the region.

1.2 Key themes: globalization, uneven development and place

In this section, we build on the Northern Rock example to examine the three main themes of the book – globalization, uneven development and place – more fully. Our selection of these themes is informed by the basic geographical concepts of **scale**, **space** and **place**. Scale refers to the different geographical levels of human activity, from the local to the regional, national and global (Figure 1.1). Space is simply an area of the Earth's surface, such as that contained within the boundaries of a particular region or country, for example. Place refers to a particular area (space), usually occupied, to which a group of people have become attached, endowing it with meaning and significance. The geographer Tim Cresswell (2004) illustrates the distinction between space and place by referring to an advertisement in a local furniture shop entitled 'turning space into place', reflecting how people use furniture and interior décor to make their houses meaningful, turning them from empty locations into personalized and comfortable homes. This domestic transformation of space into place is something with which we are all familiar, perhaps from decorating rooms in university halls of residence or shared flats.

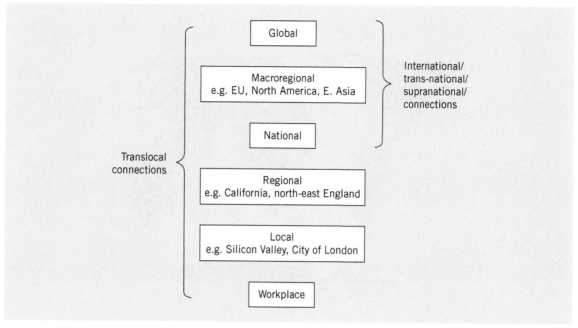

Figure 1.1 Scales of geographical analysis
Source: Castree *et al.*, 2004, p.xvix.

1.2.1 Globalization and connections across space

The first underlying theme which runs through this book is that economic activities are connected across space through flows of goods, money, information and people. These connections are not new, as trading relations between distant people and places involving the exchange of goods have existed throughout much of human history. The notion of globalization, however, emphasizes that their volume and scope of global flows has increased significantly in recent decades. Increased trade and economic interaction between distant places is dependent on technology in terms of the ease of movement and communication across space. In this context, space is understood in terms of the distance between two points and the time it takes to move between them.

A new set of transport and communications technologies has emerged since the 1960s, including jet aircraft, shipping containerization, the internet, email and mobile telephones. The effects of these 'space-shrinking technologies' have brought the world closer together, effectively reducing the distance between

places in terms of the time and costs of movement and communication (Figure 1.2). The growth of the internet from an estimated 16 million users in 1995 to around 1.7 billion in 2009 has been particularly dramatic (Internet World Statistics, 2009).

This new information and communications technology (ICT) infrastructure has made it possible for large volumes of information to be exchanged at a fraction of the previous cost, resulting in **'time–space compression'**. The term was introduced by the geographer David Harvey (1989a) who argued that the process of 'time–space compression' has been driven by the development of the economy, requiring geographical expansion in search of new markets, and supplies of labour and materials. By overcoming the constraints of geography (distance and space) through investments in transport and communications infrastructure, corporations have reduced the effects of distance as it becomes easier and cheaper to transmit information, money and goods between places. As such, time and space are effectively being compressed through the development of new technologies. This is not an entirely novel process, with a previous 'round' of time–space compression occurring towards the end of

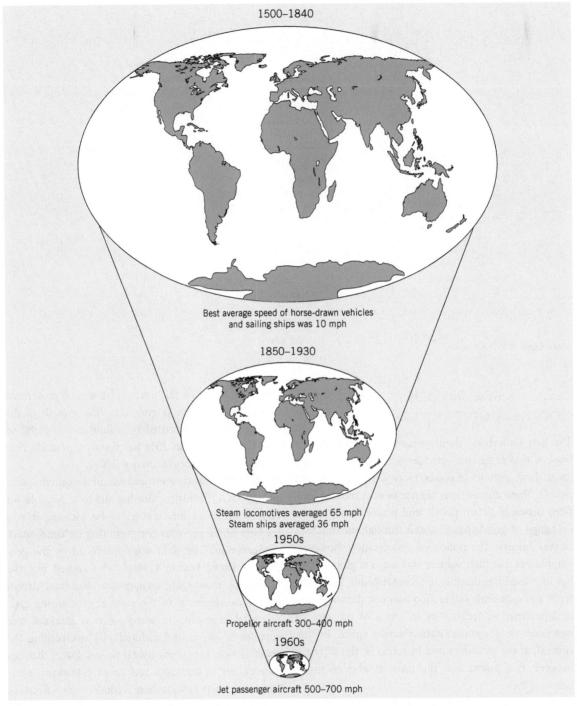

1500–1840

Best average speed of horse-drawn vehicles
and sailing ships was 10 mph

1850–1930

Steam locomotives averaged 65 mph
Steam ships averaged 36 mph

1950s

Propellor aircraft 300–400 mph

1960s

Jet passenger aircraft 500–700 mph

Figure 1.2 'A shrinking world'
Source: Dicken, 2003a, p. 92.

the nineteenth century through inventions such as railways, steamships, the telegraph and telephone (section 4.2).

While globalization has become a key buzzword of the last decade or so, it should not be viewed as a single inexorable process that people are powerless to shape

Box 1.1

Neoliberalism

A central feature of the period since the 1970s has been a changing economic policy context, with the emergence of a dominant regime known as neoliberalism in which a group of international economic organizations (Box 1.2) plays a key role by regulating the economy along neoliberal lines. Neoliberalism in simple terms is the commitment of governments and policy-makers to the principle of 'free markets' and involves the advocacy of trade liberalization, financial market deregulation, the elimination of barriers to foreign investment, the reduction of state involvement in the economy and the protection and encouragement of private property rights. Since the 1970s, it has replaced Keynesianism – based on a commitment to full employment and state intervention to stimulate growth during periods of recession – as the dominant mode of thought governing global economic policy-making (section 5.5).

(although this is how it has often been portrayed by journalists and politicians), but as something that is actively debated and contested by different groups. Two main perspectives on globalization can be identified.

The first, hyperglobalist, position contends that nation states are being bypassed by transnational business networks of production, trade and finance, leaving their governments unable to regulate and control market forces. Globalization is traced to the late 1970s when the US and UK governments became the first major powers to embrace **neoliberal** or free market policies (Box 1.1), abandoning the interventionist doctrines that had governed economic policy since the 1940s. As a result, restrictions on international investment and trade were reduced. The subsequent growth of powerful **multinational corporations (MNCs)**, allied to the increasing mobility of finance, requires states to follow orthodox pro-business policies. As this suggests, the hyperglobalist perspective closely informed government policies across the world since the 1980s, helping to bring the world that it envisions into being.

Hyperglobalists view globalization as positive, since it increases economic wellbeing by enabling a more efficient allocation of resources through the market and free trade. At the same time, it represents reduced

Box 1.2

International economic organizations

These organizations were created at the end of the Second World War as part of the Bretton Woods system, alongside the regime of fixed exchange rates. Since the early 1980s, their policies have been shaped by neoliberalism (Box 1.1).

➤ The International Monetary Fund (IMF). The role of the IMF is to promote monetary cooperation between countries, and to support economic stability and trade. The provision of financial assistance to countries experiencing budgetary problems allows the IMF to set conditions requiring countries to reform their economies. See http://www.imf.org/

➤ The World Bank (officially the International Bank for Reconstruction and Development). Its role is to provide development assistance to countries, mainly in the less-developed world. The Bank runs a range of programmes and initiatives aimed at reducing poverty and narrowing the gap between rich and poor countries. See http://www.worldbank.org/

➤ The World Trade Organization (WTO), established in 1995, took over from GATT (the General Agreement on Trade and Tariffs). The role of the WTO is to ensure a free and open trading system, working through successive 'rounds' or conferences where member countries come together to negotiate agreements. See http://www.wto.org/

government interference in our lives, leading to greater individual freedom. The role of government should be minimized to overseeing competition, although a more moderate version emphasizes the importance of investing in skills and knowledge-intensive activities, such as research and development, product design and high-tech activities, asserting that it is pointless to try to protect lesser skilled activities from low wage competition overseas (Reich, 1991). Critically, from this 'top-down' perspective, globalization is irreversible. The genie cannot be put back in the bottle and nor should it be, given its anticipated long-term benefits.

The second counter-globalization perspective has been advanced by a new global resistance movement, expressing growing resentment at global inequalities and the way the world economy is being organized and controlled. We use the term 'counter-' rather than 'anti-' globalization since it is the form that globalization is taking, rather than globalization itself, that the

movement is challenging. In particular, campaigners oppose the neoliberal or free market agenda of the hyperglobalists, associating it with increased corporate control, policies of privatization and liberalization, and particularly inequality and poverty in the developing world. While such policies had sometimes triggered local protests in different parts of the world during the 1990s, the counter-globalization movement first came to the attention of the wider public in the American city of Seattle in December 1999 (Box 1.3).

In contrast with the top-down model of economic globalization favoured by the hyperglobalists, the **counter-globalization movement** espouses a more open, participative 'bottom-up' model of globalization, evident in initiatives such as the World Social Forum which was first held in the Brazilian city of Porto Alegre in 2001, bringing together different groups and movements. Despite its success in marshalling opposition to 'corporate globalization', the movement faces some

Box 1.3

The 'Battle in Seattle'

Seattle became the focus of protest because it was there that the WTO (see Box 1.2) was holding its bi-annual 'ministerial' meeting. As the main vehicle for trade liberalization, the WTO has become a focus for what the former World Bank Chief Economist, Joseph Stiglitz, called globalization's 'discontents' (Stiglitz, 2002).

Around 70,000 protesters (Tormey, 2004, p.39) converged at Seattle to demonstrate against neoliberal globalization and to oppose the WTO agenda of launching a new round of trade liberalization, offering a wake-up call to the world's economic and political elite. As such, Seattle was not just a manifestation or 'moment' in the emergence of a broader movement. While often associated with key celebrity spokespeople such as the journalists Naomi Klein and George Monbiot, and intellectuals such as

Noam Chomsky, Walden Bello and Susan Strange, the real significance of the counter-globalization movement lies in the coming together of a huge range of different movements and campaigns, previously devoted to single-issue politics. In Seattle, protest succeeded in shutting down the WTO ministerial before a trade deal had been reached. At the time, within the conference, delegates from developing countries, led by a group of ministers from Africa, resisted the blandishments of US and European representatives, resulting in the collapse of the negotiations (Wainright, 2007). Despite repeated efforts to agree a new round of trade negotiations, the underlying dispute between US and Europe, on the one hand, and developing countries, on the other, has not yet been resolved.

There have been mass protests and movements in the recent past

– think of the anti-nuclear campaign or the environmental movement – but Seattle represented a convergence of disparate and hitherto often opposing groups against a 'common enemy'. These included trade unionists and environmentalists, NGOs and direct action networks, street dwellers and peasant movements from the developing world and middle-class consumer activists from the developed world: a so-called 'teamsters and turtles' alliance united by their opposition to neoliberal hyper-globalization. The movement has subsequently developed through other 'global days of action' in Prague, Genoa, Quebec City and elsewhere. Wherever there is a major meeting of world leaders or economic officials through institutions such as the WTO, World Bank, IMF and G20, protesters will converge to try to disrupt or contest their legitimacy.

important challenges in the years ahead. Not least among these is the need to move from global protest to forging alternatives to the current world economic order. Its diversity could prove to be a weakness in this respect. While most actors within the movement are opposed to neoliberal economic policies, there is a varied and often competing range of ideological and tactical positions. A key distinction is that between reformists – which covers a spectrum from relatively moderate liberal positions to radical social democratic positions – and genuine anti-capitalists, which includes a plethora of competing Marxisms, different strands of anarchism and 'deep greens' (Tormey, 2004). Similarly, very different positions on globalization are evident in terms of the reform/abolition divide and the geographical scale at which the economy should be regulated. Localists advocate a relocalization of the economy and the dismantling of global neoliberal institutions such as the IMF and World Bank (e.g. Hines, 2000) – the most radical option – while others call for greater intervention at the national level (e.g. George, 1999), and globalists favour new forms of global governance and regulation (e.g. Monbiot, 2003).

1.2.2 Uneven development

A basic feature of economic development under capitalism is its geographical unevenness. **Uneven development** is an inherent feature of the capitalist economy, reflecting the tendency for growth and investment to become concentrated in particular locations. These areas may be favoured by a particular set of advantages such as their geographical position, resource base, availability of capital or the skills and capabilities of the workforce. Once growth begins to accelerate in a particular area, it tends to 'suck in' investment, labour and resources from surrounding regions. Capital is attracted by the opportunities for profit while workers are drawn by abundant job opportunities and high wages. Surrounding regions are often left behind, relegated to a subordinate role supplying resources and labour to the growth area.

One key aspect of the process of uneven economic development is that it occurs at different geographical scales (see Figure 1.1). This can be illustrated with reference to three key scales of activity: the global, regional and local.

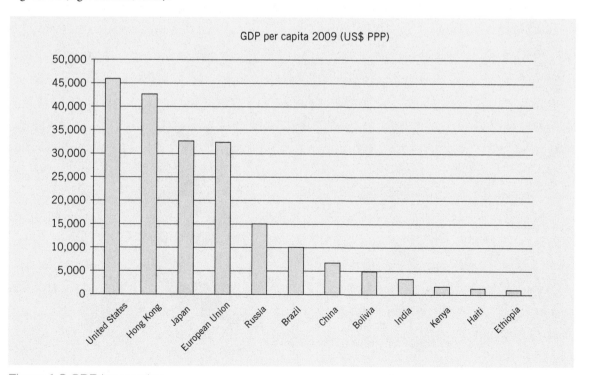

Figure 1.3 GDP by country
Source: http://www.cia.gov/library/publications/the-world-factbook/rankorder/2004rank.html

➤ At the global level, there is a marked divergence between the 'core' in North America, Japan and western Europe and the 'periphery' in the 'global south' of Asia, Latin America and Africa (see Figure 1.3). This pattern reflects the legacy of colonialism, whereby the core countries in Europe and North America produced high-value manufactured goods and the colonies produced low-value raw materials and agricultural products. While a number of East Asian countries, including China, has been able to overcome this legacy, experiencing rapid growth and rising prosperity over the past 25 years, others, particularly in sub-Saharan Africa, have been left behind, experiencing conditions of extreme deprivation and poverty.

➤ Within individual countries, too, economic disparities between regions are evident. The rapid economic development of China, for instance, has opened up a growing divide between the booming coastal provinces in the south and east and a poor, underdeveloped interior (Box 1.4). Developed countries are also characterized by regional disparities, such as the persistent north–south divide that has characterized the economic geography of the UK since the 1930s (Amin *et al.*, 2003a).

➤ Even on a local level within cities, uneven development is present in the form of social polarization between rich middle-class neighbourhoods and poorer inner city areas and public housing schemes. In the city of Glasgow in Scotland, for example, male life expectancy was just 54 years in the impoverished inner city district of Calton in 1998–2002 compared with 82 years in the affluent, middle-class suburb of Lenzie (Hanlon *et al.*, 2006).

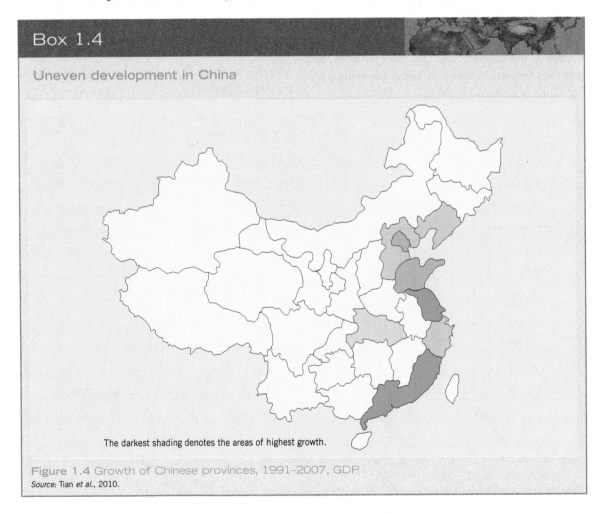

Box 1.4

Uneven development in China

The darkest shading denotes the areas of highest growth.

Figure 1.4 Growth of Chinese provinces, 1991–2007, GDP
Source: Tian et al., 2010.

Box 1.4 (continued)

Since the Communist regime opened up its economy to attract foreign investment in 1978, China has experienced rapid economic development, becoming a world manufacturing and assembly centre. The economy grew at an average rate of 9.4 per cent per year between 1978 and 2004 (Wei and Ye, 2009), a period of sustained high growth that is almost unparalleled in human history (Harvey, 2005). As a result, the per capita disposable incomes of households increased by an annual average of just over 7 per cent between 1978 and 2007, while the proportion of the population classified as undernourished fell from 30 per cent in 1981 to 12 per cent in 1997 (United Nations Development Programme (UNDP), 2008, pp.13, 14). Rapid growth has, however, created great strains in terms of environmental degradation, increased inequality and the social pressures generated by

people's rising expectations. The gap between rich and poor has widened markedly and China was found to contain 16 of the world's 20 most polluted cities in the late 1990s by the World Bank.

The experience of China since 1978 is consistent with the argument that uneven geographical development is inherent to capitalism (Smith, 1984), with the degree of spatial inequality reflecting the speed and magnitude of economic and social transformation. Economic growth has created a complex pattern of regional inequality in China, although the basic divide is between the coastal regions and the interior. The 'open door' policy was initially based on the designation of special economic zones in the coastal provinces to attract foreign investment, encouraging these regions to 'get rich quick' (Wei, 1999, p.51). Coastal provinces such as Guangdong, Jiangsu,

Zhejian, Fujian and Shandong experienced rapid growth, while much of the interior lagged far behind (Figure 1.4). In 2007, the per capita GDP of Shanghai in eastern China was 65,247 yuan, 9.55 times that of Guizhou at 6,835 yuan in the west (UNDP, 2008, p.20). According to the United Nations Development Programme (2008, p.20), the human development of the most advanced provinces such as Shanghai and Beijing is equivalent to European counties such as Cyprus and Portugal, while the worst-performing provinces such as Guizhou are comparable to Botswana and Namibia. One symptom of this polarization between coast and interior has been the migration of millions of rural workers to the coastal cities in search of employment (ibid.), generating what is estimated to be the largest mass migration in human history.

The process of economic development is highly dynamic in nature, as new technologies are developed, new forms of customer demand emerge and work practices change. Over time, patterns of uneven development are periodically restructured as capital moves between different locations, investing in those that offer the highest rate of return (profit). As a result, new growth regions emerge while established ones can experience stagnation and decline. As broader market conditions and technologies change, the specialized economic base of formerly prosperous regions can be undermined by reduced demand, rising costs, competition, and the invention of new products and methods of production. On a global scale, the most dramatic change in patterns of uneven development is the emergence of East Asia as a dynamic growth region over the last 30 years or so. Within developed countries, established industrial regions based on nineteenth- and early twentieth-century industries such as coal,

steel and shipbuilding have experienced decline while new growth centres have emerged in regions such as the south and east of the US (the so-called 'sunbelt'), southern Germany and north-eastern Italy.

1.2.3 The importance of place

The role of place is shaping economic activity is a third key theme of this book. As suggested in the previous section, processes of uneven geographical development have created distinctive forms of production in particular places. During the nineteenth and early twentieth century, **specialized industrial regions** emerged in Europe and North America. As a result:

> … distinct places are associated with sectoral and functional divisions of labour. In the United States for example, 'Pittsburgh meant steel, Lowell meant

textiles, and Detroit meant automobiles' (Clark *et al.*, 1986, p.23), while in the United Kingdom, 'one finds metal workers in the Midlands, office professionals in London, miners in South Wales, and academics in Oxford' (Storper and Walker, 1989, p.156).

(Peck, 1996, p.14)

Although some of these specific associations have been weakened by deindustrialization, the general point about distinctive forms of production being associated with particular places remains important. The City of London continues to be associated with finance and business services, for instance Silicon Valley in California with semi-conductors, Los Angeles with movies and Milan with clothing design and fashion. Such variety is continually reproduced through the interaction between wider processes of uneven development and local political, social, economic and cultural conditions. These conditions reflect the economic history of a place in terms of the particular industries found there, and the institutions and practices associated with them. In any one place, the interaction between established industries and institutions and contemporary processes of change, shapes and moulds the economic landscape (Massey, 1984).

It has become increasingly clear in recent years that globalization is a differentiated and uneven process, generating different outcomes in different places. In particular, globalization seems to be associated with a resurgence of certain regions as economic units. In particular, the success of dynamic growth regions such the City of London (financial and business services), Silicon Valley (advanced electronics), southern Germany (vehicles and electronics) and north-eastern Italy (machine tools, textiles) is rooted in the specialized production systems that have flourished there. Geographical proximity seems to encourage close linkages and communication between firms, enabling them to share information and resources. The existence of a large pool of skilled labour is a crucial feature of such regions, allowing firms to recruit easily and workers to move jobs without leaving the local area. These aspects of the local production system encourage innovation and entrepreneurship, enhancing the competitiveness of such regions within a global economy.

While globalization is not leading to the erasure of places as significant dimension of economic life, it does undermine traditional notions of places as homogenous and clearly bounded local areas. As such, there is a need to re-think place in terms of connections and relations across space (Massey, 1994). It is in this sense that the British geographer Doreen Massey's work on the development of a **global sense of place** is of particular interest. Massey develops a new conception of place as a meeting place, a kind of node or point where wider social relations and connections come together.

> … what gives place its specificity is not some long internalised history but the fact that it is constructed out of a particular constellation of social relations, meeting and weaving together at a particular locus. … Instead … of thinking of places as areas with boundaries around them, they can be imagined as articulated moments in networks of social relations and understandings … and this in turn allows a sense of place which is extroverted, which includes a consciousness of its links with the wider world, which integrates in a positive way the global and the local …

(Massey, 1994, pp.154–5)

From this perspective, place can itself be regarded as a process rather than being seen as a static and unchanging entity. Places are connected and linked through wider processes of uneven development operating through flows of capital, goods, services, information and people. Movement of particular commodities such as bananas, for example, link different parts of the UK to the economies of certain Caribbean islands (see Box 1.6).

Reflect

> Do you agree that specific places (regions) remain important within a global economy? Justify your answer with examples.

1.3 The economy and economic geography

1.3.1 The capitalist economy

'The economy' refers to the interrelated processes of production, circulation, exchange and consumption through which wealth is generated (Hudson, 2005, p.1). It is through such processes that people strive to meet their material needs, earning a living in the form of wages, profits or rent. Production involves combining land (including resources), capital, labour and knowledge – commonly known as the **factors of production** – to make or provide particular commodities. It relies on a supply of resources from nature, meaning that economic activities have a direct impact on the environment. The commodity, defined as any product or service that is sold commercially, is so basic to the workings of the economy that Karl Marx – who began his famous work *Capital* ([1867] 1976) with an examination of its properties – described it as the 'economic cell form' of capitalism. The modern economy involves the production and consumption of a vast array of commodities, spanning everything from iPods to package holidays (Boxes 1.5 and 1.6).

Human societies have tended to organize and structure their economic activities through overarching **modes of production**. These can be defined as economic and social systems that determine how resources are deployed, how work is organized and how wealth is distributed. Economic historians have identified a number of modes of production, principally subsistence, slavery, feudalism, capitalism and socialism. Each of these creates distinctive relationships between the main factors of production. **Capitalism** is clearly the dominant mode of production in the world today, operating at an increasingly global scale. It is defined by individual ownership of the means of production – factories, offices, equipment and money capital – and the associated need for most people to sell their labour power to employers or capitalists in order to earn a wage. This allows them to purchase commodities produced by other firms, creating the market demand that underpins the capitalist system. Compared with earlier modes of production, production and consumption are often geographically separate under capitalism, thus creating a need for extensive transport and distribution networks.

A key underlying point here concerns the fact that the principal features of the modern capitalist economy – such as the roles of the market, profits and competition – are not natural and eternal forces that determine human behaviour, as mainstream economists and business commentators tend to assume. Instead, capitalism is a historically specific mode of production that has emerged from its roots in early modern Europe in the sixteenth and seventeenth centuries to encompass

---

Box 1.5

Commodity chains

Commodity chains link together the production and supply of raw materials, the processing of these materials, the production of components, the assembly of finished products, and the distribution, sales and consumption of these products (see section 10.2). They involve a range of different organizations and actors, for example farmers, mining or plantation companies, component suppliers, manufacturers, sub-contractors, transport operators, distributors, retailers and consumers. Commodity chains have a distinct geography, linking together different stages of production carried out in different places (Watts, 2005). Some parts of the production process add more value or profit, creating tensions between the different participants as to who captures the value added. The role of powerful MNCs in controlling the production and distribution processes across national boundaries has been the subject of particular scrutiny in recent years (Gereffi, 1994). Exploring the production and consumption of particular commodities helps us to trace and uncover economic connections between places, linking different localities to global trading networks.

virtually the entire globe today. It has been superimposed on a complex mosaic of pre-capitalist societies and cultures, resulting in great regional variation as pre-existing local characteristics interact with broader global processes (Johnston, 1984).

While capitalism is clearly the dominant mode of production in the world today, it does not follow that all economic activity is capitalist in nature. In reality, the formal, capitalist economy, based on striving to maximize profits or earnings, co-exists with a range of other economic activities and motivations such as domestic work, volunteering, the exchange of gifts and cooperatives. In practice, non-capitalist activities interact with capitalism in a variety of ways. Think of the relationship between domestic work and paid employment, for example, or the role of gift-buying *within* capitalism. Gibson-Graham (1996) emphasizes the existence of **diverse economies**, criticizing the preoccupation with the formal, capitalist economy among economists and economic geographers. This critique has informed a number of studies of 'diverse' or 'alternative' economies such as informal work, local currencies and cooperatives (Leyshon *et al.*, 2003).

1.3.2 An economic geography perspective

Economic geography is concerned with concrete questions about the location and distribution of economic activity, the role of uneven geographical development and processes of local and regional economic development. It asks the key questions of 'what' (the type of economic activity), 'where' (location), why (requiring explanation) and 'so what' (referring to the implications and consequences of particular arrangements and processes). According to one recent definition:

> Economic geography is concerned with the economics of geography and the geography of economics. What is the spatial distribution of economic activity? How is it explained? Is it efficient and/or equitable? How has it evolved, and how can it be expected to evolve in the future? And what is the appropriate role of government in influencing this evolution?
>
> (Arnott and Wrigley, 2001, p.1)

Three key themes emerge from this:

➤ The geographical distribution and location of economic phenomena. Figure 1.5 provides an example of such a distribution, showing variations in employment in financial and business services – a key growth sector in developed countries in the 1990s and early 2000s – across the UK. Describing and mapping distributions of economic activity in this way can perhaps be seen as the first task of economic geography, addressing the basic questions of 'what' and 'where'.

➤ The next stage is to explain and understand these spatial distributions and patterns of economic activity. This is a more demanding task which brings in theory and requires us to have some appreciation of history. It involves addressing the more advanced questions of 'why' and 'how'. In trying to explain why financial and business services employment is clustered in South-east England and one or two other areas, then, we would need to draw on theories of the spatial concentration of economic activity and have some understanding of the basic economic history of the UK since the industrial revolution.

➤ A third issue is that of engagement with policymakers in government and the private sector, making recommendations and offering advice about particular geographical issues and problems. As well as describing and explaining the distribution of certain economic phenomena, geographers have also sought to outline how the economic geography of particular countries and regions should be organized. This role has sparked periodic debates about the social 'relevance' of the subject (Peck, 1999).

Reflect

➤ How would economic geographers explain the concentration of financial and business services employment in the south-east of England?

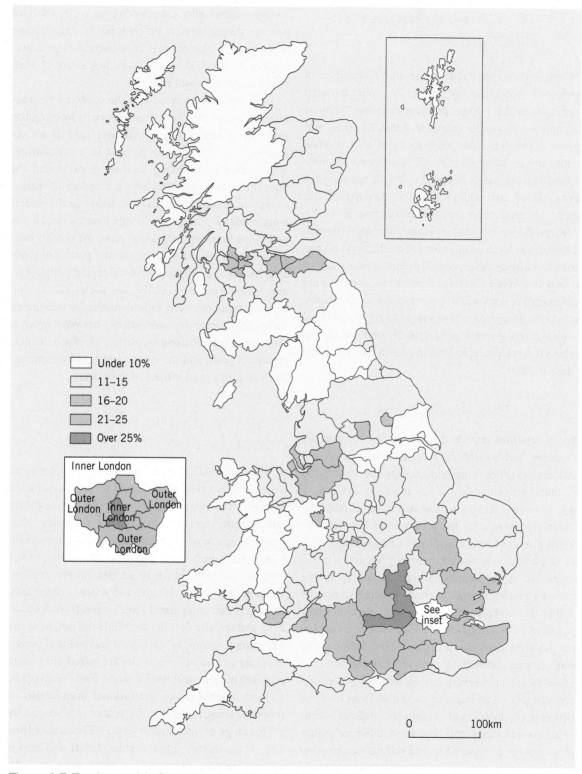

Figure 1.5 Employment in financial and business services in Britain, 2005
Source: ONS, 2006.

1.4 A political economy approach

In this section, we build on the above definition of economic geography by introducing our favoured approach to the subject: **political economy**. Political economy analyses the economy within its social and political context, rather then seeing it as a separate entity driven by its own set of rules based on individual self-interest. It is concerned with not only the exchange of commodities through the market, but also with production and the distribution of wealth between the various sections of the population (Barnes, 2000b). This focus on questions of production and distribution distinguishes political economy from modern economics. While the classical political economy of the nineteenth century was not concerned with geography, economic geographers have sought to apply a political economy framework to geographical questions such as regional development and urban growth (Harvey, 1982; Massey, 1984).

1.4.1 Social relations

Social relations provide the general link between the economy and society. In contrast with mainstream economics, which is underpinned by the assumption of individual rationality, political economists believe that economic activity is grounded in social relations. These simply refer to the relationships between different groups of people involved in the economy, such as employers, workers, consumers, government regulators, etc. A key defining social relationship is that between employers and workers, structuring activity within the workplace. The two groups are commonly regarded as different classes within society, a position set out with particular force by Marx, although for some commentators the growth of a large middle class of 'white collar' workers has blurred the distinction considerably. Other important social relations are those between producers and consumers, different firms (e.g., manufacturers and suppliers), different groups of workers (e.g., supervisors and ordinary employees) and government agencies and firms. As our case study of the banana indicates (Box 1.6), the production of

a simple commodity creates complex social relations between people based in different places. Even if some of the parties – for example consumers at large supermarkets in the UK or US – are not aware of such relationships, they still exist.

Once we have accepted that the economy is structured by social relations, it is important to recognize that these relations will change over time as society evolves. We have already pointed to a transformation occurring in western societies from around the fifteenth century onwards, where a system of market capitalism gradually replaced feudalism as the underlying mode of production. In other words, feudal ties and values between the peasantry and the nobility have given way to market competition, the pursuit of profit and the wage relationship between capital (employers) and labour (workers) as the key social relationship shaping economic and indeed human development. Increasingly intensive competition between firms is also a critical relationship, driven by the relentless pursuit of profit and the subsequent need to eliminate rivals to gain a greater share of the market.

1.4.2 Institutions and the construction of markets

A second crucial element in our political economy perspective is a view of markets as socially constructed entities that require social and political regulation rather than representing naturally occurring phenomena capable of self-regulation. Unlike mainstream economics, which holds that markets, if left to their own devices, will return to an equilibrium position where supply equals demand and waste is eliminated, we believe that unregulated ('free') markets are destabilizing and socially destructive. While the notion of the free market retains its ideological and political power, in reality virtually all economies are mixed, containing substantial public sectors. Following Karl Polanyi (Box 1.7), we emphasize the **institutional foundations of markets**, recognizing that the economy is shaped by a wide range of institutional forms and practices (Box 1.8). These include cultural rules, habits and norms which structure the social relations between individuals, helping to generate the trust that underpins

Box 1.6

'The secret life of a banana' (Vidal, 1999)

The banana is the world's most popular fruit, with consumers spending £10 billion a year on the fruit globally (Fairtrade Foundation, 2009). This simple statistic, however, masks a complex geography of production and distribution that links different people and places together (Watts, 2005). While individual consumers may remain unaware of such linkages, they create real social relationships between people in different places, for example consumers in countries such as the UK and banana farmers in tropical countries in Africa, the Caribbean and Latin America.

Bananas are grown by either individual small farmers, sometimes working under contract, or on large industrial plantations run by MNCs. Either way, the fruit must be grown, picked and packed and transported to the nearest port, from where it is shipped in special temperature-controlled compartments to another port in the destination country (Cook, 2002). The fruit is then ripened in special ripening centres before being sent to the supermarket by truck. Bananas exported from the tiny West Indian island of St Vincent to the UK, for instance, are transported to Southampton by the Geest line shipping company, taking roughly two weeks (Vidal, 1999).

The complex chain of linkages involved in the production, distribution and exchange of any particular commodity creates real conflicts of interest between groups of people over who captures the most added value from the product (Watts, 2005). Figure 1.6 shows how the price of a 30 pence banana is distributed between the various actors in the production chain. In the UK, supermarket price wars (reflecting the fact that bananas are supermarkets' third best-selling goods after petrol and lottery tickets) have seen the price of loose bananas in the UK fall from £1.10 a kilo to 64 pence a kilo between 2002 and 2007 (Fairtrade Foundation, 2009). Although the price recovered to 95 pence by the end of 2008, Asda cut it again to 38 pence by October 2009 (Doward, 2009).

The costs of cheap bananas are typically passed down the supply chain to small farmers and plantation workers, leading to deteriorations in pay and conditions. In 1999, for instance, Del Monte sacked all 4,300 of its workers on a large plantation in Costa Rica, re-employing them on wages reduced by 30–50 per cent and on longer hours with fewer benefits (Lawrence, 2009). These cuts underpinned a global deal with Wal-Mart, allowing Del Monte to grant the global retailer a low price. In response, development organizations and activists have promoted Fairtrade which now accounts for one in every four bananas sold in the UK and which has been adopted by large retailers like Sainsbury's and Waitrose (Robinson, 2009).

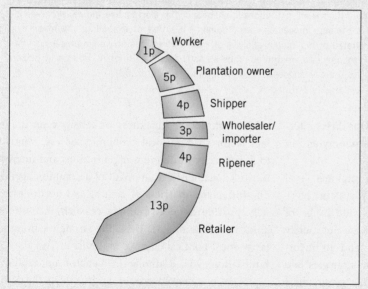

Figure 1.6 Banana split: who gets what in the banana chain
Source: International Development Education Association of Scotland website 'Banana Link', at http://www.ideas-forum.org/images/Bsplit.org: Jan Nimmo.

Box 1.7

Karl Polanyi: the economy as an instituted process

The work of Karl Polanyi (1886–1964) on the institutional foundations of economic processes is well known within the social sciences. Polanyi's broad conception of the economy and his insights into the underlying nature of market society have come to inform debates about contemporary processes of globalization and the social 'embeddedness' of the economy (Block, 2003).

In *The Great Transformation*, published in 1944, Polanyi explores the origins and development of market society during the nineteenth century. He identified land, labour and money as 'fictitious commodities' because they are not actually produced for sale on the market like true commodities – a crucial point neglected by orthodox economic analyses, which assume that the price mechanism will balance supply and demand in the normal fashion. The identification of these fictitious commodities

is important to Polanyi's argument, indicating that a pure self-regulating market economy is impossible since state intervention is required to match the supply and demand for land, labour and money (Polanyi, 1944). The book demonstrates how the construction of competitive markets depended upon state action through the upholding of property rights and contracts, the introduction of labour legislation, the establishment of measures to ensure a stable food supply and the regulation of the banking system.

In the 1950s, Polanyi turned away from modern market economies to the analysis of primitive and archaic economies. This substantive focus was closely informed by his enduring interest in the scope of the economy and the role of institutions in shaping economic processes, as demonstrated by his famous analysis of 'the economy as an instituted process'. As this phrase suggests, Polanyi views

institutions as constitutive of the economy:

> The instituting of the economic process vests that process with unity and stability; it produces a structure with a definite function in society; it shifts the place of the process in society, thus adding significance to its history; it centres interest on values, motives and policy.
>
> (Polanyi, [1959] 1982, p.250)

Institutions are both economic and non-economic, with the inclusion of the latter regarded as vital since religion or government, for example, may be as important in underpinning the operation of the economy as the monetary system or the development of new labour-saving technologies. From this perspective the research agenda is one of examining the manner in which 'the economic process is instituted at different times and places' (ibid., p.250).

legal and contractual relationships, and the direct intervention of the state in managing the economy and in running systems of social welfare.

Key forms of **institution** at the national level include firms, markets, the monetary system, business organizations, the state, and a wide range of state agencies and trade unions. These are not merely organizational structures; they also tend to incorporate and embody specific practices, strategies and values that evolve over time. Economic geographers are particularly interested in how local and regional economies are shaped by distinctive institutional arrangements (Martin, 2000). Important forms of institutions in this respect include local authorities and development agencies, employers' organizations, business associations and chambers of commerce, local political groupings, trade union branches and voluntary agencies. According to Amin and Thrift

(1994) institutional 'thickness' or density is an important factor shaping local economic success, referring to the capacity of different organizations and interests to work together in the pursuit of a common agenda. The **industrial districts** of central and north-eastern Italy provide a good example of such institutional 'thickness' (Figure 1.7). Here, the strongly communitarian local political cultures (socialist in Tuscany and Emilia-Romagna, Catholic in Veneto) underpinned a process of renewed economic growth in the 1970s and 1980s through the roles of political parties, trade unions, craft associations, local authorities, chambers of commerce and cooperatives of small entrepreneurs in providing support and a range of services to small firms (Amin, 2000). This created a sophisticated reservoir of knowledge and skills, and generated high levels of trust between firms, facilitating collaborative forms of innovation.

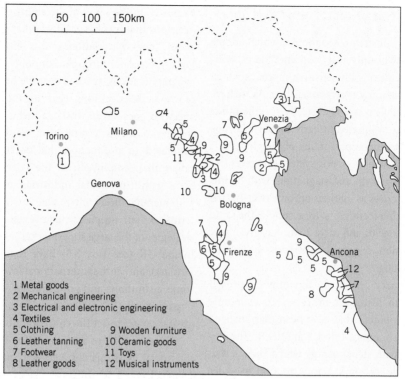

Figure 1.7 Industrial districts in Italy
Source: Amin, 2000, p.155.

1 Metal goods
2 Mechanical engineering
3 Electrical and electronic engineering
4 Textiles
5 Clothing
6 Leather tanning
7 Footwear
8 Leather goods
9 Wooden furniture
10 Ceramic goods
11 Toys
12 Musical instruments

Box 1.8

The construction of housing markets in the UK

Markets have traditionally been taken for granted in economic analysis, but this is changing as social researchers (including sociologists, anthropologists and geographers) challenge conventional approaches. This recent research stresses how markets are actively made, not given, by economic actors who construct and perform markets in various ways. This process has been examined in a fascinating study of housing markets in the Scottish city of Edinburgh by Susan Smith, Moira Munro and Hazel Christie (2006). The study focused on key 'market intermediaries' – solicitors, estate agents, surveyors and property developers – who shape the flow of information between buyers and sellers. It was conducted in the late 1990s, when the UK housing market was booming, with Edinburgh experiencing the second steepest rise in prices after the south-east of England.

One of the key themes of the research was how housing professionals constructed the housing market as a separate and self-contained economic object, which was invested with a life of its own (described, variously, as 'hot', 'active', 'exciting', 'amazing') (Smith *et al.*, 2006, p.86). They performed housing markets through a process of rational economic calculation, as prices were set according to their detailed knowledge of market forces. Yet, while market professionals characterized housing markets according to established economic models, this framing became increasingly incompatible with the actual performance of markets during the boom. Increasingly, asking prices and property values became detached from actual sale prices as some prospective buyers were prepared to pay up to 25 to 30 per cent over valuation in order to secure a property. This created great uncertainty, disrupting the smooth operation of the market as decision-making became confusing and idiosyncratic. Interestingly, Smith *et al.* (2006) suggest that professionals' detachment from the market may itself be a contributory factor to price instability with the belief that they are powerless to influence the market perhaps helping to place that market beyond control (p.92). As this indicates, 'far from *being* the economy, markets have to be made "economic" through a complex interplay of cultural, legal, political and institutional arrangements' (ibid., p.95, original emphasis).

1.4.3 History and evolution

In contrast with mainstream economics, which adopts a timeless purview of universal economic forces, political economy is a historically sensitive approach which stresses that the economy evolves over time. As we have already seen, capitalism is a historically specific mode of production that originated in early modern Europe before spreading to encompass much of the globe. It is a dynamic and unstable economic system that gives rise to periods of both growth and stagnation, with booms often followed by crises as markets become over-heated and profits crash. In recent years, for example, the rapid growth of the late 1990s and early 2000s has given way to a global recession, triggered by a crisis in housing markets and banking.

Our evolutionary political economy approach is concerned with historical processes of economic change and development that have been marginalized by mainstream economics (Hodgson, 1993). The key concept here is 'path dependence', which means that the ways in which economic actors respond to wider processes of economic change are shaped and informed by past decisions and experiences (Walker, 2000). Typically, geographers have 'spatialized' this concept, adapting it to explore patterns of regional growth and decline, emphasizing the role of social and cultural factors within regions in encouraging growth or decline through, for example, the development of an open and collaborative industrial system (Saxenian, 1994) or a succession of defensive and reactive responses to change (Grabher, 1993). Our approach emphasizes the interaction between pre-existing regional characteristics and wider processes of change. Thus, for example, the local craft traditions and skills of the Italian industrial districts of central and north-eastern Italy were revived in the 1970s and 1980s as part of a wider shift to more flexible production methods, leading to sustained growth in the region (Amin, 2000).

1.4.4 Power

Having accepted that the economy is structured by social relationships, it is also important to recognize the role of **power** in underpinning these relationships. Ultimately, all human social relations are underpinned by power, in the sense of the ability or capacity to take decisions that involve or affect other people (see Allen, 2003). Economic relations in this sense are no different. Power percolates through economic relationships at all geographic levels: from that of the household in terms of who makes decisions regarding the domestic budget, and who 'goes out to work' and who 'stays at home'; at the level of the firm in terms of the share of the wealth generated in production that accrues to employers rather than employees; in the relationships between firms within particular industries, with large retailers and manufacturers often able to dictate prices and terms to their suppliers (for example, the cost-cutting strategies of the large supermarket chains such as Tesco in the UK (see Box 3.5) have reduced earnings for farmers); and at the international level, in the way that some institutions and governments – the World Trade Organization (WTO) or the US, for example – have greater power to set the rules of trade than others.

'The secret life of a banana' (Box 1.6) indicates how the social relationships between different groups of people located in different places are structured by power. At a very basic level, it is clear that some actors in the chain, particularly the supermarkets and the multinational firms that coordinate the production and distribution processes, are in a more powerful position than the small Caribbean farmers or the labourers in the large Central American plantations. Unequal power relations in this sense are central to understanding the concept of uneven development.

> ### Reflect
>
> ➤ In what ways do institutions shape the process of economic development?

1.5 Outline of the book

In this first part of the book, entitled 'Foundations', we develop our conceptual approach and identify some of the main underlying features of capitalism and its geographies. In Chapter 2 we consider the main sets of approaches that have been adopted by economic geographers, focusing particularly on political economy, and the institutional and cultural perspectives that have been developed over the last decade. The following chapter introduces the main groups of actors and processes shaping the development of the capitalist economy, focusing on capital, labour, consumers and the state in turn. While Chapter 3 focuses upon the basic workings of the capitalist economy, Chapter 4 is directly concerned with its geography, outlining the development of key spaces of production and consumption since the nineteenth century. Chapters 3 and 4 together provide an important historical context for the remainder of the book.

Informed by these foundations, we turn to explore the geographies of the contemporary economy in Part 2, entitled 'Key Actors and Processes'. Chapter 5 examines the changing role of the state in the economy, assessing the changing geography of economic governance and the evolution of states. We go on, in Chapter 6, to consider the role of MNCs as key agents of globalization, outlining their growth over time and assessing claims about the emergence of truly global corporations in the last couple of decades. Chapter 7 focuses on the geography of employment and the changing role of labour within a more integrated world economy. The active role of workers in shaping processes of economic development is stressed (Herod, 1997). This is followed by a chapter on development in the Global South, which identifies the main approaches to development, examines broad patterns of inequality and outlines contemporary policy debates on trade, aid and debt.

The third part of the book covers 'Contemporary Issues in Economic Geography'. Chapter 9 examines the uneven geographies of finance in the wake of financial globalization, deregulation and the financial crisis of 2007–08. Chapter 10 focuses on the organizational geographies and governance of global industries, covering the global commodity chain and global production network approaches. This is followed by a chapter on knowledge, creativity and regional development, which considers debates on innovation and agglomerations and the 'creative classes' (Florida, 2002). Chapter 12 is concerned with alternative economic geographies that extend beyond and challenge the capitalist mainstream (Gibson-Graham, 2006), covering the counter-globalization movement, various

alternative local initiatives and broader geographies of responsibility and justice, as expressed in the notion of fair trade, for instance. Finally, a brief conclusion pulls the main themes together and considers the wider political economy of development in the wake of the recent financial crisis and global recession.

Exercise

Think of a commodity that you have recently consumed. This could be something you ate for lunch or breakfast, or an item of clothing that you have recently bought. A jar of coffee or a pair of trainers could be an example.

When you purchased these, were you primarily concerned with the price and physical qualities of these goods? Was there anything, a label possibly, to indicate the geographical origin of this good? Why would it have been produced in that particular region or country? Under what conditions do you think it would have been produced, for example, in a factory, by craft workers, on a farm? What main actors would have involved in its production, MNCs, small firms, farmers, etc.? How might the profits be distributed among these main actors?

Key reading

Barnes, T.J. and Sheppard, E. (2000) 'The art of economic geography', in Sheppard, E. and Barnes, T.J. (eds) *A Companion to Economic Geography*, Oxford: Blackwell, pp.1–8.
A brief introduction to economic geography as an academic subject area. Highlights the key questions that economic geographers address and the changing perspectives than inform their work.

Leyshon, A. (1995) 'Annihilating space?: the speed-up of communications', in Allen, J. and Hamnett, C. (eds) *A Shrinking World?: Global Unevenness and Inequality*, Oxford: Oxford University Press, pp.11–54.
Introduces the key concept of time–space compression, emphasizing the speed-up of global communications since the early 1980s. Explains this in terms of the geographical expansion of capitalism over time, facilitated by successive advances in transport and communications technologies.

Taylor, P.J., Watts, M. and Johnston, R.J. (2002) 'Geography/globalisation', in Johnston, R.J., Taylor, P. and Watts, M. (eds) *Geographies of Global Change: Remapping the World*, 2nd edn, Oxford: Blackwell, pp.1–17, 21–8.
A stimulating introduction to the relationships between geography as a discipline and the issue of globalization. Outlines the rise of globalization as a key topic of interest, discusses the uneven geography of globalization and highlights recent political debates about the nature of globalization.

Watts, M. (2005) 'Commodities', in Cloke, P., Crang, P. and Goodwin, M. (eds) *Introducing Human Geographies*, 2nd edn, London: Arnold, pp.527–46.
A review of the commodity as a key topic of interest to geographers. Highlights the economic importance of the commodity within capitalism and the role of commodities in linking production together in different places through commodity chains and networks.

Useful websites

http://www.polity.co.uk/global/default.asp
Provides a very accessible and comprehensive introduction to key aspects of globalization. The companion website for the *Global Transformations* textbook (Held *et al.*, 2004).

http://www.exchange-values.org/
The website of a social sculptures project by Shelley Sacks in collaboration with the banana growers of the Windward Islands and a range of representative organizations. Contains a range of useful articles under the 'texts, debates, discussion' link and images. The articles by the geographers Ian Cook and Luke Desforges are particularly recommended along with the ones on 'banana wars', 'banana lives' and 'unfair trade'.

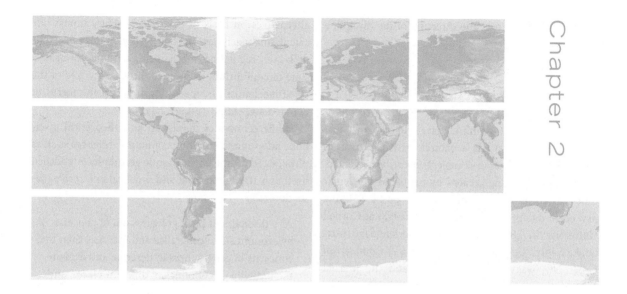

Approaches to economic geography

Chapter map

Having defined economic geography in section 1.3, in this chapter we set out the main approaches that economic geographers have adopted over time. This serves to frame and position the political economy perspective that informs this book. We begin by considering the relationship between economic geography and the neighbouring discipline of economics. In the reminder of the chapter, we examine the different approaches that have been adopted in economic geography, gaining favour at different times. Four main approaches are identified: traditional economic geography, which was dominant from the late nineteenth century to the 1950s; spatial analysis, embraced in the 1960s and 1970s; political economy, which became popular in the 1970s and 1980s; and a set of cultural and institutional frameworks which economic geographers have turned towards since the early 1990s. The latter two approaches are considered in greater detail because of their importance in shaping debates over the direction of the discipline in recent years.

2.1 Introduction

A key starting point for this chapter is to recognize that no academic subject has a natural existence. Instead, as Barnes (2000a) argues, subjects must be 'invented' in the sense of being created by people at particular times. According to Barnes (2000a, pp.14–15), the first economic geography course was taught at Cornell University in 1893, the first English-language textbook, George G. Chisholm's *Handbook of Commercial Geography*, was published in 1889 and the journal *Economic Geography* established in 1925. The neighbouring discipline of economics was also established in the late nineteenth century, along with a number of other social sciences. From the start, however, the two disciplines assumed different characteristics.

Economics views the economy as governed by market forces that basically operate in the same fashion everywhere, irrespective of time and space. The market is composed of a multitude of buyers and sellers – the forces of demand and supply – who dictate how scarce resources are allocated through their decisions about what to produce and consume. Mainstream neoclassical economics is underpinned by the idea of 'economic man', assuming that people act in a rational and self-interested manner, continually weighing up

alternatives on basis of cost and benefits, almost like calculating machines. The market is viewed as an essentially self-regulating mechanism, tending towards a state of equilibrium or balance through the role of the price mechanism in mediating between the forces of demand and supply (Figure 2.1).

While economics developed as a theoretical discipline adopting the methods of natural sciences such as physics and chemistry, economic geography established itself as a strongly factual and practical enterprise (see section 2.2):

> As a discipline it [EG] grew less out of concerns by economists to generalise and theorise, than the concerns of geographers to describe and explain the individual economics of different places, and their connections one to another.
>
> (Barnes and Sheppard, 2000, pp.2–3)

In general, a clear contrast can be drawn between the formal and theoretical approach of economics and geography's more open-ended ethos and substantive concerns. While geography can be seen as synthetic in nature, focusing on the relationships between, rather than the separation of, processes and things (Lee, 2002, p.333), economics is analytic, seeking to separate the economy from its social and cultural context. Key features of each of the four approaches to economic geography examined in this chapter are set out in Table 2.1, providing an important backdrop to the ensuing discussion.

2.2 Traditional economic geography

What we can term the **traditional approach to economic geography** held sway from the late nineteenth century to the 1950s. It was factual and descriptive in nature, focusing on the compilation of information about economic conditions and resources in particular regions. Economic geography or **commercial geography** was highly prominent from the 1880s to the 1930s, a period in which geography assumed a key role as the 'handmaiden' of empire, providing useful knowledge for colonial administrators and traders (Livingstone, 1992). Commercial geography was

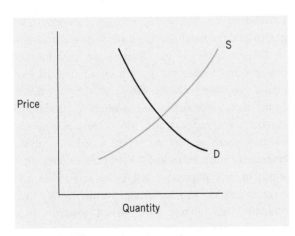

Figure 2.1 Demand and supply curves
Source: Lee, 2002, p.337.

Table 2.1 Approaches to economic geography

	Traditional	Spatial analysis	Political economy	Institutional/cultural
Supporting philosophy	Empiricism[a]	Positivism (Box 2.1)	Dialectical materialism[b]	Postmodernism (Box 2.6) & institutionalism
Main source of ideas	Classical German geography, anthropology, biology	Neo-classical economics	Marxist economics, sociology and history	Cultural studies, institutional economics, economic sociology
Conception of the economy	Closely integrated with the natural resources and culture of an area	Driven by rational choices of individual actors	Structured by social relations of production. Driven by search for profit and competition	Importance of social context. Informal conventions and norms shape economic action
Geographical orientation (place or wider processes)	a) Commercial geography stressed global trading system; b) regional geography highlighting unique places (regions)	Wider forms of spatial organization	Wider processes of capitalist development. Places passive 'victims' of these wider processes	Emphasis on individual places in context of globalization
Geographical focus	a) Colonial territories; b) distinctive regions, mainly in Europe and North America, often rural and geographically marginal	Urban regions in North America, Britain and Germany	Major cities in industrial regions in Europe and North America. Cities and regions in developing countries, especially Latin America	Growth regions in developed countries. Global financial centres. Key sites of consumption
Key research topics	Effects of the natural environment on production and trade; identifying distinctive regional economies	Industrial location; urban settlement systems; spatial diffusion of technologies; land use patterns	Urbanization processes; industrial restructuring in developed countries; global inequalities and underdevelopment	Social and institutional foundations of economic development; consumption; work identities; financial services; corporate cultures
Research methods	Direct observation and fieldwork	Quantitative analysis based on survey results and secondary data	Reinterpretation of secondary data according to Marxist categories. Interviews	Interviews, focus groups, textual analysis, ethnography, participant observation

[a] Emphasis on gaining knowledge directly from the senses, particularly through observation, and 'the facts'.

[b] Dialectical means that social change is seen in terms of a struggle between opposing forces (Box 2.2). Materialism stresses the real social and economic conditions of existence (production, labour, class relations, technology, resources) over ideas and culture, effectively privileging matter over mind.

particularly useful in providing knowledge about the colonial territories of the European powers in Africa, Asia and Latin America, detailing the key resources and identifying the type of crops that could be grown there (Figure 2.2).

Commercial geography was based on the 'great geographical fact' that different parts of the world yield different products, underpinning the system of global commerce (Barnes, 2000a, p.15). The *Handbook of Commercial Geography* (Chisholm, 1889) emphasized factual detail over abstract theory, with George G. Chisholm credited as having wished the 'love of pure theory to the devil', displaying, instead, a 'meticulous mastery of detail' (MacLean, 1988, p.25). Chisholm provided detailed statistics on trade, compiling information about the production and exchange of a wide

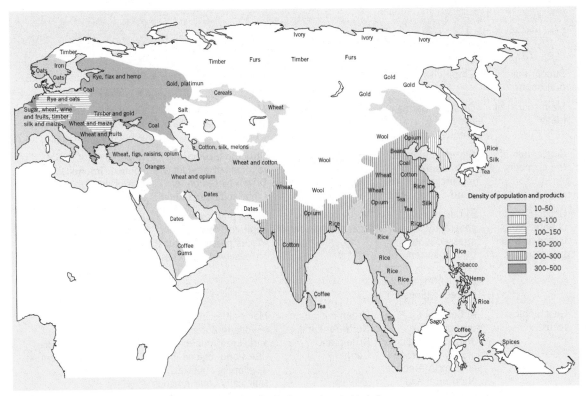

Figure 2.2 Regional economic specialization in Asia under colonialism
Source: Chisholm, 1889, pp.302–3.

range of commodities and resources. His numerous maps, figures and tables played an important role in offering a concrete expression of imperialism as an economic system, rendering the complex flow of goods and resources more visible and tangible (Barnes, 2000a, p.15). The development of commercial geography was crucial in establishing economic geography as a distinct sub-discipline, helping to define many of its enduring characteristics, such as an avoidance of theory, an emphasis on factual detail, a celebration of numbers and a reliance on geographical categories made visible by the map (ibid., p.16).

By the 1930s, however, the empires of the major powers were in decline, with policy debates becoming increasingly focused upon internal national issues such as unemployment and poverty. Partly in response, the focus of economic geography 'shifted from the general commercial relations of a global system to the geography of narrowly bounded, unique regions, especially those close to home' (ibid., p.18). This was the era of **regional geography**, defined by Hartshorne

(1939, p.21) as a project of 'areal differentiation' which describes and interprets the variable character of the Earth's surface, expressed through the identification of distinct regions. These were classified and described through works of synthesis that presented the key characteristics of an area in a logical order, beginning with the physical landscape and proceeding through to human settlement and culture (Livingstone, 1992). Economic geography was largely subsumed within this broader regional enterprise, emphasizing regional uniqueness, the compilation of regional typologies and the need for first-hand fieldwork (Barnes, 2000a, pp.20–1).

Reflect

➤ What were the main links between commercial geography and imperialism between the 1880s and 1930s?

2.3 Spatial analysis in economic geography

By the mid-1950s, considerable dissatisfaction was being directed towards the traditional approach. A new generation of researchers increasingly came to reject the idea that regional synthesis was the proper goal of geography, seeking to develop a more scientific approach. The attack on the traditional establishment found particularly cogent expression in a 1953 paper by Fred K. Schaefer, who called for geographers to employ scientific methods in a search for general theories and laws of location and spatial organization (Scott, 2000, p.485). This argument drew directly on the established **positivist** philosophy of science (Box 2.1)

Schaefer's philosophical arguments fitted with a new style of practical research being developed at the Universities of Iowa and Washington, Seattle where younger geographers were using statistical and mathematical methods to analyse problems of industrial location, distance and movement (Barnes, 2000a). A vibrant spatial analysis research programme developed at Seattle, for example, which focused on issues of industrial location and land use patterns, urbanization and central place theory, transport networks and the geographical dynamics of trade and social interaction. The new scientific geography spread to other departments in the US while the Universities of Cambridge and Bristol became key centres in the UK during the 1960s. Progress was such that Burton (1963) could proclaim the 'quantitative revolution' complete, defined 'as a radical transformation of the spirit and purpose of geography'. Economic geography was at the forefront of this movement, viewed as an area that was particularly suited to the application of quantitative methods.

Real world conditions were increasingly favourable to this new approach in the late 1950s and 1960s, as policy-makers focused on economic and urban problems in developed countries providing funds for research and a demand for academic analysis and advice. A period of sustained economic growth and an underlying faith in science and technology created an optimistic 'can do' attitude with urban and regional planning embraced as the means of addressing problems of location, land use management and transportation (Barnes, 2000a). In this context, regional geography appeared increasingly backward and anachronistic, with its focus on rural backwaters and concern with description and classification offering little of practical value to the planner or developer.

Neo-classical economic theory provided a ready source of concepts for quantitative economic geography in the 1960s. Geographers sought to apply the same

---

Box 2.1

Positivism

This is a philosophy of science originally associated with the French philosopher and sociologist Auguste Comte (1798–1857) and developed further by the Vienna Circle of thinkers in the 1920s and 1930s (Gregory, 2000). It seems to have gained broad acceptance as an account of the goals of natural science in the post-war period, although specific aspects of it sparked debate. Since the 1970s, particularly, some social scientists (including many human geographers) have rejected positivism.

Positivism holds that a real world exists independent of our knowledge of it. This real world has an underlying order and regularity which science seeks to discover and explain. Facts can be directly observed and analysed in a neutral manner. The separation of fact and value is a central tenet of positivism; personal beliefs and positions should not influence scientific research. The aim of science is to generate explanatory laws which explain and predict events and patterns in the real world. In the classic deductive method (moving from theory to practical research), scientists formulate hypotheses – formal statements of how a force or relationship is thought to operate in the real world – which are then tested against data collected by the scientist through experiment or measurement. Hypotheses that are supported by initial testing must then be verified or proved correct through objective and replicable procedures. If verification is successful, they gain the status of scientific laws.

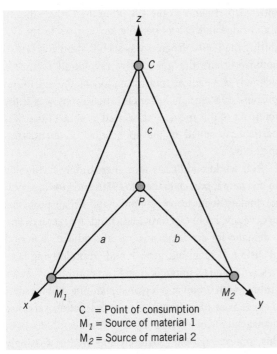

C = Point of consumption
M_1 = Source of material 1
M_2 = Source of material 2

Figure 2.3 Weber's locational triangle
Source: Knox and Agnew, 1994, p.77.

style of deductive theorizing and analysis, beginning by simplifying assumptions to the development and testing of hypotheses and models against numerical data representing real world conditions. As one of the pioneers of quantitative economic geography, Harold McCarty of the University of Iowa, put it, 'economic geography derives its concepts largely from the field of economics and its method largely from the field of geography' (McCarty, 1940, quoted in Barnes, 2000a, p.22).

The tradition of **German location theory** provided a body of economic theory applied to geography, which the new economic geography of the 1950s and 1960s could draw upon. The work of theorists such as Von Thunen, Weber, Christaller and Losch was applied to the circumstances of North America and the UK in the 1960s, being used to explain and predict land use patterns, the location of industry and the organization of settlements and market areas. Weber's theory of industrial location emphasized the importance of transport costs in determining where a factory or plant would be located in relation to the sources of raw

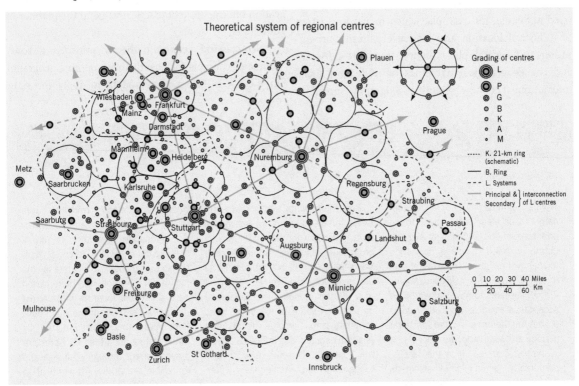

Figure 2.4 Central places in southern Germany
Source: Christaller, Walter, Baskin, Carlisle Wir, *Central Places in Southern Germany*, 1st edn © 1966, pp.224–25. Reprinted by permission of Pearson Education, Inc., Upper Saddle River, NJ.

materials and the market area, represented in terms of a locational triangle (Figure 2.3). Point P is where the costs of transporting the material to the factory and the finished goods to market are minimized. If the raw materials lost weight during manufacturing, the factory would be drawn towards the material sources. If, on the other hand, distribution costs are higher than the costs of transporting materials, the industry would be drawn towards its market. Weber formulated this model in 1929, a time in which heavy industries based in coal-field regions dominated the economic landscape.

Perhaps the best known of the German locational models is Christaller's **central place theory** (Johnston, 2000). Based on the assumption of economic rationality and the existence of certain geographical conditions, such as uniform population distribution across an area, central place theory offers an account of the size and distribution of settlements within an urban system. The need for shop owners to select central locations produces a hexagonal network of central places, organized into a distinct hierarchy of lower- and higher-order centres (Figure 2.4). More recently, a new version of spatial analysis has been developed by mainstream economists such as the Nobel prize-winner Paul Krugman. This **new geographical economics** – termed 'new economic geography' by Krugman himself – applies mathematical modelling techniques to analyse issues of industrial location (Box 2.2).

Box 2.2

The 'new geographical economics'

The starting point for the 'new geographical economics' (NGE) is the basic geographical fact that economic activity is unevenly distributed. It addresses questions such as why, and under what conditions, do industries concentrate? How do centripetal (centralizing) forces favouring the geographical concentration of industries in a particular location interact with opposing forces favouring geographical dispersal to a range of locations?

In addressing these issues, the NGE applies the methods of mainstream economics, devising models based on a number of simplifying assumptions, described as 'silly but convenient' (Krugman, 2000, p.51). It retains much of the basic architecture of neo-classical economics, requiring explanations that are based in the rational decisions of individual actors. Some of the specific assumptions are different, though, incorporating notions of imperfect (monopolistic) competition and economies of scale where

additional investment in production capacity brings firms increased profits (neo-classical economics assumes that there are no economies of scale).

Centripetal forces favouring geographical concentration are identified as the effects of market size, a large and specialized labour market, and access to information from other firms located there. Conversely the main centrifugal forces encouraging dispersal are immobile factors of production such as land and, to a considerable extent, labour, and the costs of concentration such as congestion (Krugman, 1998, p.8). Much actual research has been devoted to assessing the interaction between these forces under different conditions. Small changes in, for example, transport costs or technology can 'tip' the economy from a pattern of dispersal to one of concentration. The typical outcome of concentration processes will be a simple core-periphery pattern, while dispersal will create a number of specialized

and relatively evenly sized centres of industry.

The new geographical economics, therefore, focuses on the same basic questions that have long interested economic geographers, but adopts a distinctive approach based on the methods of economics. For many economic geographers, it is simply a more sophisticated version of the spatial analysis popular in the 1960s, sharing its underlying limitations and generating 'a dull sense of déjà vu' (Martin, 1999a, p.70). A particular weakness of the new geographical economics is the characteristic tendency to 'focus on what is easier to model rather than on what is probably most important in practice' (Krugman, 2000, p.59). At the same time, while this approach can never hope to capture the complexity and richness of the real economic landscape, its analytical clarity and sense of purpose does perhaps carry some lessons for economic geographers (Martin, 1999b).

The so-called quantitative revolution transformed the nature of economic geography 'from a field-based, craft form of inquiry to a desk-bound technical one in which places were often analysed from afar ...' (Barnes, 2001, p.553). Instead of directly observing and mapping regions in the field, economic geographers now tended to use secondary information and statistical methods to analyse patterns of spatial organization from their desks. While not all practioners of economic geography adopted the new methods, spatial analysis came to occupy centre-stage with those who refused to follow its approach increasingly relegated to the sidelines (ibid.). By the late 1960s, however, the mood was changing again, with a growing number of geographers beginning to question 'the spirit and purpose' of this new quantitative geography.

> ### Reflect
>
> Do you think that economic geographers should adopt similar methods and perspectives to economists or should they seek to differentiate themselves?

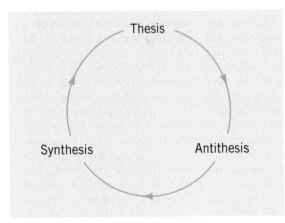

Figure 2.5 Dialectics
Source: Johnston and Sidaway, 2004, p.231.

2.4 The political economy approach in economic geography

2.4.1 The origins of Marxist geography

In the late 1960s and early 1970s, quantitative geography was subject to criticism for its lack of social relevance and concern. To a new generation of geographers, it seemed as if the discipline was narrowly focused on technical issues of urban and regional planning to the neglect of deeper questions about how society was organized. The key issues here were racial divisions in US cities, the Vietnam War (symbolizing the imperialism of US foreign policy), gender inequalities and the rediscovery of poverty in inner-city ghettos. Geography remained largely silent about such questions, leading Harvey to call for a revolution in geographic thought:

> The quantitative revolution has run its course, and diminishing marginal returns are apparently setting in ... There is an ecological problem, an urban problem, an international trade problem, and yet we seem incapable of saying anything of depth or profundity about any of them.
>
> (Harvey, 1973, pp.128–9)

In response to these pressing social issues, a group of geographers in the US particularly sought to fashion a new **radical geography**. This movement began at Clark University in Massachusetts, where a group of postgraduate students, led by Richard Peet, launched *Antipode: A Radical Journal of Geography* in 1969.

This radical new geography turned to political economy for its intellectual foundations. The work of Marx in particular provided a framework for the critical analysis of advanced capitalism. Once again, economic geography was at the forefront of these developments. In general, **Marxism** emphasizes processes and relationships rather than fixed things, adopting a dialectical perspective which sees change as driven by the tensions between opposing forces, usually in the form of thesis – antithesis – synthesis (Figure 2.5). Capitalist society in particular is characterized by continual change and flux, driven by the search for profits in the face of competition:

> The bourgeoisie [capitalists] cannot exist without constantly revolutionising the instruments of production, and thereby the relations of production, and with them the whole relations of society ... Constant revolutionising of production, uninterrupted disturbance of all social conditions, everlasting uncertainty and agitation distinguish the bourgeois epoch from all earlier ones. All fixed, fast-frozen relations ... are swept away, all new formed ones become antiquated before they can ossify. All that is solid melts into air, all that is holy is profaned ...
>
> (Marx and Engels, [1848] 1967, p.83)

From this perspective, particular geographical objects, for example cities or a transport system, exist as expressions of wider relationships and are subject to transformation through the movement of broader forces of change (Swyngedouw, 2000).

Box 2.3

The political economy tradition and the origins of a Marxist approach

It is important to set Marx's contribution to political economy within its historical context. Born in 1817, Marx's early academic career was as a student of German philosophy, influenced by Hegel's dialectical approach, in which human relations are conceived in terms of the struggle between opposing forces Figure 2.5. During the 1840s, however, he became increasingly concerned with economic issues, fascinated by the rise of industrial and urban society. His big theoretical project, culminating in the three volumes of *Capital* ([1867] 1976) became one of revising the classical political economy of Smith and Ricardo to understand the dramatic changes that were sweeping society in the mid-nineteenth century (Dowd 2000).

Smith, whose great work *The Wealth of Nations* was published in 1776 and Ricardo, whose *Principles of Political Economy and Taxation* appeared in 1817, were great advocates of the market and free trade. Smith's guiding philosophy was that greater wealth would accrue to the community of nations if governments desisted from intervening in the economy. Ricardo argued that a system of international trade without tariffs (import taxes) would encourage much greater efficiency since each country would specialize and produce goods that they were relatively more efficient at making, therefore benefiting the broader commonwealth of nations (Box 4.1).

Both, however, were writing during the very early stages of industrial capitalism when feudalism allied to mercantilism – national protectionism – were viewed as impediments to social progress. For Smith, the development of market relations and competition would produce a more progressive and equal society than feudalism, liberating the individual from traditional social bonds. In this sense, he was very much in tune with the eighteenth-century Enlightenment which sought to overthrow the *ancien régime* and which influenced the French Revolution and American War of Independence. While Marx, writing over half a century after Smith, also recognized that capitalism was more progressive than feudal society, unlike Smith he was able to witness the negative consequences of industrial capitalism. The growth of the factory system, mass urbanization and the development of a large industrial working class, many of whom were living in conditions of appalling poverty and squalor, had transformed Smith's capitalist utopia into William Blake's 'dark satanic mills'.

According to Marx, class struggle is the basic driving force of history, from the relations of master and slave, to feudal lord and serf, to capitalist and labourer. Under the capitalist mode of production, class relations are structured by the private ownership of capital (money, factories and equipment), creating a class of capitalists, and a class of labourers who must sell their labour in exchange for a wage. The difference between the value of what the labourer actually produces and what he/she is paid in wages is retained and accumulated by the capitalist as surplus value or profit, representing the basis for class exploitation. The development of factories under capitalism brings large numbers of workers together in industrial cities, providing them with the means of organizing against the system that exploits them. When coupled with the inevitable overproduction that would result from the continual expansion of production, driven by competition, the growing power of the working class would eventually bring down the system and usher in a new socialist era. In this way, capitalism acts as its own gravedigger (Dowd, 2000, p.87).

2.4.2 The development of Marxist theory

While the writings of Marx provided only a few scattered comments and insights into the geography of capitalism, Marxist geographers such as David Harvey have sought to build on these by developing a distinctively Marxist analysis of geographical change (Box 2.4). From this perspective, the economic landscape is shaped by the conflict-laden relationship between capital and labour, mediated by the State, providing a stark contrast to the harmonious equilibrium state of regional balance posited by neo-classical economic theory (Sheppard *et al.*, 2004, p.4).

The first phase of Marxist geography concentrated on establishing how capitalism produces specific

---

Box 2.4

Neil Smith and the theory of uneven development

One of the most notable contributions to Marxist economic geography is Neil Smith's **theory of uneven development**. Processes of uneven development, according to Smith, are the result of a dialectic of spatial differentiation and equalization that is central to the logic of capitalism, transforming the complex mosaic of landscapes inherited from pre-capitalist systems. Capital moves to the areas that offer highest profits for investors, resulting in the economic development of these areas. The geographical concentration of production in such locations results in differentiation as they experience rapid development, while other regions are left behind (see section 4.3). As a result, living standards and wage rates vary markedly between regions and, especially, countries. At the same time, the tendency towards equalization reflects the importance of expanding the market for goods and services, implying a need to develop newly incorporated colonies and territories so as to generate the income to underpin consumption.

Over time, the process of economic development in a particular region tends to undermine its own foundations, leading to higher wages, rising land prices, lower unemployment and the development of trade unions, thus reducing profit rates. In other regions, underdevelopment leads to low wages, high unemployment and the absence of trade unions, creating a basis for profit that attracts capital investment. Over time, capital will 'see-saw' from developed to underdeveloped areas, 'jumping' between locations in its efforts to maintain profit levels. It is this movement of capital that creates patterns of uneven development. In this sense, 'capital is like a plague of locusts. It settles on one place, devours it, then moves on to plague another place' (Smith, 1984, p.152).

According to Smith, the production of space under capitalism leads to the emergence of three primary geographical scales of economic and political organization: the urban, the national and the global. The dynamic nature of the uneven development process is most pronounced at the urban scale at which capital is most mobile, resulting in, for example, the rapid gentrification (upgrading through the attraction of investment and new middle-class residents) of previously declining inner-city areas (Figure 2.6). Conversely, patterns of uneven development exhibit most stability at the global scale, where the divide between developed and developing counties remains as wide as ever, although East Asia has risen to the core of the world economy through sustained economic growth since the 1960s.

Figure 2.6 Urban gentrification in Glasgow
Source: RCAHMS Enterprises: Resource for Urban Design Information (RUDI). Licensor www.scran.ac.uk.

geographical landscapes (Smith, 2001). In *The Limits to Capital*, Harvey (1982) identified a central contradiction between the geographical fixity and motion of capital. There is a need, on the one hand, for fixity of capital in one place for a sustained period, creating a built environment of factories, offices, houses, transport infrastructures and communication networks to enable production to take place. Such fixity is countered by the need, on the other hand, for capital to remain mobile, enabling firms to respond to changing economic conditions by seeking out more profitable locations (Box 2.4). This may require them to withdraw from existing centres of production in which they have invested heavily. Capital is never completely mobile, but must put down roots in particular places to be effectively deployed. Nonetheless, its relative mobility lends capital an important spatial advantage over labour that is more place-bound (see section 3.3.1).

Harvey (1982) argues that capital overcomes the friction of space or distance through the production of space, in the form of a built environment that enables production and consumption to occur. Indeed, such investment in the built environment can act as a **'spatial fix'** to capitalism's inherent tendency towards overproduction by absorbing excess capital, performing an important displacement function. As economic conditions change, however, these infrastructures can themselves become a barrier to further expansion, appearing increasingly obsolete and redundant in the face of more attractive investment opportunities elsewhere. In these circumstances, capital is likely to abandon existing centres of production and establish a new 'spatial fix' involving investment in different regions (Box 2.4). The deindustrialization of many established centres of production in the 'rustbelts' of North America and Western Europe since the late 1970s and the growth of new industry in 'sunbelt' regions and the newly industrializing countries of East Asia can be understood in this light.

The second phase of Marxist geography began in the early 1980s, concentrating on developing Marxist analyses of specific situations and circumstances (Smith, 2001). One key research question was how particular places were affected by wider processes of economic restructuring. In a landmark text, Massey (1984) investigated the changing location of industry

in Britain, developing the concept of 'spatial divisions of labour' (see section 4.3.2). In contrast with conventional location theory which emphasizes the influence of resources, markets and transport costs, labour – its availability, cost and skills – is the key location factor from a Marxist perspective.

2.4.3 The regulation approach

Another strand of Marxist-informed research that was very influential in economic geography during the 1990s is the regulation approach. Derived from the work of a group of French economists in the 1970s, the regulation approach stresses the important role that wider processes of social regulation play in stabilizing and sustaining capitalist development. These wider processes of regulation find expression in specific institutional arrangements that mediate and manage the underlying contradictions of the capitalist system (see Chapter 3), expressed in the form of periodic crises, enabling renewed growth to occur. This occurs through the coming together and consolidation of specific **modes of regulation**, referring to the institutions and conventions that shape the process of capitalist development. Regulation is focused on five key aspects of capitalism: labour and the wage relation, forms of competition and business organization, the monetary system, the state and the international regime (Boyer, 1990). When these act in concert, a period of stable growth, known as a **regime of accumulation**, ensues (Jones *et al.*, 2004, p.59).

Two regimes of accumulation can be identified since the 1940s (Table 2.2). After the global economic depression of the 1930s, a new regime of **Fordism** emerged, named after the American car manufacturer, Henry Ford, who pioneered the introduction of mass production techniques. Fordism was based on a crucial link between mass production and mass consumption, provided by rising wages for workers and increased productivity in the workplace. The state adopted a far more interventionist approach, informed by the theories of the British economist John Maynard Keynes, which stressed the important role of government in managing the overall level of demand in the economy to secure full employment. At the same time,

Table 2.2 Fordist and post-Fordist modes of regulation

	Fordism	Post-Fordism
The wage-relation	Rising wages in exchange for productivity gains. Full recognition of trade union rights. System of national collective bargaining	Flexible labour markets based on individual's position in market. Limited recognition of trade unions
Forms of competition and business organization	Dominance of large corporations. Nationalization of key sectors such as utilities	Key role of small- and medium-sized enterprises (SMEs) alongside large corporations. Privatization of state enterprises and general liberalization of the economy
The monetary system	Monetary policy focused on demand management and maintaining full employment. Use of interest rates to facilitate economic expansion and contraction	Focus on reducing inflation, relying on high interest rates if necessary. International monetary integration and coordination, for example the European single currency
The state	Highly interventionist. Adoption of Keynesian policies of demand management. Provision of social services through welfare state. Goals of full employment and modest income redistribution	Reduce state intervention in the economy. Abandonment of Keynesian policies for neoliberalism. Efforts to reduce welfare expenditure and privatize services
The international regime	Cold War division into capitalist and communist blocs. Bretton Woods system of fixed exchange rates, anchored to the US dollar. Promotion of free trade, monetary stabilization and 'Third World' development	Increased global economic integration. System of floating exchange rates. Renewed emphasis on free trade and openness to foreign investment. Imposition of neoliberal adjustment policies on developing countries

governments provided a wide range of social services through the welfare state.

Fordism experienced mounting economic problems from the early 1970s, however, and elements of a new 'post-Fordist' regime of accumulation based on flexible production became prominent in the 1980s. Theories of **post-Fordism** emphasize the role of small firms, advanced ICT, and more segmented and individualized forms of consumption (section 3.4.3). Whether these features amount to a distinctive and coherent regime of accumulation remains questionable, however, with mass production remaining important in some sectors. One key trend has been the abandonment of Keynesian policies since the 1970s in favour of a **neoliberal approach** which seeks to reduce state intervention in the economy and embrace the free market, stressing the virtues of enterprise, competition and individual self-reliance (Harvey, 2005).

2.4.4 The relevance and value of Marxist political economy

By the late 1980s, Marxist geography was becoming subject to increasing criticism, informed by the emergence of postmodern thought (Box 2.6). Marxism had also become rather out of touch with the 'new times' of the 1980s, marked by the dominance of neoliberal ideas, particularly in the UK and US. In the realm of left-wing politics too, the focus was shifting from the traditional 'politics of distribution' concerned with work, wages and welfare to a 'politics of identity' concerned with asserting the rights of various groups such as women, ethnic minorities and gay people to recognition and justice (Crang, 1997). Such claims were channelled through broader social movements rather than the traditional labour movement.

Three main criticisms of Marxism in geography can be identified:

➤ Its apparent neglect of human agency in terms of an impoverished view of individuals and a failure to recognize human autonomy and creativity. Instead, Marxists tend to privilege wider social forces such as class and see people as bearers of class powers and identities, reading off their behaviour from this, not as unique individuals.

➤ Its emphasis on economic forces and relations. While the Marxist concept of production is, as we have seen, much broader than conventional notions of the economy, Marxists have been criticized for

stressing the determining role of economic forces. Culture and ideas are often viewed as products of this economic base.

➤ Its overwhelming emphasis on class, and neglect of other social categories such as gender and race. Leading Marxists such as Harvey were attacked by feminist geographers in the early 1990s for their neglect of gender issues, accusing them of subsuming these within a class-based Marxist analysis.

These are important criticisms, although it is questionable whether Marxist geography really was as economically

---

Box 2.5

Is Marxist political economy still relevant?

Since the early 1980s, there has been a wholesale abandonment of Marxist and socialist thinking in the face of the upsurge of neoliberal thinking in the West and the collapse of communism in the Soviet Union and eastern Europe. The fall of the Berlin Wall in 1989 was heralded as bringing about the 'end of history' by the conservative American writer Francis Fukuyama in the sense that liberal democracy and the market had won the battle of ideas over socialism. The spirit of market triumphalism associated with globalization in the 1990s has, however, been punctured in recent years as the limitations of global capitalism have become increasingly apparent (Gray, 1999), highlighted by the anti-globalization movement and the recent financial crisis.

As part of the more sober climate of the late 1990s and 2000s, Marx has been rediscovered. Triggered by the 150th anniversary of the publication of The Communist Manifesto (Marx and Engels, [1848] 1967) in 1998 and the financial crises then engulfing East Asian and Russia, 'impeccably bourgeois magazines' such as the Financial Times and New Yorker published articles heralding Marx's thought (Smith, 2001,

p.5). Writing in the New Yorker, John Cassidy praised Marx as the 'next big thinker', citing his relevance to the workings of the global financial system and stating that his analyses will be worth reading 'as long as capitalism endures' (quoted in Rees, 1998). Although Wall Street's discovery of Marx was predictably short-lived, listeners voted Marx the greatest philosopher of all time in a BBC Radio 4 poll in 2005. The British Marxist historian, Eric Hobsbawm, explained this in terms of the 'stunning prediction of the nature and effects of globalisation' found in The Communist Manifesto (quoted in Seddon, 2005).

We believe that Marxist political economy is still relevant because of its value as a framework for understanding the evolution of the global capitalist system. Marx's primary contribution to knowledge was as an analyst of capitalism not as an architect of communism. While he offered only a few scatted comments about geography, this has been rectified by Marxist geographers such as Harvey and Smith who have developed theories of uneven development (see Box 2.4). In the absence of other approaches which can match

its historical-geographical reach and analytical purchase, Marxism provides the most suitable framework for analysing the 'big questions' concerning the economic geography of global capitalism (Swyngedouw, 2000). Marxism also retains a strong sense of social and political commitment, emphasizing issues of inequality and injustice and the need to change the world as well as interpret it.

At the same time, we recognize that Marxism does not hold all the answers, containing several limitations and weaknesses. This points to the need for a modified Marxist political economy that can incorporate insights from other perspectives, particularly the cultural and institutional approaches that economic geographers have turned towards in recent years (section 2.5). In particular, these approaches can provide a stronger sense of agency, sensitivity towards the cultural construction of the economy and a concern with the role of institutions and evolution (Hudson, 2006). In short, we favour a kind of 'open' Marxism that does not claim to have a monopoly on truth, is receptive to insights from other perspectives and evolves in line with capitalism as its object of analysis.

Reflect

➤ Do you agree that Marxism is still relevant to the analysis of global capitalism?

determinist as its critics allege (Hudson, 2006). We do not think that they mean that Marxist political economy is no longer relevant or useful and should be abandoned (Box 2.5). As recent research on geographical scale and labour has shown, basic Marxist categories such as capital and labour remain highly relevant to an understanding of contemporary capitalism. In particular, such categories help us to address the 'big questions' of uneven development, social justice and environmental degradation on an increasingly global scale (Swyngedouw, 2000).

Partly in response to the above criticisms and reflecting its encounters with other philosophical perspectives such as postmodernism, the political economy approach has become increasingly complex and diffuse in the 1990s and beyond. At the same

time, the culturally and institutionally informed work of the past 15 or so years has raised important new questions about knowledge, identity and consumption. What seems to be needed is a framework for bringing together ideas from the two sets of perspectives, directing us towards an **'open' political economy approach** that is receptive to the insights offered by other perspectives (Box 2.5).

2.5 Cultural and institutional approaches in economic geography

2.5.1 The cultural 'turn'

A new set of approaches has emerged since the early 1990s, emphasizing the institutional and cultural foundations of economic processes. The key context here is the **cultural 'turn'** that took place in human geography and the social sciences in the late 1980s and 1990s,

Box 2.6

Postmodernism

Postmodernist ideas have attracted widespread interest since the 1980s, coming to exert considerable influence in architecture, the humanities and the social sciences. Ley (1994, p.466) defines postmodernism as 'a movement in philosophy, the arts and social sciences characterised by scepticism towards the grand claims and grand theory of the modern era, and their privileged vantage point, stressing in its place openness to a range of voices in social enquiry, artistic experimentation and political empowerment'.

As this quote indicates, pluralism is a key characteristic of postmodern thought in terms of embracing the knowledge claims of different social groups. Grand theories or 'meta-narratives' (big stories) claiming to

uncover the changing organization of society are rejected as a product of the privileged position and authority of the observer rather than being accepted as objective representations of the realities that they purport to explain. Instead of functioning as a set of universal truths, knowledge should be regarded as partial and situated in particular places and times. Postmodernists reject conventional notions of scientific rationality and progress, favouring an open interplay of multiple local knowledges.

Rather than assuming that social life has an underlying order and coherence, postmodernists celebrate difference and variety. Difference and variety are held to be a basic characteristic of the world, applied to a range of different phenomena,

including human groups and cultures, buildings, urban neighbourhoods, texts and artistic products. This basic attention to difference quickly attracted the interest of geographers, reflecting the fact that 'the discipline has always ... displayed a sensitivity to the specific kinds of differences to be found between different (and 'unique') places, district, regions and countries' (Cloke et al., 1991, p.171). For Gregory (1996), post-modernism provides an opportunity for geographers to return to the notion of a real differentiation, emphasized by traditional regional geography, but armed with a new 'theoretical sensitivity' derived from work in cultural studies.

creating a 'new cultural geography'. Rather than the traditional view of culture as a possession of upper- and middle-class groups in society, the 'new cultural geography' has developed a broad and dynamic concept of culture. It is seen as a process through which individuals and social groups make sense of the world, often defining their identity against 'other' groups regarded as different according to categories such as nationality, race, gender and sexuality (Jackson, 1989). Meaning is generated through language which, instead of simply reflecting an underlying reality, actively creates that reality through **discourses** – networks of concepts, statements and practices that produce distinct bodies of knowledge (Barnes, 2000a). The cultural turn has been closely tied to the rise of **postmodernism** (Box 2.6).

2.5.2 Economy–culture links

While economic geography was, along with urban geography, regarded as the leading area of human geography during the 1960s, 1970s and 1980s when spatial analysis and Marxist political economy held sway, the 'new' cultural geography has come to be seen as the most exciting, 'cutting edge' area of the discipline since the late 1980s (Thrift, 2000). In response, some economic geographers have sought to adapt their interests, approaches and methods, incorporating notions of difference, identity and language into their research (Lee and Wills, 1997). The links between economy and culture are of central importance here, with many observers agreeing that the economy has become increasingly cultural in terms of the growing importance of sectors such as entertainment, retail and tourism, while culture has become increasingly

---

Box 2.7

Linking the economy and culture: the example of Christmas

The annual festival of Christmas offers an instructive example of the close links between the economy and culture, representing 'a cultural event of vast economic significance – or an economic event of vast cultural significance' (Thrift and Olds, 1996, pp.331–32). For most people, in developed countries particularly, Christmas is a time of leisure and consumption, associated with a holiday from work, usually spent with friends and family (Figure 2.7). As a cultural festival, Christmas draws together traditions from different countries: the Christmas tree from Germany, the practice of filling stockings from the Netherlands, the idea of Santa Claus or Father Christmas from the US and the Christmas card from Britain (Miller, 1993, cited in Thrift and Olds, 1996, p.332). As part of the global spread of a Western consumer culture, these practices are spreading to emerging economies such as China.

The social customs associated with Christmas support the practice of gift-buying. This provides the economy with a crucial stimulus in the final quarter of the retail year, with the Christmas shopping season typically accounting for between 26 to 40 per cent of a store's annual sales (*Moneyweek*, 2004). In 2008, the British population was predicted to spend a total of £82.3 billion on Christmas foods and gifts – an average of £1,363 per person (Rigby, 2008). Until the recent 'credit crunch', much of this was paid for by savings and debt, with around 15 million people getting further into debt to fuel their spending in the year 2003–04 (*Moneyweek*, 2004).

Such expenditure benefits not only retailers but also manufacturers, often located in areas distant from the main centres of global consumption. The Chinese city of Shenzhen, for example, where about 75 per cent of the world's toys are made, benefits hugely from Christmas, with 1,000 containers a day leaving for Southampton, Rotterdam and New York (ibid.).

Figure 2.7 Christmas consumption
Source: Franco Zecchin, Getty Images.

economic, viewed as a set of commodities that can be bought and sold in the market (Box 2.7).

This has led to talk of a **'new' economic geography** that examines the links between economic action and social and cultural practices in different places (Thrift and Olds, 1996). As Wills and Lee (1997, p.xvii) put it, 'the point is to contextualise rather than undermine the economic, by locating it within the cultural, social and political relations through which it takes on meaning and direction'. Four main strands of culturally informed research in economic geography can be identified (Barnes, 2003):

> Consumption, with studies focusing, for instance, on the creation and experience of particular landscapes of consumption such as shopping malls, supermarkets and heritage parks (see sections 3.4 and 4.4).

> Gender, performance and identity in the workplace and labour market. This work focuses on how employers perform particular roles and tasks at work, often informed by cultural and gendered norms. One of the most notable studies on this is Linda McDowell's research on work cultures in

merchant banks in the City of London (see Box 2.8).

> Research on the importance of personal contact and interpretative skills in financial and business services. This has focused on the cultural practices that underpin communication and interaction, the sites in which this takes place and the consequences for our understanding of financial markets. Research on financial centres such as the City of London, has shown the importance of social networks and trust, encouraging geographical concentration through the need for regular face-to-face contact (Thrift, 1994).

> Corporate cultures and identities. Research has focused on how managers and workers create distinctive corporate cultures through particular discourses and day-to-day practices. Of particular interest here is Schoenberger's (1997) work on large American corporations such as Xerox, DEC and Lockheed, showing the limitations of such cultures in dealing with a turbulent and unpredictable economic environment.

Box 2.8

Capital culture: gender and work in the City

Linda McDowell's research on the construction and performance of gender relations in the City of London represents one notable example of new economic geography research, linking economic issues to wider social and cultural relations and focusing on questions of discourse, identity and power. McDowell aims to understand how an international financial centre such as the City of London actually operates, 'viewing it through the lens of the lives and careers of individual men and women working in the City's merchant banks at the end of this period of radical change [the late 1980s and early 1990s] both in the global economy and in the city' (McDowell, 1997, p.4). In focusing on gender segregation and identity

at work, the research assesses men's experiences in addition to women's. At the same time, the extent to which the City has become more open to women and the career prospects it offers them remain key issues.

Capital Culture illustrates another key feature of the new economic geography by employing a range of research methods (see above), going beyond the conventional reliance on questionnaire and statistics to employ qualitative methods involving direct fieldwork. McDowell focused on the merchant bank sector, utilizing postal questionnaires sent out to all such banks in the City and detailed face-to-face interviews with employees of three of them.

This multi-method approach meant

that information on changing employment trends and relations within companies could be combined with first-hand accounts of how individual employees have managed and negotiated processes of employment change within the workplace. Through the case study interviews, in particular, the research examined people's everyday work experience, assessing how they assumed and performed particular gender roles. The importance of image, bodies and the presentation of self are major themes. A key conclusion is that changing work relations and power structures in the City still favour men as 'the cultural construction of the banking world remains elitist and masculinist' (McDowell, p.207).

2.5.3 Institutions, embeddedness and path dependence

Economic geographers have drawn upon concepts from institutional economics that emphasize the social context of economic life and the evolutionary nature of economic development. Institutions are important because they link 'the economic' and 'the social' through a set of habits, practices and routines (Hodgson, 1993). Beyond the specific types of organization identified in section 1.4.2, institutions are defined broadly as informal conventions and norms that shape and influence the behaviour of economic actors. For example, modern capitalist societies are characterized by a set of norms that emphasizes the importance of consumption in terms of personal fulfilment and identity (section 3.4). At the same time, of course, increased consumption is vital in enabling retailers and manufacturers to generate profits, fuelling the process of economic growth (Box 2.7). Amin and Thrift (2004) cite the example of the sports utility vehicle (SUV) which has become particularly prominent in North America, addressing middle-class families' desires for 'safety', 'security' and 'status' (Figure 2.8).

A related set of ideas from economic sociology has also been influential in stressing that economic processes are grounded or 'embedded' in social relations (Granovetter, 1985). In contrast with the orthodox economic conception of the economy as a separate domain driven by the rational decisions of individual actors, economic sociologists, following Polanyi, argue that the economy is socially constructed with social norms and institutions playing a key role in informing and shaping economic action (see Box 1.7). One key concept that has been incorporated into economic geography in recent years is **territorial 'embeddedness'**, emphasizing how particular forms of economic activity are rooted in particular places.

In a study of advanced manufacturing technologies in southern Ontario, Gertler (1995) shows the importance of 'being there' in facilitating adoption of the technologies by ensuring close links between producers (organizations that actually develop, distribute and sell aspects of the technology, for example robots) and users (manufacturing firms). A shared 'embeddedness' in a distinctive industrial culture enabled users and producers to develop appropriate training regimes and industrial practices, involving the sharing of information and knowledge. This does not mean, however, that

Figure 2.8 The sports utility vehicle (SUV)
Source: © David Cooper/Toronto Star/Corbis.

industrial cultures are always defined geographically or that learning and interaction cannot occur over longer distances, utilizing information and communication technologies. An awareness of the importance of this issue has fostered an interest in the development and organization of knowledge within large firms, raising questions about how such global and local knowledges are combined (see section 11.2).

Institutionalist ideas have encouraged the rise of a **'new regionalism'** in economic geography that examines the effects of social and cultural conditions within regions in helping to promote or hinder economic growth (Storper, 1997). In particular, inherited institutional frameworks and routines are held to be of considerable importance in influencing how particular regions respond to the challenges of globalization (Amin, 1999). Individual places have, as such, attracted renewed attention, as economic geographers have attempted to identify the social and cultural foundations of economic growth and prosperity in successful regions such as 'Silicon Valley' in California, and 'Motor Sport Valley' in Cambridge, England. In contrast with the political economy approaches of the 1980s, institutionalist perspectives emphasize the importance of internal conditions within regions in shaping their experience of economic development as opposed to external processes, and treat localities and regions as active participants in economic development rather than as passive arenas that are exploited by capital (for example, large MNCs).

The idea that the process of economic development is **'path dependent'** is another key institutionalist idea adopted by economic geographers. This means that the ways in which economic actors respond to wider processes of economic change are shaped and informed by past decisions and experiences:

> One of the most exciting ideas in contemporary economic geography is that industrial history is literally embodied in the present. That is, choices made in the past – technologies embodied in machinery and product design, firm assets gained as patents or specific competencies, or labour skills acquired through learning – influence subsequent choices of methods, design and practices.
>
> (Walker, 2000, p.126)

The past is expressed and sustained through technology, machinery and equipment, and organizations as well as a broader set of attitudes and habits that informs current practice. Regional culture is a product of the past in this sense.

The process of 'path dependence' can be illustrated with reference to 'old' industrial regions such as the Ruhr Valley in Germany and north-east England, in which cultural factors have been closely associated with decline. The industrial cultures of these areas seemed to have become rigid and fossilized, meaning that firms and institutions were tied to obsolete production systems and methods, militating against more positive responses to economic change in terms of generating new products and methods. These declining areas can be seen as:

> … the hard-luck cases: once-successful places where local cultures fostered ties so strong, structures so rigid and attitudes so unbending that newcomers and new ways of doing things encountered insurmountable barriers to entry.
>
> (Gertler, 2003, p.134)

In this sense, declining regions tend to become 'locked in' to out-moded practices and habits, preventing them from acquiring new knowledge in an economically effective manner. In some cases, 'old' industrial regions have attempted to address these deficiencies by trying to transform their cultures through initiatives that encourage learning and collaboration between firms. Parts of the American Midwest, South Wales and the Basque Country are examples of such 'reclamation projects' (ibid.).

2.5.4 Relational approaches

In recent years, economic geography has been influenced by the rise of relational thinking in geography. This builds on Doreen Massey's 'global sense of place', which defines place as constructed out of the coming together of wider social relations (section 1.2.3). According to Massey (2005, pp.9–12), the **relational approach** to space is grounded in three basic propositions. First, space should be seen as the product of interrelations, meaning that places and their identities are created through such relations, rather than pre-existing in an *a priori* fashion. Second, space emphasizes the possibility of multiplicity, created out of the different relationships that link places together, resonating with the postmodernist concern for difference. Third, space is always changing rather than being static or fixed. Relational economic geography focuses on the relationships between economic actors and broader processes of change, informed by the institutional and cultural approaches discussed above (Bathelt and Gluckler, 2003; Yeung, 2005). It is concerned with economic action and interaction which it views as being embedded in wider social and economic relations (Bathelt and Gluckler, 2003). From this perspective, globalization marks a new era of space/place relations, focusing attention on economic networks and wider circuits of knowledge generation and exchange rather than bounded regions (Amin, 2002).

This conception of space has fostered a new concern with '**the relational region**' (Allen *et al.*, 1998) which re-imagines regions as open and discontinuous spaces, defined by the wider social relations in which they are situated (Box 2.9). They are created for particular purposes by, for instance, policy-makers, social movements or academic analysts, meaning that they have no essential character or identity outside of these acts of creation and becoming. Instead of being internally coherent and unified, regions are discontinuous and divided (ibid.). John Allen, Massey and Allan Cochrane, based at the Open University, illustrate these arguments by an analysis of south-east England, the emblematic growth region of neoliberal Britain in the 1980s and 1990s, defined by its close connections to the global economy through the City of London in particular. Within the south-east, areas of economic decline and deprivation

Box 2.9

Trans-national communities and regional development

Anna-Lee Saxenian's work on trans-national communities of engineers and entrepreneurs provides a useful illustration of recent relational approaches to regional development. Saxenian is renowned for her previous work on Silicon Valley in California (Box 4.4), and she has maintained this focus in her recent research, while modifying her approach to take account of the trans-national links and flows.

Saxenian (2006) argues that the past couple of decades have seen the rise of what she calls the 'new Argonauts' in the form of engineers and entrepreneurs who operate internationally, often having moved overseas for education or employment and having gone on to launch new firms in the countries they moved to. The Argonaut metaphor is based on the Greeks who accompanied Jason overseas to find the Golden Fleece in Greek mythology, with the common denominator being that of travelling to seek their fortunes in distant lands.

The main example of such Argonauts discussed by Saxenian are first-generation immigrants to the US from Asian countries such as India, China and Taiwan who gained technical skills in engineering and became entrepreneurs in the US. Increasingly, these individuals form communities of technically skilled immigrants with work experience and connections to Silicon Valley and related American technology centres.

In the 1960s and 1970s, the relationship between the home countries of these Asian engineers and the US as the receiving country conformed to the classic pattern of migration between the rich and poor worlds. From the late 1980s, however, Saxenian argues, a new trend emerged, as US-educated en-gineers began to return to their home countries as their economies started to grow and as the development of ICTs created new opportunities for firms and individuals to operate internationally. Initially, this reverse flow was

confined to Taiwan – and other more developed countries such as Israel, which has traditionally exported engineers to the US – before spreading to India and China from the late 1990s. This trans-national movement of entrepreneurs and engineers in the IT sector underpins Saxenian's key argument that the traditional pattern of 'brain drain', for the originating countries, is being replaced by a mutually beneficial one of 'brain circulation'. As such, the 'new Argonauts' are replacing the old pattern of one-way flows of skilled people from the periphery to the core with new, two-way linkages, fuelling growth and employment creation in the (former) periphery through the infusion of skills, capital and technology incubated in the core. It remains unclear, however, as to how widespread such Argonaut-based 'brain circulation' is compared with traditional 'brain drain' beyond the specific countries she discusses.

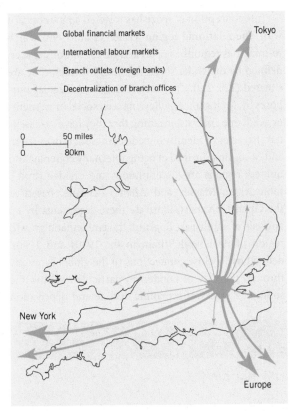

Figure 2.9 International financial links of south-east England
Source: Allen *et al.*, 1998, p.49.

exist, complicating and confounding the overarching image of growth and prosperity, while the boundaries of the region can be seen as open and porous (Figure 2.9), moving both 'outwards' and 'upwards' in the 1980s and 1990s through the westward movement of wealthy groups for residence and leisure, and its enhanced status as a strategic node within the world economy (Amin and Thrift, 1992). This relational theory of the region has been accompanied by a critique of 'new regionalist' thinking, rejecting its portrayal of regions as internally coherent and externally bounded (Allen *et al.*, 1998; Amin, 2004).

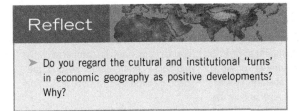

Reflect

➤ Do you regard the cultural and institutional 'turns' in economic geography as positive developments? Why?

2.6 Summary

Prevailing approaches to economic geography have changed considerably over time, as we have demonstrated, mirroring the development of geography more broadly. The traditional framework based on regional classification and description gave way to a more quantitative approach in the late 1950s and early 1960s, based on the development of concepts and techniques of spatial analysis. This was, in turn, replaced by Marxist political economy in the 1970s and 1980s before the 'cultural turn' led to the introduction of new cultural and institutional perspectives in the 1990s. These periodic shifts in intellectual orientation should be seen as changes in broad focus and direction rather than as wholesale transformations, with research in the spatial analysis and political economy traditions remaining important.

Our favoured approach in this book can be described as 'new' political economy, signalling that it has moved beyond the rather clunky and deterministic theories of the 1970s and early 1980s to become more flexible and open to the importance of context, difference and identity, partly as a result of encountering cultural and institutional approaches. As such, our position brings together concepts from the two traditions, but from a standpoint that is anchored in political economy. It represents, as such, a 'culturally sensitive political economy' rather than a 'politically sensitive cultural economy' (Hudson, 2005, p.15). Our approach can be described as broadly regulationist in nature, stressing the important role that wider processes of social regulation play in stabilizing and sustaining capitalism.

From a base in political economy, we examine the uneven development of capitalism, 'the individual economics of different places, and their connections one to another' (Barnes and Sheppard, 2000, pp.2–3) in this book. In adopting a broad, synthetic conception of the economy, we recognize the importance of the link between the economy and culture, and assess some connections between the two in specific geographical settings. Our approach also takes difference and variety seriously, emphasizing the distinctive character of individual places, but insisting that we need to consider how geographical difference is produced and reproduced through wider processes of economic development and state intervention.

Exercise

A major supermarket chain is proposing to open a new out-of-town superstore in the town or city where you live.

How would each of the three latter perspectives covered in this chapter – spatial analysis, political economy and the cultural/institutional approach – examine and understand this issue? What aspects of the development would each approach focus attention on; for example, people's experience of shopping, the broader policies and practices of the corporation in question, finding the optimum location, the links between consumption and identity, the impact on local shops, analysing the characteristics of the local market, assessing why customers shop in major superstores, competition among supermarket chains and relationships with suppliers?

On this basis, assess the strengths and weaknesses of each approach. Which, if any, offers the 'best' understanding of the issue? Why? Is it appropriate to try and bring elements of the different approaches together? If so, which ones?

Key reading

Barnes, T.J. (2000) 'Inventing Anglo-American economic geography', in Sheppard, E. and Barnes, T.J. (eds) *A Companion to Economic Geography*, Oxford: Blackwell, pp. 11–26.
An engaging account of the formation and early development of economic geography. Stresses how academic subjects are 'invented as specific projects at particular ties and in particular places by groups of people. Covers the traditional approach and spatial analysis.

Lee, R. (2008) 'Economic geography', in Johnston, R.J., Gregory, D., Pratt, G. and Watts, M. (eds) *The Dictionary of Human Geography*, 5th edn, Oxford: Blackwell, pp.195–8.
Defines economic geography as 'the geography of people's struggles to make a living' (p.261). Outlines how the subject has evolved over time and highlights a range of examples of contemporary economic geography research.

Scott, A.J. (2000) 'Economic geography: the great half century', in Clark, G., Feldmann, M. and Gertler, M. (eds) *The Oxford Handbook of Economic Geography*, Oxford: Oxford University Press, pp.18–44.
An upbeat review of the development of economic geography in the post-war period. Focuses particularly on the spatial analysis and political economy approaches, highlighting specific research topics such as localities and the rediscovery of regions since the 1980s.

Swyngedouw, E. (2000) 'The Marxian alternative: historical–geographical materialism and the political economy of capitalism', in Sheppard, E. and Barnes, T.J. (eds) *A Companion to Economic Geography*, Oxford: Blackwell, pp.40–59.
A compelling account of Marxist political economy from one of its leading exponents. Identifies key aspects of the Marxist approach and its development within geography. Emphasizes its commitment to social and political change and relevance in addressing big questions associated with globalization, uneven development and environmental degradation.

Thrift, N. (2000) 'Pandora's box? Cultural geographies of economies', in Clark, G., Feldmann, M. and Gertler, M. (eds) *The Oxford Handbook of Economic Geography*. Oxford: Oxford University Press, pp.689–704.
A review and summary of the cultural 'turn' in economic geography. Emphasizes how it has broadened and enlivened the field, opening up an array of new questions. Identifies the main areas of research and speculates on future developments.

Useful websites

http://www.egrg.org.uk/index.html
The website of the Economic Geography Research Group of the Royal Geographical Society (with the Institute of the British Geographers). This site is mainly used by academic researchers in the field, so much of the material is likely to be difficult. It is worth exploring the site, however, to get a feel for the types of issues and topics that economic geographers conduct research on.

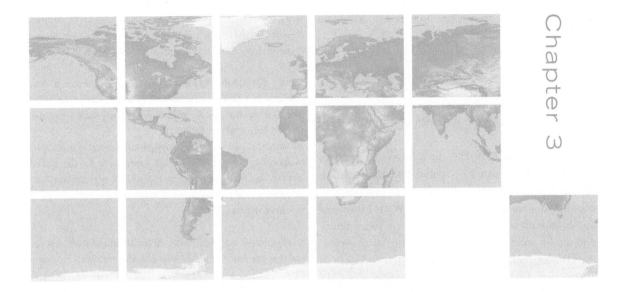

Chapter 3

Shaping the capitalist economy: key actors and processes

Chapter map

The development of the capitalist economy is shaped by the actions of groups of people such as capitalists, workers, consumers and state officials. In this chapter, we consider the role of these key sets of actors and assess the main historical processes that have structured their actions. Following a brief introduction, section 3.2 examines the role of capital in the economy, focusing on the circuit of capital, the nature of the firm, the organization of finance and investment and processes of innovation and technological change. We then turn to labour, considering the position of labour in capitalism and assessing the division of labour within production. Section 3.4 is concerned with consumers and the process of consumption, focusing attention on the convergence between economy and culture (section 2.5). The role of the state in the economy is briefly examined in the penultimate part of the chapter, highlighting its importance in regulating and coordinating economic activities.

3.1 Introduction

Our approach in this book emphasizes the social relations structuring the economy, rejecting the individualism of mainstream economics. While we deal with each of the main groups of economic actors – capital, labour, consumers and the state – in turn within this chapter, it is the (social) relations between these groups that are crucial. Foremost among these is the relationship between capital and labour within the production process. The decisions of consumers about which goods and services to purchase also underpins the economic process. Increased consumption is crucial in allowing the economy to grow and expand. The state plays an important role in regulating the economy, mediating between the interests of capital and labour in an effort to create stable conditions that facilitate economic growth. We believe that the position of each of these groups within capitalism gives them a particular set of material interests that shapes their actions. At a general level, we can say that capital seeks to generate profits; labour strives to maximize its wages and living conditions; consumers seek to further their interests and aspirations by purchasing goods and services; and the state aims to promote economic growth within its territory.

While people are not the wholly rational and self-interested actors that underpin conventional economic theory, they are knowledgeable individuals who have their own interests and aspirations (Hudson, 2005, p.3). The fact that capitalism operates through the actions of individuals whose knowledge of alternatives and outcomes is always imperfect, introduces an important element of uncertainty and openness into the economic process. At the same time, individual actors operate within the broader structures of the capitalist economy, meaning that they are confronted by pressures such as the need to earn a wage or generate profits. The pressing need for people to earn an income to support themselves and their families has led one commentator to describe economic geography as 'the study of people's struggle to make a living' (Lee, 2000, p.195). The specific ways in which people respond to wider economic imperatives will vary according to a wide range of factors, including their economic status, gender, family situation, age and cultural values. These responses can be expected to vary spatially, given the variety of local and regional environments that exists within the contemporary world economy, reflecting the interaction between capitalism and a range of pre-existing societies and cultures (Johnston, 1984).

3.2 Capital

Capital is a complex term, used in different ways by different people. For our purposes here, however, it can be simply defined as money that is invested in production or financial markets. Capitalists are those people who have acquired capital, allowing them to own the means of production: land, materials, factories, offices and machines. The pressures of competition mean that the capitalist has to seek to expand his stock of capital by reinvesting it in production to generate higher profits. This reinvestment in order to generate more profits – which are, in turn, reinvested – is often referred to as the process of **capital accumulation** (Barnes, 1997). It lies at the heart of the capitalist system with profit-seeking representing the basic driving force for economic growth and expansion.

3.2.1 The circuit of capital

The basic economic process under capitalism can be understood in terms of the **circuit of capital** through which profits are generated (Harvey, 1982). The first stage is that capital in its money form (M) is transformed into commodity form (C) by purchasing the means of production (MP) – factories, machines, materials, etc. – and labour power (LP) (Figure 3.1). The means of production (MP) and labour power (LP) are then combined in the production process, under the supervision of the owners of capital or their managers and representatives, to produce a commodity for sale (C*) – for example, a car, house or, even, a haircut. This commodity is sold for the initial money outlay plus a profit Δ, representing what some economists call surplus value.

Part of the money (M) realized from the sale of the commodity is reinvested back into the production process, which recreates the circuit anew. Following each circuit of capital there is an expansion in the total

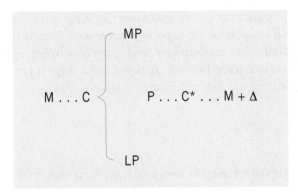

$$M \ldots C \left\{ \begin{array}{c} MP \\ \\ \\ LP \end{array} \right. \quad P \ldots C^* \ldots M + \Delta$$

Figure 3.1 The process of production under capitalism
Source: Castree *et al.*, 2004, p.28.

amount of capital, forming the basis of capital accumulation. Clearly, the distribution of income at the end of each circuit becomes an important political question and governments play a regulatory role in diverting capital into welfare spending and other state functions (investment in transport infrastructure, housing, etc.) through taxation.

3.2.2 The firm as grounded capital

The main organizational form that capital takes is the **firm**. Firms are legal entities owned by individual capitalists or, more commonly now, a range of shareholders. Along with the commodity, the firm could be described as a basic 'economic cell form' of capitalism (see section 1.3.1). Our approach in this book is informed by the **competence- or resource-based theory** of the firm. Derived from the work of the economist Edith Penrose (1959), the competence perspective views firms as bundles of assets and competencies that have been built up over time. A firm becomes distinct from its competitors through the specific way in which it combines the resources of land, labour and capital in the production of certain commodities. Competencies can be seen as particular sets of skills, practices and forms of knowledge. The need for firms to focus on identifying and developing their 'core competencies' is a key theme of recent management theories. For example, the core competencies that have enabled Dell to acquire such a strong position in the personal computer market include the online 'bespoking' of each

computer according to customer requirements and the high quality of its manufacturing and distribution, offering reliable products at competitive prices (Fields, 2006).

The knowledge and skills developed by a firm become embedded within its organizational rules and routines (Taylor and Asheim, 2001, p.323). Firms acquire and develop these knowledge and skills through processes of learning associated with the repetition of particular tasks. As such, the competence-based approach draws attention to the dynamic processes of learning and knowledge creation that occur within firms, providing much of the basis for understanding firm strategy (Nelson and Winter, 1982). It is well suited to economic geography, indicating that firms can derive some of their competitiveness from factors present in the broader regional environment (Maskell, 2001, p.339), combining them with internal resources to create distinctive competencies.

In addition to this abstract theory of what firms are and how they operate, it is important to appreciate the diversity of firms in terms of size, ownership and structure. In practice, the nature of **business organization** is very dynamic, taking new forms over time. The heroic small firms and entrepreneurs of the nineteenth century have given rise to large trans-national corporations or what some Marxist economists have termed monopoly capital (Baran and Sweezy, 1966). While mainstream economics had assumed the existence of a world of small firms with no power to influence prices in the marketplace, the emergence of these large multi-plant corporations required more sophisticated analyses (Galbraith, 1967). This work looked at the internal organization of business, emphasizing how the growth of the large joint stock company had resulted in a separation of ownership (shareholders) from control (managers) and the emergence of vast internal labour markets through which employment is organized (Doeringer and Piore, 1971). In the large public limited firm or joint stock company, shares are typically distributed between a diverse range of interests that represent different sections of society (see Table 3.1). At the same time, small firms (of less than 100 employees) continue to exist and employ more people than large corporations, even in advanced national economies, as suppliers and competitors to larger firms.

Table 3.1 Ownership structure of share capital in UK's privatized utilities, 1993

Investor	% of share in privatized utilities
Central government	9.2
Pension funds	27.6
Insurance companies	12.4
Unit trusts	4.5
Individuals	24.5
Overseas	8.3
Others	12.1

Source: Sawyer and O'Donnell, 1999

3.2.3 Finance and investment

Two key functions of money within the capitalist economy can be identified. First, it acts as a measure and store of value, with the value of particular commodities expressed in term of their price (section 9.2). Similarly, the wealth of individuals is expressed in terms of the monetary value of their income and assets, while share prices measure the fluctuating fortunes of large firms. Second, money is also the medium of exchange through which commodities are bought and sold, enabling the seller to buy other commodities with the money received. Money is advanced by banks and other organizations to firms and individuals as credit – effectively loans that must be paid back later – allowing them to invest in production or buy goods that they would not otherwise have been able to at that particular time (effectively bringing investment and consumption forward in time).

In order to grow and expand, firms need capital for investment. Some of this capital can be generated internally through the profits made from selling commodities in the marketplace, but many firms require access to external sources of finance such as loans from banks, the financial markets and venture capital. For large public corporations, the stock market provides an additional source of funds through the issuing of shares that are purchased by individuals and institutional investors, referred to as **equity finance**. In recent decades, the pension funds have become increasingly

important as investors in the stock market, as indicated in Table 3.1 (Clark, 1999). **Venture capital** is equity finance that investors provide to firms not quoted on the stock exchange. Such firms are generally small and have a high growth potential, attracting venture capitalists who aim to make high returns by selling their stake in the firm at a later date. Such investment is high risk, with investors typically expecting returns of over 30 per cent (Tickell, 2005, p.249). The investors can be either wealthy individuals, often former entrepreneurs, known as 'business angels' or institutions in the form of professional investment companies that raise funds from financial institutions such as banks (Mason and Harrison, 1999). While the venture capital industry is relatively small compared with other financial markets, it is of considerable economic significance because of its importance in funding high-growth firms that generate employment and wealth.

Finance has tended to become more **centralized** and integrated over time, with local and regional banks giving way to national and, increasingly, global systems (Dow, 1999). The UK, for example, has a highly centralized financial system, with the London region accounting for 45 per cent of the gross value-added generated from financial intermediation in 2008 (Office for National Statistics, 2010). 'Classic' venture capital, which concentrates on funding new businesses, is highly concentrated in south-east England and, to a lesser extent, Scotland (Mason and Harrison, 2002; Mason, 2007). In the US, more than half of listed venture capital offices were located in only three metropolitan areas: San Francisco, Boston and New York (Chen *et al*., 2009). Regions that are deficient in 'classic' venture capital, including most UK regions outside south-east England, lack access to funds for supporting dynamic new businesses, leading to low levels of entrepreneurship and innovation. Furthermore, given that venture capitalists raise funds from financial institutions such as pension funds and banks, savings are drained from such regions and invested in prosperous locations such as south-east England (Mason and Harrison, 2002).

Capital is the most mobile of the factors of production, in contrast with labour and land, which are relatively place-bound. As we have seen, it tends to become concentrated in regions that offer a high rate of return or profit on money invested, leaving underdeveloped countries and regions bereft of the capital required to invest in production facilities and development projects. At the same time, investment is not contained indefinitely within core regions, with firms and investors also responding to investment opportunities in other regions. Over time, there is a 'see-saw' effect as capital flows back and forth between different sectors and regions (Smith, 1984). Financial deregulation in most of the major economies seems to have exacerbated this situation in recent decades, unleashing a far more rapid and volatile phase of

accumulation (Harvey, 1989a), and culminating in the 'credit crunch' and global recession of recent years. **Capital switching** has important geographical dimensions, often being transferred from regions dominated by declining sectors to 'new industrial spaces' in distant regions offering more attractive conditions for investment. An example of the geographical effects of capital switching and uneven development is provided by the experience of the north-east of England over the twentieth century (Box 3.1).

Box 3.1

Capital switching and its geographical effects: the case of the north-east of England

As the first country to industrialize, the UK enjoyed an advantage over the rest of the world in the production of heavy engineering and manufacturing goods throughout the nineteenth century. The north-east of England became a heartland region of industrial capitalism during the second half of the nineteenth century on account of its iron ore and coal reserves (Figure 3.2). It became particularly dominant in the growing shipbuilding market – producing around 40 per cent of the world's ships at one point – but also became an important centre for rail and bridge engineering, and arms production. This growth was associated with the emergence of a local bourgeoisie, the 'coal combines' (capitalist class), a group of families that had originally made their fortunes through the coal trade but had diversified into the new industries to take advantage of the investment opportunities.

With the growth of competition from other countries – especially the US and Germany – and a worldwide

downturn in the 1920s, the region's industries faced an economic crisis reflected in declining markets and rising unemployment. At the peak of the Depression in 1933, over 80 per cent of the region's shipbuilding and repair workforce were unemployed. Faced with this situation, the region's capitalists had two alternatives: reinvesting in the traditional industries, introducing new production methods and technologies, or finding alternative avenues for investment. For the most part they chose the latter, switching capital from what were increasingly perceived as declining industries to new growth sectors such as car and aircraft production, the public utilities and financial services. The nationalization of the coal industry in the 1940s provided further opportunities for the region's capitalists when 'the coal owners' capital, locked in the fixed assets of mines and machinery, became suddenly transformed under the compensation terms into highly liquid government bonds' (Benwell Community Project, 1978, p.58).

Diversification out of the heavy industrial sector, was accompanied by a geographical expansion of the 'coal combines' interests both at the national and international level. Through the establishment of special investment trusts in the inter-war years, local capitalists were able to exploit the growing market in trading and financial transactions to the extent that by 1930, one of the largest of these trusts, the 'Tyneside', had only 23 per cent of its investment in north-eastern industry. Despite efforts to modernise the region by both state-led restructuring and the attempt to encourage new inward investment, the legacy of this withdrawal of capital from the north-east has been the region's transition from a core region of capitalism in the nineteenth and early twentieth centuries to an increasingly peripheral position from the 1930s. This is manifest in lower levels of economic development and higher levels of social deprivation than the national average.

Box 3.1 (continued)

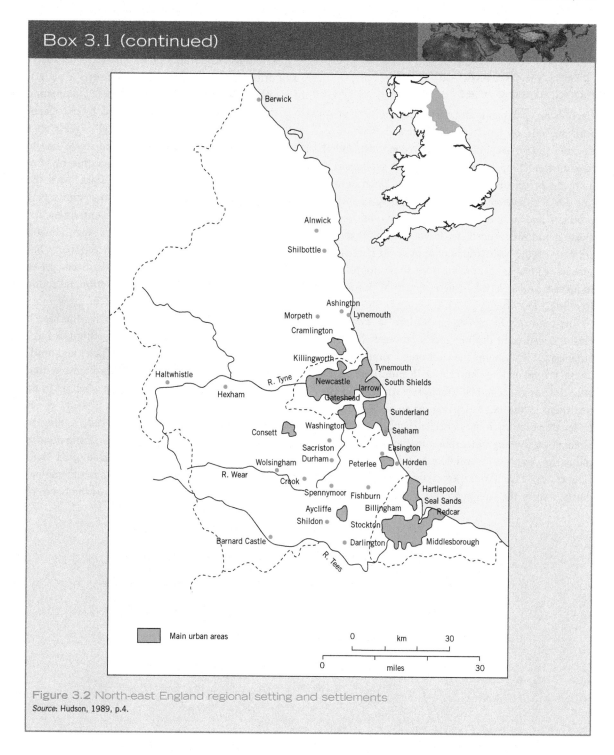

Figure 3.2 North-east England regional setting and settlements
Source: Hudson, 1989, p.4.

3.2.4 Innovation and technological change

'Creative destruction' and Kondratiev cycles

Capitalism is a dynamic and unstable economic system, with periods of rapid growth followed by ones of decline and crisis. The Austrian economist Joseph Schumpeter (1943) famously described capitalism as a process of '**creative destruction**' based upon innovation and the development of new technologies. As new products and technologies emerge, they often render existing industries obsolete, leaving them unable to compete on the basis of quality or price. As part of this process, capital is withdrawn from these unprofitable industries and invested in new centres of production (see Box 3.1).

The tendency for major innovations to 'swarm' or cluster together in distinct cycles or waves has been emphasized by Schumpeter and other commentators. These are sometimes known as '**Kondratiev cycles**' after the Soviet economist Kondratiev who first identified them in the 1920s. Lasting for some 50–60 years in length, each cycle is associated with a distinctive system of technology, incorporating the key propulsive industries, transport technologies and energy sources (Figure 3.3). Five Kondratiev cycles are usually distinguished since the late eighteenth century. Each cycle

consists of two distinct phases: one of growth (A) and one of stagnation (B) (Taylor and Flint, 2000, p.14).

The notion of Kondratiev cycles is based on the analysis of price trends, which show a characteristic pattern of steady increases for about 20 years, culminating in a rapid inflationary spiral, followed by a collapse with prices reaching a trough some 50–5 years after the start of the cycle. This has sparked much debate about the specific mechanisms behind this pattern. At a basic level, though, each cycle starts with the bunching together of key inventions that create new economic opportunities for firms and entrepreneurs. At this stage, technology is expensive and demand high, resulting in rising prices; after the technology has matured, becoming routine and standardized, prices fall. An inherent problem is **overproduction**, reflecting a tendency for the volume of output to grow more rapidly than market demand. Driven by the search for profits and the pressures of competition, a large number of firms invests in new technologies and products during periods of growth. Accordingly, output increases rapidly until it reaches a point when it can no longer be absorbed by the market. As a result, prices drop and profits and wages are reduced, leading to bankruptcies and unemployment. Even during periods of growth, new technologies make established skills obsolete, leading to marginalization and unemployment for groups of workers (Box 3.2). In the boom of

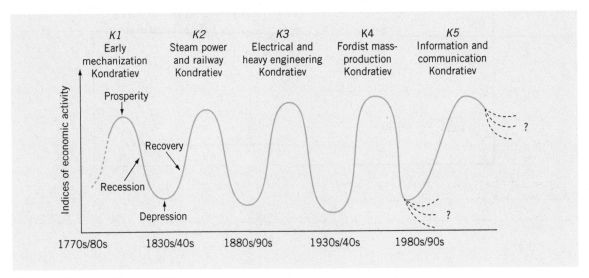

Figure 3.3 Kondratiev cycles
Source: Dicken, 2003a, p.88.

the late 1990s and early 2000s, the tendency towards overproduction spread from manufacturing to financial services, as banks and other financial institutions invested in complex new products and instruments, borrowing heavily on the markets to do so (Blackburn, 2008). High levels of debt and the pattern of complex inter-bank trading ultimately led to a crisis of confidence in credit ('the credit crunch'), prompted by defaults on 'sub-prime' loans in the US.

Successive waves of industrialization

Rather than representing a single abrupt change – the industrial revolution – industrialization should be seen as an uncertain and uneven process, occurring in successive waves of transformation (Figure 3.3). In the narrow sense, industrialization refers to a series of changes in manufacturing technology and organization that radically transformed the economic landscape of capitalism. It can be defined, in narrow terms, as the 'application of power-driven machinery to manufacturing' (Gremple, undated, p.1). While technological innovations provided a base for large rises in productivity and output, it was the development of the factory system that enabled this possibility to be exploited to the full (Knox et al., 2003, p.244). The factory was at the core of industrialization, replacing the existing system of domestic labour and independent craftsmen, representing perhaps its most enduring legacy. Basically, it allowed capitalists to gather and order a large workforce under one roof, enabling them to control the labour process and exploit the new technologies and emerging division of labour to the full (section 3.3.2).

The first Kondratiev cycle involved early mechanization based on water power and steam engines from the 1770s. This was focused on cotton textiles, coal and iron working, facilitated by the development of river systems, canals and turnpike roads for transporting raw materials and finished products. The cotton textiles industry was the main focus of the industrial revolution, acting as the 'pacemaker' of change and providing the basis for the first industrial regions (Hobsbawm, 1999, p.34). The industrial revolution greatly increased output while reducing production costs massively. This occurred through the creation of **economies of scale**, referring to the tendency for firms' costs for each unit of output to fall when production is carried out on a large scale, reflecting greater efficiency. By 1812, for example, the cost of making cotton yarn had dropped by nine-tenths since 1760 and by 1800 the number of workers needed to turn wool into yarn had been reduced by four-fifths (Grempel, undated, p.2). By the 1830s and 1840s, however, markets were not growing quickly enough to absorb output, leading to reduced profits, falling wages and unemployment (Box 3.2) (Hobsbawm, 1999, pp.54–5).

Box 3.2

The decline of the handloom weavers

The changing fortunes of the handloom weavers of England in the late eighteenth and early nineteenth centuries illustrate how the processes of 'creative destruction' associated with industrialization actually operate, particularly in terms of the impact of new technologies on labour. The cotton textiles industry – based in Lancashire in northern England (Figure 3.4) – in particular was transformed by a sequence of innovations in the late eighteenth century. These included Arkwright's water frame (1771), Hargreave's spinning jenny (1778) and Compton's mule (1779), which combined the jenny and the water frame. The effect was to revolutionize the spinning process, allowing yarn to be produced in large quantities, impacting upon weaving which had been speeded up by the flying shuttle (1733).

Weaving expanded hugely in this period, absorbing the growing quantity of yarn through the multiplication of handloom weavers. In retrospect, the years from 1760 to 1810 appear as something of a 'golden age' for handloom weaving with the number of handloom weavers in cotton rising from about 30,000 to over 200,000 (Thompson, 1963, p.327). Demand for cloth was high and wages rose steadily, attracting many new entrants to the trade. Yet the general climate of prosperity disguised a loss of status as the weavers' ancient craft protections were eroded,

Box 3.2 (continued)

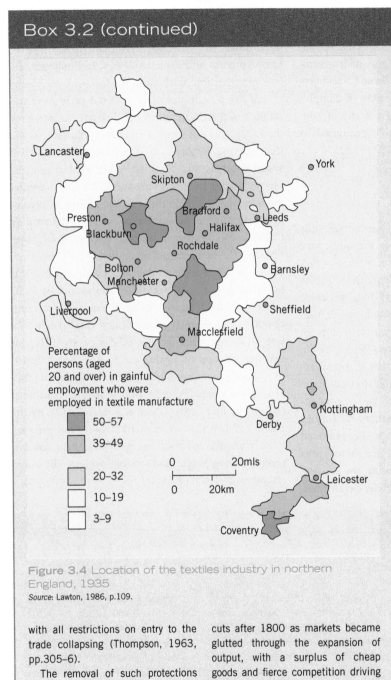

Figure 3.4 Location of the textiles industry in northern England, 1935
Source: Lawton, 1986, p.109.

Percentage of persons (aged 20 and over) in gainful employment who were employed in textile manufacture

- 50–57
- 39–49
- 20–32
- 10–19
- 3–9

pre-dated the spread of power-looms in the cotton textile industry, but this became a key force which rendered handloom weaving uncompetitive. As a result, the average weekly wage of the handloom weaver in Bolton fell from 33s. in 1795 to 14s in 1815 to 5s.6d in 1829-34 (Hobsbawm, 1962, p.57). Production moved from the cottage to the factory as the way of life and culture of the handloom weavers were undermined.

Widespread distress and desperation saw the weavers resort to expressions of impotent 'rage against the machine' by smashing the new power-looms and factories. Alongside such Luddite acts, strikes were organized and petitions to the government launched. The doctrines of political economy that held sway at the time, however, prohibited any interference with the 'natural' operation of market forces, and the government responded with repression.

By the 1820s, the great majority of weavers were living on the borders of starvation (Thompson, 1963, p.316). In the town of Blackburn, at the heart of Lancashire's textile industry, over 75 per cent of handloom weaves were unemployed in the early 1820s. One week in April 1826, 76 people entered the Blackburn Workhouse, bringing the total 'crammed together' to 678 (Turner, undated). The labour of the handloom weavers had been rendered obsolete by the industrial revolution, demonstrating how technological innovation could displace skilled labour from established craft industries.

with all restrictions on entry to the trade collapsing (Thompson, 1963, pp.305–6).

The removal of such protections exposed the weavers to savage wage cuts after 1800 as markets became glutted through the expansion of output, with a surplus of cheap goods and fierce competition driving wages down. The decline actually

A second cycle of industrialization began in the 1840s, lasting until the 1890s, just as the first textile-based phase had reached its limits. The cycle was based on the industries of coal, iron and steel, heavy engineering and shipbuilding, creating a much firmer foundation for economic growth. Advances in transportation were of central importance here, particularly the development of the railways. Again, however, a

Table 3.2 The growth of the service sector

	% employed in services				
	1976	1986	1996	2004	2008
France	52.2	61.4	69.4	75.3	76.5
Germany	48.7	54.7	61.6	66.6	67.8
Japan	52	57.1	61.2	67.1	68.5
UK	57.5	63.1	70.7	76.4	77.2
US	65.3	68.8	73.3	78.4	79.5

Sources: OECD, 2005b, 2009.

phase of expansion from the 1840s to the early 1870s gave way to depression as markets became saturated and prices fell.

The development of the internal combustion engine and electricity was central to the third wave, lasting from the 1880s to the 1920s, associated with automobiles, oil, heavy chemicals and plastics. Britain lost its lead to Germany and the US who pioneered the development of the new technologies. The characteristic pattern of a growth phase being succeeded by stagnation was again apparent with the boom years from the mid-1890s to 1914 matched by the inter-war depression.

The fourth cycle occurred between the 1930s and 1970s based upon key industries such as electronics, automobiles, aerospace engineering and petrochemicals. This is the period of Fordism, involving the mass production of consumer goods. It also saw the spread of road transport and aircraft. A fifth cycle seems to have begun in the late 1970s, centred upon the 'knowledge economy' sectors of information technology, telecommunications and biotechnology (Hall, 1985). The spread of computer networks has generated a revolution in communications, while the containerization of ship cargos and the development of jet aircraft have transformed the long-distance transport of freight and people. The service sector has become the dominant source of employment in developed economies, accompanied by a process of **deindustrialization** as manufacturing industry declines in importance (Table 3.2).

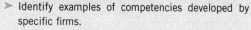

Reflect

➤ Identify examples of competencies developed by specific firms.
➤ In what ways does the mobility of capital between regions contribute to the process of uneven development?

3.3 The labour process

3.3.1 The position of labour under capitalism

As we have emphasized, the position of labour is a central defining feature of the capitalist mode of production. Unlike previous and more primitive societies where the great mass of the population, as slaves or peasants, was in thrall and they were effectively owned by their masters, under capitalism workers are released from feudal ties and are free to sell their own labour. Indeed, labour has to sell its own 'labour power' to earn wages to pay for the essentials of life (e.g. food, clothing or shelter) because it does not own the means of production and therefore the means for its own sustenance. Capital certainly needs labour in order to produce things to sell, but labour's needs are more urgent; it has to engage in production to secure a wage to sustain itself. In this sense, there is a fundamental imbalance at the heart of capitalist social relations. While labour can withdraw its labour power (i.e. strike) as a sanction against capital, its immediate needs are greater than those of capital – which can afford an empty factory/office for a certain amount of time, although not of course for a sustained period.

Labour can be regarded as a '**fictitious commodity**', which takes the appearance of a commodity, but cannot be reduced to one (Polanyi, 1944). In the sense that it is purchased by capital, and combined with the means of production to produce commodities for sale, labour functions as a commodity. As such, there is a labour market where labour is purchased by capital, and sold by workers, for a certain price (wage). Yet, as Storper and Walker argue, echoing Marx:

> Labour differs fundamentally from real commodities because it is embodied in living, conscious human beings and because human activity (work) is an irreducible, ubiquitous feature of human existence and social life.
>
> (Storper and Walker, 1989, p.154)

Labour is not produced for sale on the market like other commodities, as Polanyi (1944) observed (Box 1.7), but emerges into the labour market from society, through the education and training system. In this sense, the supply of labour is relatively autonomous from demand. Unlike other commodities, when capital buys labour it does not buy a specified amount but rather the time and labour power of the worker. After the working day is over, workers are free to pursue their lives beyond the confines of the factory gates or office.

This brings us to another crucial point: labour must be reproduced outside the market (Peck, 1996). **Reproduction**, in this context, refers to the daily processes of feeding, clothing, sheltering and socializing that support and sustain labour, processes which rely on family, friends and the local community. As Harvey (1989a, p.19) has memorably expressed, 'unlike other commodities, labour power has to go home every night'. The notion of labour reproduction focuses attention on the connection between work, home and the community. In many conventional accounts of work, the focus on full-time waged employment, usually that of men, has obscured the importance of domestic labour in the home, often performed by women (Gregson, 2000). Domestic labour includes activities such as childcare, cooking, cleaning, washing, ironing, etc. It is typically unpaid, but is vital to the reproduction of the paid workforce.

The key geographical expression of the nature of labour as a 'fictitious commodity' is its **relative immobility**. Labour is reproduced in particular places, something that is determined by the need for work and home to be located in fairly close proximity for the vast majority of people. The result is a patchwork of local labour markets, the geographical range of which is determined by the distance over which people are able to commute. While this has expanded over time, with the growth of suburbs and the dominance of the private car, most people still live and work within local labour market areas:

> It takes time and spatial propinquity for the central institutions of daily life – family, church, clubs, schools, sports teams, union locals, etc. – to take shape … Once established, these outlive individual participants to benefit and be sustained by generations of workers. The result is a fabric of distinctive, lasting local communities and cultures woven into the landscape of labour.
>
> (Storper and Walker, 1989, p.157)

The relative immobility of labour, and its concern with sustaining and defending its communities, contrasts with the mobility of capital (although this distinction should not be taken too far, as firms' investment in particular places does constrain their mobility to a certain extent). This point can be broadened to suggest that economic landscapes are formed out of the interaction between the conflicting forces of capital seeking profits and labour seeking to defend and promote its interests (Peck, 1996).

3.3.2 Divisions of labour

As we have already indicated, a key feature of the industrial revolution was the reorganization of production into the factory system, bringing large numbers of workers together under the control of capitalists. More broadly, a central principle of the factory system and industrial society is the division of labour which has technical, social and geographical dimensions (Sayer, 1995). We deal with the former two dimensions here, integrating the latter into our discussion of the geography of industrial development in Chapter 4. The technical division of labour can be defined as the

Box 3.3

Adam Smith and the division of labour

The concept of the division of labour was first formalized and developed by Adam Smith, writing in the late eighteenth century, just as the first shoots of industrialization were starting to appear. Smith argued that the division of labour in production is limited by the extent of the market. In pre-industrial times, localized markets were associated with small-scale domestic industry, employing craftsmen and artisans who undertook a number of different tasks. The rapid extension of markets on a global scale that took place in this period (section 3.4), by contrast, created the conditions for large-scale industrial production to flourish, employing an elaborate division of labour.

Smith famously used the example of a pin factory to demonstrate that it is far more efficient for an individual worker to concentrate on one particular task rather than to try to perform a number of activities:

One man draws out the wire, another straightens it, a third cuts it, a fourth points it, a fifth grinds the top for receiving the head; to make the head requires two or three distinct operations ... and the important business of making a pin is, in this manner, divided into about eighteen distinct operations.

(Smith, [1776] 1991, pp. 14–15)

The key principle here is specialization. In a small factory employing ten workers, an untrained worker could make less than 20 pins a day. If workers became specialized through an increased division of labour, though, Smith observed that ten people could make 48,000 pins in a day, even when the small size of the workforce meant that one worker had to perform two or three operations. In this way an increased division of labour resulted in huge rises in productivity.

There are three specific reasons why an enhanced division of labour increases productivity and efficiency in the workplace (Smith, [1776] 1991, p.13):

➤ It improves the dexterity of workers who become highly adept in performing the same routine task thousands of time in the same day.

➤ Lost time, referring to the time wasted by workers moving between tasks, tools and machines, is sharply reduced.

➤ It facilitates the replacement of labour with machines. This is due to the fact that specialization involves breaking the labour process down into a large number of standardized and routine tasks that can be performed by machines (fixed capital).

---

process of dividing production into a large number of highly specialized parts, so that each worker concentrates on a single task rather than trying to cover several (Box 3.3).

An increased division of labour results in the **deskilling of labour** as more rewarding aspects of work such as design, planning and variation are removed. The aim of the eighteenth-century pottery owner, Josiah Wedgwood, for instance, was to increase the division of labour so as to convert his employees into 'such machines of men that cannot err' (quoted in Bryson and Henry, 2005, p.315). The sub-division and fragmentation of the labour process increases employers' control of production, reducing workers to small cogs in the system.

Following the initial development of the factory system, further developments took place in the early twentieth century. These can be described as Fordist, after the American automobile manufacturer Henry Ford who introduced the key innovations (section 2.4.3). Fordism is based on an **intensification of the labour process**, developing techniques of mass production. This was to be balanced by mass consumption with increased wages for workers giving them additional purchasing power in the market, creating the consumer demand needed to underpin mass production.

The first key element of the Fordist organization of production occurred with introduction of **scientific management or Taylorism** – after its key advocate F.W. Taylor – which involved the reorganization of work according to rational principles designed to maximize productivity. Three key elements of scientific management can be identified (Meegan, 1988):

➤ A greatly increased technical division of labour, based on the complete separation of the design and planning of work, undertaken by management, from its execution by workers who became increasingly focused on simplified and repetitive tasks.

- ➤ The sub-division of operations was matched by the reintegration of the production process, involving increased coordination and control by management who were to exercise complete authority over the planning and direction of work, removing the power of foreman and workers.

- ➤ The performance and organization of workers was subject to very close monitoring and analysis by management, employing techniques such as time-and-motion studies.

This organizational revolution was matched by experimentation with new techniques to increase productivity. The most famous and widely adopted of these is the moving assembly line, first established at Ford's Highland Park automobile plant in Detroit, Michigan between 1911 and 1913 (Figure 3.5). This revolutionized production methods in the automobile industry. Rather than the worker assembling cars by moving around a factory to pick up different parts, the parts now came to the worker who would be placed at a fixed position with typically just one dedicated task:

'The man who places the part does not fasten it', said Henry Ford. 'the man who puts in a bolt does not put on the nut; the man who puts on the bolt does not tighten it.' Average chassis assembly time fell to ninety-three minutes. The lesson was obvious. Within months Highland Park was a buzzing network of belts, assembly lines, and sub-assemblies ... The entire place was whirled up into a vast, intricate and never-ending mechanical ballet.

(Lacey, 1986, cited in Meegan, 1988, p.142)

The saving of time and restructuring of work tasks had a dramatic effect on productivity; between 1911 and 1914 annual production of cars quadrupled from 78,000 to around 300,000, while the workforce only doubled in size over the period and even fell between 1913–14 (Meegan, 1988, p.143). Box 3.4 gives a brief flavour of the experience of 'working for Ford' (Beynon, 1984).

Figure 3.5 The Ford assembly line
Source: Mary Evans Picture Library.

Box 3.4

Working for Ford

The industrial sociologist Huw Beynon (1984) undertook a detailed study of working conditions at Ford's Halewood plant in Liverpool, north-west England in the late 1960s and early 1970s, conveying workers' experiences of assembly line work:

> It's the most boring job in the world. It's the same thing over and over again. There's no change in it, it wears you out. It makes you awful tired. It slows your thinking right down. There's no need to think. It's just a formality. You must carry on. You just endure it for the money. That's what we're paid for, to endure the boredom of it all.
>
> If I had a chance to move I'd leave right way. It's the conditions here. Ford class you more as machines than men. They're on top of you all the time. They expect you to work every minute of the day.
>
> (Ford workers at Halewood, quoted in Beynon, 1984, p.129)

According to another worker, while the white-collar staff that worked in the 'office' were part of Ford's, the men on the shop floor were just regarded as 'numbers'. Getting used to the incessant demands of the production process was difficult for many new workers with no experience of working in a car plant. The effects of assembly line work could sometimes extend into domestic life as well. According to one night-shift worker,

> My wife always used to insist that I had my breakfast before I went to bed. And I would get into such a state that I would sit down to bacon and egg and the table would be moving away from me. I thought, 'crikey, how long am I going to have to put up with this?' But the pay was good. It was a case of really getting stuck in and saying 'to hell with it, get it while it's here'. And this is the way it went, but the elderly chaps couldn't stand the pace.
>
> (Joe Dennis, quoted in Meegan, 1988, p.144)

It is important to recognize that the workers were not just passive, meekly accepting the dictates of management. The role of the trade unions was central, in particular the plant-level shop stewards' committee, in representing the interests of the workers. One of the key things that shaped workers' day-to-day experience was the speed of the line, with Beynon (1984, p.148) stating that 'the history of the line is a history of conflict over speed-up'. After considerable disruption and struggle at Halewood, the workers won an important victory when management conceded the right for the shop stewards to hold the key that locked the assembly line (Beynon, 1984, p.149), giving them more control over their work conditions.

The increased technical division of labour in the workplace is matched by what is termed the **social division of labour** in society. This refers to the vast array of specialized jobs that people perform in society, from doctors and lawyers to plumbers, painters and construction workers (Sayer and Walker, 1992). Modern industrial societies are characterized by a highly complex division of labour in this respect. In the course of their work, individuals enter into a range of social relations with other people who occupy roles such as colleagues, supervisors and clients or competitors (section 1.4). Students, for example, enter into social relations with academics in their role as university teachers during their degree programmes. The different jobs that people do have acquire different values in society, using that term in its broadest sense to incorporate the social status and prestige that particular jobs confer on people (compare, for example, an investment banker with a hairdresser). Such varying levels of status and prestige play an important part in determining pay rates for different occupations, alongside patterns of supply and demand in the labour market (Peck, 1996).

Reflect

➤ Why does labour tend to be relatively immobile geographically, compared with capital?
➤ Outline the distinction between technical and social divisions of labour.

3.4 Consumers and consumption

3.4.1 Understanding consumption and 'the consumer'

Consumption can be defined as those processes involved in the sale, purchase and use of commodities (Mansvelt, 2005, p.6). For most people, it is a central, taken-for-granted part of life in contemporary society. Watching television, eating a burger, going to the cinema, shopping and clubbing are all acts of consumption. While such everyday activities may seem mundane and trivial, they are of considerable economic and cultural significance (Crang, 2005, p.360). In economic terms, the sale of commodities is crucial in enabling firms to generate revenue and profits, fuelling the process of economic growth. The importance of consumption as a driver of economic growth is emphasized through regular media references to patterns of consumer spending and levels of consumer confidence

---

Box 3.5

The rise of Tesco

Tesco is the most successful supermarket chain in the UK, posting profits of just over £1.4 billion for the first six months of 2009 (Finch, 2009) and controlling 30.9 per cent of the UK grocery market in November 2009, more than any other chain (TNS Worldpanel, undated). It is the third largest global food retailer by revenue, behind Wal-Mart from the US and France's Carrefour, reflecting an aggressive policy of international expansion in recent years, focusing on selected countries in Europe, Asia and the US, which it entered in 2007 (Box 6.3). The company was founded by Jack Cohen in the East End of London in 1919; he combined the initials of his tea supplier, TE Stockwell, with the first two letters of his surname, to create the brand name. Cohen's 'pile it high, sell it cheap' philosophy led to gradual expansion over the decades, but endowed the company with a dowdy, downmarket image, leaving it lagging behind competitors such as Sainsbury's by the 1970s. After Cohen retired in 1973, Tesco adopted a new approach, attempting to become an 'aspirational mass retailer' through price wars with rivals, advertising campaigns, the development of

large out-of-town stores and a range of other innovations, resulting in substantial sales growth in the 1980s. Following a brief downturn in the early 1990s, growth resumed under a new Chief Executive, Terry Leahy, in the 1990s with Tesco overtaking its traditional rival, Sainsbury's, as the UK's largest supermarket in 1995 (Corporate Watch, 2004).

The rise of Tesco has generated much controversy regarding the company's business practices, sparking the formation of a number of oppositional groups and campaigns. It is viewed as emblematic of the power of the major UK food retailers – a grouping that also includes Sainsbury's, Asda (owned by Wal-Mart) and Morrisons – which together account for around 80 per cent of the UK grocery market. Three main areas of concern can be identified. First, relations with suppliers whereby the market power of Tesco and other leading retailers allows them to push down prices to suppliers and farmers, reflecting the retailers' effort to capture much of the value generated within specific **commodity chains** (Box 1.5). A UK Competition Commission report on supermarkets in 2000 found that Tesco consistently

paid suppliers 4 per cent below the industry average. Other manifestations of retailer power uncovered in a Friends of the Earth (2003) survey for suppliers included: being asked to pay rebates on agreed prices, waiting over 30 days for invoices to be paid, incurring additional transport and packaging costs, and having to meet the costs of unsold or wasted products. Second, the dominance of Tesco over the retail markets of particular towns has sparked protests against so-called 'Tescopoly', with opponents highlighting the impact on independent retailers which are vulnerable to being undercut and forced out of business by large supermarket chains (Prynn, 2006). Third, concerns also exist over the environmental impact of Tesco, with data showing that the production, transport, consumption and disposal of food is the number one contributor to climate change for British households, accounting for nearly one-third of each person's impact (Arbuthnott, 2009). Despite a number of initiatives, a recent report criticized the efforts of Tesco, Morrisons and Asda to engage with more environmentally sustainable practices (ibid.).

(*The Economist* 2005a, 2005b). In the UK, for example, consumer spending increased by 68 per cent between 1997 and 2007 (Office for National Statistics, 2009), outstripping the growth of gross household income (55 per cent) (Wallop, 2007), and underpinning the growth of the economy, partly through increased personal indebtedness (Blackburn, 2008).

Developed countries such as the UK and US have experienced a shift from an industrial economy to a **post-industrial** service-based economy in recent decades as the manufacturing sector has contracted in size and service industries such as retail and financial services have grown. As a result, the circulation of knowledge, information and images has become increasingly important relative to the production of manufactured goods (Lash and Urry, 1994). Major corporations and 'brands' such as Nike, McDonald's and Coca-Cola have focused increasingly on marketing and sales activity, out-sourcing actual production to a range of suppliers. The spread and influence of such brands has become an important dimension of the process of globalization, raising concerns about the creation of a global consumer culture erasing the distinctiveness of local cultures and places (Ritzer, 2004). The retail sector is an important part of this post-industrial economy, accounting for a sizeable share of paid employment. In the US, for example, 'retail salesperson' was the largest occupation recorded in official statistics, accounting for 11.5 per cent of total employment in May 2009 (Bureau of Labor Statistics, 2010). Particular retailers such as the major supermarket chains, like Tesco in the UK, exercise great power over manufacturers and suppliers through their supply chains, with efforts to reduce consumer prices translating into lower prices for suppliers (Box 3.5).

At the same time, 'the consumer' or 'the market' is often invoked as the reason for producing goods and organizing services in particular ways. For example, bananas must be of a particular size and quality to satisfy consumer expectations, while services such as banking should be provided through the internet rather than face-to-face because this is what 'the consumer demands' (Crang, 2005, p.126). The implications of this can be serious, leading to the impoverishment of Caribbean banana growers who cannot meet these standards (Box 1.6) or the closure of bank branches. In

this sense, 'the consumer' has become a kind of 'global dictator' (Miller, 1995), with the demands of affluent Western consumers in particular determining how goods are produced and services delivered throughout the world economy. Consumer demand cannot be understood in narrowly economic terms with consumption playing a central role in people's wider social and cultural lives (going shopping, visiting a bar or a restaurant). At the same time, consumption is itself shaped by 'those wider contours of society and culture' (Crang, 2005, p.126) with people's desire for goods bound up with their identities and positions in society.

3.4.2 Consumption, culture and identity

Consumption has become a major focus of interest for geographers and other social scientists over the last couple of decades. Much of this is a response to the previous neglect of consumption, seen as very much secondary to, and derivative of, the more fundamental process of production. Three broad premises underpin this recent concern with consumption (Slater, 2003). First, it is seen as central to the reproduction of social and cultural life, referring to people's everyday actions in supporting themselves and their families, involving feeding, clothing, sheltering, socializing, etc. Second, modern market societies are said to be characterized by a **consumer culture** organized around the logic of individual choice in the marketplace. Third, studying consumption enables us to better comprehend the importance of culture in shaping economic processes and institutions, representing 'the site on which culture and economy most dramatically converge' (Slater, 2003, p.149).

Two main perspectives on consumption can be identified. First, a number of influential social critics, including Karl Marx and Herbert Marcuse, have viewed it as signalling the triumph of market exchange and industrial society over deeper human qualities and meanings. As such, consumption marks the process through which culture is colonized by economic forces (Slater, 2003, p.150). In capitalist societies, needs and wants are artificially created and manufactured, inducing people to consume far more than they actually need. A number of studies has focused on the role of advertising in stimulating demand for products. While there is certainly a sense in which market demand is induced through processes such as advertising, this approach invariably tends to cast the consumer in a passive role as a 'cultural dupe' or 'dope', manipulated and controlled by corporations and the media. Some versions of postmodernism (Box 2.6) have reproduced this theme of passive consumers, emphasizing their powerlessness in the face of an infinite universe of abstract signs and meanings.

A second view emphasizes the active role of consumers in utilizing things for their own ends. Earlier work in this vein viewed consumption and leisure practices as a source of social status and distinction, as captured in the institutional economist Thorstein Veblen's notion of 'conspicuous consumption' and the sociologist Pierre Bourdieu's more recent work on cultural capital and taste. Contemporary studies have moved away from this concern with social status and distinction to stress how consumers creatively re-work the products they buy, generating new meanings in the process (Crang, 2005, p.363). Rather than reading off the process of consumption from production and corporate strategies, as critical approaches have tended to do, one has to understand the social and cultural relations in which it is entangled. From this perspective,

consumption is seen as a relatively fertile arena for the expression of individuality and creativity, compared with the world of work that involves considerable drudgery and monotony for many people.

Examples of the wider social and cultural relations in which active consumption is rooted are those of family and friendship. According to the anthropologist Danny Miller, instead of being driven by individual greed and hedonism, consumption is based on acts of love and devotion involving the purchase of commodities for partners, children and friends (Miller, 1998b). Of particular significance is gift-buying, focused around rituals such as Christmas (Box 2.7) and birthdays. Much of this work on active and creative consumption has been ethnographic in nature, attempting to understand the meaning and significance that people attach to consumption through detailed fieldwork and observation.

3.4.3 Changing patterns of consumption

One of the key expressions of modern consumer culture is the department store, which came to prominence in the second half of the nineteenth century.

Figure 3.6 Macey's: a famous New York department store with nineteenth-century origins
Source: D. MacKinnon.

Box 3.6

Gender and consumption in nineteenth-century New York City

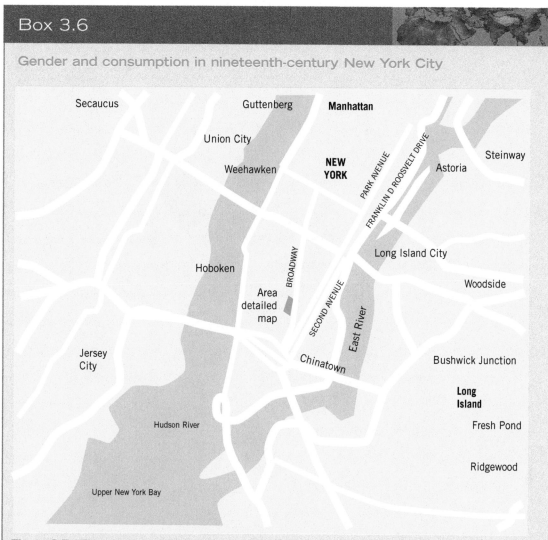

Figure 3.7a The original retail district in nineteeth-century New York City
Source: Constructed from area map from Multimap, at http://www.multimap.com/.

While the rapid and massive expansion of New York's retail district in the nineteenth century reflected the spectacular growth of the city's economy, its form was shaped by the shopping habits of the middles classes, particularly women. Shopping in the city's burgeoning department stores was an important activity for middle-class women in New York, becoming almost a daily ritual for some (Domosh, 1996). The very wealthy defined the styles and fashions that middle-class women sought to emulate, allowing

the latter to define their own tastes and, by extension, their social status (ibid., p.261). Rather than being entirely confined to the domestic sphere, women made frequent trips from their suburban homes to downtown Manhattan, thus playing an important role in shaping this and other American downtowns. While shopping was a frequent activity, they also met for church prayer groups, lectures and concerts, and to pay bills or make social calls (ibid., p.258).

A distinct retailing area developed

in late nineteenth-century New York, focused on Fifth Avenue between Union and Madison Squares, extending west to Sixth Avenue and east to Broadway (Figure 3.7). This was an 'urban landscape designed specifically for consumption', made up of 'ornamental architecture and grand boulevards, of restaurants and bars, and of small boutiques and large department stores' (ibid., pp.263–64). Improvements in urban transportation through the development of a rapid transit network

Box 3.6 (continued)

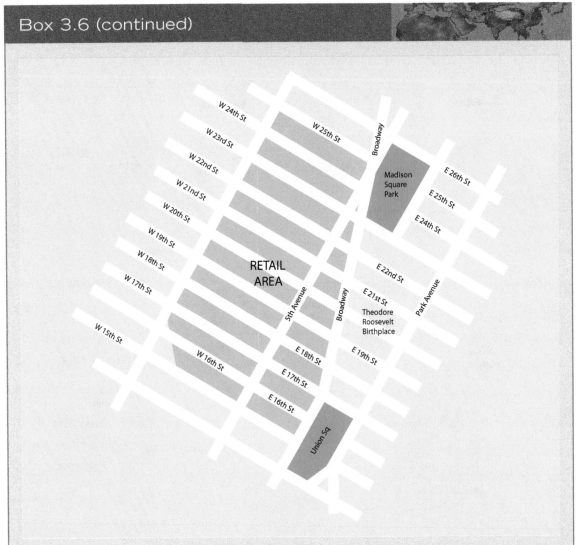

Fugure 3.7b Downtown Manhattan retail area

allowed middle-class women to travel downtown to shop in the morning, returning to their uptown residences for lunch before proceeding back downtown in the afternoon. Wide, paved streets, lit by gas and electricity, made the retailing district feel safe and congenial for the women.

The city's first department store, Stewart's, opened on Broadway in 1846. Its four storeys, devoted entirely to retailing, and its white marble façade were unprecedented in the city (ibid., p.264). The interior of the store was arranged in order to maximize the display of goods and create an appropriate atmosphere for women. It included large mirrors, chandeliers, a gallery and parlour. A later store was opened by Stewart further uptown on Broadway in 1862, becoming an important tourist attraction. Its six storeys contained a main floor and five encircling balconies, lit by a skylight that provided natural light. Reflecting the analogies drawn between consumption and religion, a commentator compared Stewart's with the nearby Grace Church. By the end of the nineteenth century, the domestic sphere had become more fully incorporated into stores in the form of tea rooms, restaurants, art galleries and grand architectural displays.

Described as 'the quintessential consumption site of the late nineteenth and early twentieth century' (Lowe and Wrigley, 1996, p.18), department stores presented 'the most visible, urban manifestation of consumer culture and the economics of mass production and selling' (Domosh, 1996, p.257) (Figure 3.6). A vast array of goods was placed on public display in the store and prices fixed for standardized goods. While we take this for granted today, prior to the development of modern stores, goods were not displayed, with customers having to ask to view them, and prices were negotiated between the customer and shop owner or clerk (Domosh, 1996, p.264). In the new department stores, shoppers were directly able to compare the prices and qualities of different goods as shopping became a knowledgeable and skilled activity, encouraged by the retailers and advertisers.

Shopping also became highly gendered as the department stores targeted middle-class women in particular (Box 3.6). In this way, a crucial set of links was forged between gender, class and culture in the late nineteenth century, which continue to shape retail and consumption today. Shopping became a major part of women's work, with the purchase of goods replacing the domestic production of food and clothing, thus creating a market for manufactures and retailers. The realm of consumption was defined as feminine, associated with leisure and self-indulgence, in contrast with the masculine domain of production, governed by the work ethic and associated notions of self-denial and self-discipline (ibid., p.262). Department stores provided spaces in which women could be taught to shop through the display of goods as a spectacle, the use of advertising and demonstration, and the assistance of specialist staff (Hudson, 2005, p.147). The introduction and manipulation of fashion was a key mechanism for increasing demand, meaning that frequent changes in style were required to keep up. Shopping became an important female duty, with store owners cultivating associations between women, fashion and religion, referring to their stores as 'cathedrals' and goods as 'objects of devotion' (Domosh, 1996, p.266).

While mass consumption became established in the nineteenth century, it was consolidated and reinforced during the period of Fordism from the 1940s to the 1970s. As we have emphasized (sections 3.3.2),

Fordism was a system of industrial organization based on a balance between mass production and **mass consumption**. The key link was higher wages for workers, received in exchange for increased productivity. Fordism involved the mass production of consumer durables such as automobiles, fridges and washing machines, produced in standard forms. Rising wages meant that more and more workers were able to afford such goods, ensuring a larger market for manufacturers and retailers. A key trend in the geographical organization of society was the growth of suburbs, particularly in North America, facilitated by state investment in infrastructure such as roads and electricity. Suburban lifestyles become closely associated with mass consumption patterns, with every household requiring its car, washing machine and lawnmower (Goss, 2005).

Mass markets for standardized goods became increasingly saturated from the late 1960s, as economic growth slowed and the Fordist system experienced growing problems. As a result, mass consumption has been eclipsed by the rise of **post-Fordist** patterns of **consumption** since the 1970s. These are defined by flexibility, as markets have become fragmented into distinct segments and niches. Accordingly, patterns of consumption are defined individually rather than collectively. Consumer choice and identity have become increasingly important, with individual consumers regarding the purchase and consumption of commodities as expressions of their lifestyles and aspirations (Mansvelt, 2005, p.44). Individualized patterns of consumption oriented towards identity and lifestyles, are a key component of postmodern culture, characterized by an emphasis on flexibility, difference and diversity (Box 2.6). The rapid circulation of ideas, images and signs, fuelled by the advertising industry and the media, is another central aspect of postmodern consumer culture.

Producers and retailers have become increasingly consumer-oriented, striving to tailor goods and services to the demands of individuals and specific groups of consumers. It is in this sense that the consumer has become a kind of 'global dictator' (Miller, 1995), determining how production is organized across a wide range of economic sectors and geographical locations. The growth of information and communication technologies has allowed retailers to store, process and convey data about changing patterns of consumer demand. The introduction of point-of-sale terminals in the 1980s allowed retailers to rapidly transmit information about consumption trends from shops to company headquarters, from where it was passed on to designers and suppliers. The clothing company Benetton is one of the best examples of a company that has grown through a strategy of niche marketing and product differentiation, employing advanced communication technologies to the fullest possible extent in the spheres of design, manufacturing, procurement and marketing (Knox *et al.*, 2003, pp.188–9).

Reflect

➤ How does the idea of the active consumer differ from previous theories adopted by social scientists?

➤ To what extent do you view consumption as an important means of expressing your individuality and identity?

3.5 The state

In this book, we adopt a broadly regulationist perspective (section 2.4.3), rejecting the notion of an autonomous self-regulating economy. Rather, the economy is regulated through a wide range of forms of social regulation, including social habits, administrative rules and cultural norms, with the **state** playing a key role in harnessing and coordinating these different mechanisms (Aglietta, 1979). By the state, we are referring to a set of public institutions that exercise authority over a particular territory, including the government, parliament, civil service, judiciary, police, security services and local authorities. As this suggests, the state is a complex entity, encompassing a number of institutions beyond what is normally referred to as government. The role of the state in social regulation is directed towards the objective of ensuring social and economic stability, thereby creating the conditions that allow the national economy to expand.

The state assumes two key functions in this respect (Johnston, 1986). First, the **'accumulation' function**

means that a key task of the state is supporting and promoting economic development within its territory, ensuring that business can accumulate capital for investment and growth. Such an objective underpins many of the activities listed in Table 3.3, involving the state in designing economic strategies, promoting innovation and 'competitiveness' in certain strategic industries and administering education and training programmes for workers. In recent decades, governments have viewed their performance in managing and overseeing the economy as vital to the retention of power, striving to build a reputation for economic competence. In order to gain power in Britain in 1997, for instance, the leaders of the (new) Labour Party had to drop its old 'tax and spend' policies, fostering a new image of themselves as responsible and prudent custodians of the public finances. This required them to develop good relations with business and the City of London.

The second key function is **'legitimation'**, referring to the range of activities undertaken to maintain social order, ensuring that the capitalist system and the associated social order are regarded as legitimate and 'natural' by the majority of citizens. As well as managing the legal system, this has seen states construct elaborate welfare systems over the course of the twentieth century to try to spread the benefits of growth and offer social protection to their citizens against the vagaries of the market, including unemployment and ill health. This **welfare state** has come under attack from neoliberal commentators and politicians since the 1970s, but much of its basic architecture remains in place in the majority of advanced industrial countries. While the 'legitimation' function requires the state to sponsor collective projects aimed at increasing the wellbeing of its citizens, it remains dependent on the expansion of the capitalist economy (accumulation) to ensure its continued financial viability through the revenue provided by taxation.

The key regulatory activities of the state are listed in Table 3.3. These include ensuring macroeconomic stability through fiscal and monetary policy, helping to secure the reproduction of labour through, among other things, education and training programmes, health and safety legislation and unemployment benefit, and the provision of basic infrastructure. More broadly, states

Table 3.3 The economic roles of the state

1. Maintenance of property rights and governance frameworks

Maintenance of law and order

Maintenance of private property rights

Recognition of institutional property rights

Rules of ownership and use of productive assets

Rules for the exploitation of natural resources

Rules for the transfer of property rights

Protection of intellectual property rights

The governance of economic relations between: family members, employers and workers, landlords and tenants, buyers and sellers

2. Management of territorial boundaries

Provision of military force

Regulation of money flows, goods flows, service flows, labour flows, knowledge flows

3. Control of macro-economic trends

Fiscal policy

Monetary policy

4. Governance of product markets

Regulation of the market power of firms

The selection and regulation of natural monopolies

The provision of public goods and goods unlikely to be supplied fairly

5. Governance of financial markets

Rules for the establishment and operation of financial institutions

Designation of the means of economic payment

Rules for the use of credit

Maintenance of the lender of last resort

6. Provision of basic infrastructure

Includes transportation and communication systems, energy and water supply, waste disposal systems

Assembly and conduct of communications media

Assembly and conduct of public information

Land-use planning and regulation

7. Selection and development of economic growth strategy

Promotion and maintenance of strategic industries

Development of science and technology

Urban and regional development

Identification of key outcomes and targets (employment, growth, innovation, etc.)

Selection and implementation of financial incentives

8. Production and reproduction of labour

Demographic planning

Provision of universal education and training

Governance of workplace conditions

Wages policies

Social wage provision

Supply and governance of childcare

Governance of retirement pensions

9. Other legitimation activities

Maintenance of public health

Citizenship rights

Income and wealth distribution

Reduction of poverty

Cultural development

Socialization

Environmental protection/enhancement

Source: Adapted from O'Neill, 1997, p.295.

are charged with safeguarding property relations, regulating trade and overseeing the operation of financial markets. The importance of the latter was illustrated by the financial crisis of 2007–08 when the state stepped in to rescue certain well-known high-street banks in the UK and US, injecting capital and purchasing shares in order to prevent the failure of the banks leading to a wider economic collapse (French et al., 2009). The state is also of course, an important economic entity in its own right, employing large numbers of people, purchasing a wide range of goods and services and harnessing public and private resources behind major strategic initiatives and projects (Painter, 2000).

States are explicitly geographical in form, having clearly marked borders which separate the territory of one state from that of others. Within these territorial limits, states have sought to define and promote national economies, building integrated national markets through the creation of common legal standards and financial rules, the expansion of transport and communication systems and the careful regulation of flows of goods, money and people across their borders.

At the same time, states contain considerable internal diversity in terms of economic conditions, cultural values and political allegiances. The problems of territorial management created by such internal diversity have generally been recognized through the creation of a tier of local government which administers state programmes and represents local interests (Duncan and Goodwin, 1988).

The notion of a coherent national economy has been seriously disrupted by the process of economic globalization in recent years as flows of goods, money and information across national borders have grown rapidly (section 1.2), making it more difficult for governments to regulate their economies in the conventional fashion. Despite this, states remain important actors within globalization (Dicken *et al.*, 1997), not least in prosecuting the interests of their own multinationals in overseas markets and in international trade forums such as the WTO.

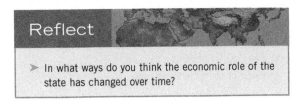

Reflect

➤ In what ways do you think the economic role of the state has changed over time?

3.6 Summary

This chapter has examined the role of the main economic actors and outlined the basic underlying processes that have shaped the development of capitalism since the late eighteenth century. The basic process of production was described in terms of the circuit of capital, in which capital, labour and the means of production are combined to generate commodities that are sold in competitive markets for a profit. The firm can be seen as a distinctive bundle of competencies and assets, expressed in terms of skills, practices and knowledge. While capital tends to be concentrated in prosperous core areas such as south-east England, it remains highly mobile, subject to being 'switched' from declining regions to new growth regions. The process of innovation and technological change is an inherent feature of capitalism, reflecting the pressures

of profit-seeking and competition between firms. Economic historians have observed the tendency of major innovations to 'swarm' together, underpinning 'long waves' or Kondratiev cycles of economic development. This chapter also considered the role of labour, noting that, although workers are required to sell their labour to capitalists for a wage, labour should be viewed as a 'fictitious commodity' on account of its irreducible human and social qualities (Polanyi, 1944). At the same time, the evolving technical division of labour in production has shaped the development of industrial capitalism, finding perhaps its fullest expression in the Fordist mass production industries that typified the middle decades of the twentieth century.

The role of consumers and consumption is central to modern economic life, providing the market demand that underpins economic growth. Recent work on consumption has been based on notions of the 'active consumer', rejecting the view of the consumer as a 'dupe' manipulated by corporations and the media. Modern mass consumption grew in the nineteenth century, finding particular expression in the department store, described as 'the quintessential consumption site of the late nineteenth and early twentieth century' (Lowe and Wrigley, 1996, p.18). Middle-class women were targeted as a key market by the retailers, forging an enduring set of links between gender, class and consumer culture. Mass consumption practices were consolidated and reinforced under Fordism in the twentieth century. Since the 1970s, a new post-Fordist mode of consumption has emerged, characterized by a focus on different market segments and niches with consumption defined individually rather than collectively. Companies have become increasingly consumer-oriented, with the focus on meeting consumers' demands and expectations leading to the figure of the western consumer being described as a 'global dictator' (Miller, 1995). We also considered the role of the state in regulating the economy, focusing on its 'accumulation' and 'legitimation' functions. While the role of the state in the economy has been attacked by conservative politicians since the 1970s, much of the welfare state remains in place and the state is an important actor within contemporary processes of globalization (Chapter 5).

Exercise

Select a major innovation introduced during the twentieth century (e.g. motor car, moving assembly line, commercial jet aircraft, television or personal computer) and assess whether the actual benefits of this innovation outweighed the negative effects.

Consider the following issues: the growth and spread of the product or method; who owned or controlled the technology; its appeal to consumers; how production was organized; the impact on existing industries and workers; and its effects of experiences of time and space.

Key reading

Gregory, D. (2008) 'Capitalism', in Johnston, R.J., Gregory, D., Pratt, G. and Watts, M. (eds) *The Dictionary of Human Geography*, 5th edn, Oxford: Blackwell, pp.56–9.
An introduction to capitalism as a distinctive mode of production based on the private ownership of the means of production and the commodification (commercialization) of labour power. Discusses different approaches to the analysis of capitalism and outlines the circuit of capital.

Hobsbawm, E.J. (1999) *Industry and Empire*, Revised edn, London: Penguin, pp.34–56, 87–111.
A highly readable account of the development of the British economy since the industrial revolution by the renowned Marxist historian. Particularly informative on the Industrial Revolution and links with empire.

Harvey, D. (1985) 'The geopolitics of capitalism', in Gregory, D. and Urry, J. (eds) *Social Relations and Spatial Structures*, London: Macmillan, pp.128–63.
A forceful and stimulating essay on the key processes of capital accumulation by the leading Marxist geographer. Difficult but worth persevering with for the valuable insights it offers into the role of capitalism in shaping the geographical landscape.

Meegan, R. (1988) 'A crisis of mass production', in Allen, J. and Massey, D. (eds) *The Economy in Question*, London: Sage, pp.136–83.
Probably the best geographical introduction to Fordism as an economic system based on mass production and mass consumption. Examines the growth of Fordism and the experience of working in Fordist industries.

Peck, J. (2000) 'Places of work', in Sheppard, E. and Barnes, T.J. (eds) *A Companion to Economic Geography*, Oxford: Blackwell, pp.133–48.
A very useful summary of the geography of labour and labour markets. Relates the changing organization of labour to wider processes of industrial restructuring, emphasizing the distinctiveness of local labour markets and the shift towards flexible production.

Slater, D. (2003) 'Cultures of consumption', in Anderson, K., Domosh. M., Pile, S. and Thrift, N. (eds) *Handbook of Cultural Geography*, London: Sage, pp.146–63.
A wide-ranging survey of recent work on consumption which stresses the importance of consumption as a crucial link between the economy and culture. Identifies different approaches to the study of consumption and discusses recent debates about new shopping patterns and the emergence of a global consumer culture.

Useful websites

http://en.wikipedia.org/wiki/Capitalism
A description of the basic principles and operations of capitalism in accessible terms from Wikipedia, the free online encyclopedia. Defines and discusses key elements of capitalism such as private ownership of the means of production, the pursuit of self-interest and the operation of free markets in an engaging manner.

http://en.wikipedia.org/wiki/Industrial_Revolution
An account of the Industrial Revolution from the same source. Identifies the main causes of the Industrial Revolution and the key innovations, as well as outlining how these were applied in the key industries such as coal mining and textiles. The development of factories and machines is emphasized alongside the changing organization of labour.

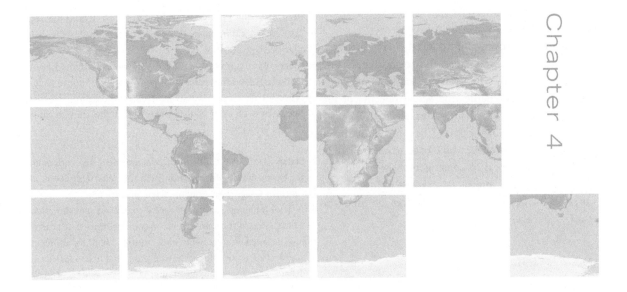

Spaces of production and consumption

between the Industrial Revolution and the geographical expansion of capitalism, facilitated by the development of the 'space shrinking' technologies of transport and communications (Leyshon, 1995). In section 4.3, the uneven geography of production is examined in relation to successive waves of industrialization from the late eighteenth century (section 3.2.4). We focus particular attention on the growth of 'new industrial spaces' since the 1970s. We turn our attention to consumption in section 4.4, examining key spaces of consumption at the global and local scales. The notion of a single global consumer culture is assessed and the role of specific sites of consumption such as large mega-malls, streets, car boot sales and the home is highlighted. Finally, we summarize the main points of the chapter.

Chapter map

In the introduction to this chapter, we emphasize the geographical expansion and historical evolution of the capitalist economy. We then outline the relationship

4.1 Introduction

As we indicated in section 3.2, the contemporary economy is dominated by a capitalist mode of production in which the pursuit of profit has been the key driving force behind development. Over the last 400 or so years, this has led to the emergence of a single world economy, as capitalism has spread geographically from its origins in western Europe across the globe to incorporate virtually all countries and regions (Knox *et al.*, 2003). This process of expansion has transformed pre-capitalist societies and systems, not least through the process of colonialism, which reached its high point in the late nineteenth century. Despite periodic set-backs and reverses, including serious challenges from alternative social systems such as forms of communism in the Soviet Union and China during the twentieth century, capitalism has continued to expand. Indeed, the latest stage involves the incorporation of formerly socialist countries through China's 'open door' policies and following the collapse of communist regimes in the former Soviet Union and eastern Europe.

The growth of capitalism has not only been uneven geographically but also historically, sparking periodic crises such as the Great Depression of the 1930s, the 'stagflation' (a combination of economic stagnation and rising inflation) of the 1970s, and the recent financial crisis and recession. Such crises seem to reflect, among other things, a tendency towards over-expansion, as favourable economic conditions encourage firms to produce more and more goods, leading ultimately to a slump when the market cannot absorb all this output. Such internal contradictions of capitalism were recognized by commentators as diverse as Karl Marx, Joseph Schumpeter and John Maynard Keynes, leading them to forecast or fear its demise. Adopting a longer-term view, however, capitalism has proved remarkably durable and adaptable in finding short- and medium-term solutions to its internal crises, although it has never been able to completely solve the underlying problem of over-production. As we have seen, the regulation approach explains this in terms of the role of historically specific regimes of accumulation such as Fordism in sustaining and stabilizing capitalism (section 2.4.3). Since the early 1980s, the spread of neo-liberal policies has shaped the development of a new phase of capitalist globalization (Harvey, 2005).

4.2 The Industrial Revolution and the geographical expansion of capitalism

The growth of industrial production from the late eighteenth century was supported and facilitated by the geographical expansion of capitalism. The great increases in output and productivity associated with the application of new technology and the develop-ment of an elaborate division of labour within factories required, as Adam Smith observed, a corresponding increase in the extent of the market. At the same time, industrialization was dependent on an increased volume and range of raw materials, drawn from different parts of the globe. Through the growth of capi-talist production and trade, an **international division of labour** was created during the nineteenth century. This involved the developed countries of Europe and North America producing manufactured goods, while the underdeveloped world specialized in the produc-tion of raw materials and foodstuffs. This global trading system was supported and justified by the doctrine of **comparative advantage**, classically expressed by David Ricardo, the great English economist, in 1817. It emphasizes the benefits of international trade, stating that countries should export the goods which they can produce with greater relative efficiency. In recent years, the theory of comparative advantage has provided an important intellectual foundation for the dominant neo-liberal ideology of globalization.

While the growth of a world economy can be traced back to the voyages of exploration and discovery in the sixteenth century, **geographical expansion** gained new momentum as the industrial revolution took off, cul-minating in the 'age of empire' between 1875 and 1914 (Hobsbawm, 1987). The inherently expansive nature of capitalism as an economic system has underpinned a search for new markets, new sources of raw materials and new supplies of labour, forging economic relation-ships between territories on a global scale.

The need of a constantly expanding market for its products chases the bourgeoisie over the whole

Box 4.1

Trade and comparative advantage

Comparative advantage is the principle that a country should specialize in producing and exporting goods in which it has a comparative or relative cost advantage over others and import goods in which it has a cost disadvantage. For example, while the developed country in Table 4.1 has an absolute advantage in producing both wheat and cloth (it can produce them more efficiently), the developing country has a comparative advantage in wheat, and the developed country in cloth. Since wheat is relatively cheaper to produce than cloth in the developing country, it only needs to sacrifice 1 metre of cloth to produce 2 kilos of wheat. Conversely, cloth is relatively cheaper to produce than wheat in the developed country, with the production of 8 metres entailing the sacrifice of 4 kilos of wheat. In the developing country, by contrast, producing 1 metre of cloth involves giving up 2 kilos of wheat. Thus, the principle of comparative advantage states that countries should specialize in the goods that lead them to give up least in terms of the production of other goods. Through specialization, both countries gain by focusing on the good which they can produce most efficiently and by importing the other.

One key question concerns the sources of comparative advantage. How is it that certain countries can produce some goods more efficiently than others? Ricardo explained this in terms of countries' different endowments of the factors of production: land, labour and capital. For example, Canada has a lot of land, China a lot of labour and the US is rich in capital. This means that relative costs of producing goods vary between countries, providing a basis for trade. Basically, countries should specialize in producing goods that use the factors they have in abundance, enabling them to produce these goods cheaply (grain in Canada, textiles and footwear in China, pharmaceuticals in the US). In this sense, Ricardo believed that trade patterns had a natural basis. The doctrine of comparative advantage supported the colonial trading system where the European countries exported capital-intensive manufactured goods and the colonies produced raw materials and agricultural goods which were labour (mines, plantations, etc.) and land intensive.

In recent decades, however, it has become apparent that most trade takes place between developed countries with similar factor endowments. This has helped to stimulate the development of a **new trade theory** by the Nobel prize-winning economist Paul Krugman (Box 2.2) and others, which recognizes that comparative advantage does not simply reflect pre-existing factor endowments. Rather, it is actively created by firms through the development of technology, human skills and economies of scale, something which is often referred to in terms of **competitive advantage**. This helps to account for patterns of trade and regional specialization at a more detailed level, for example aircraft in Seattle, cars in southern Germany and finance and business services in London or New York.

Table 4.1 Quantities of wheat and cloth production

	Kilos of wheat	Metres of cloth
Developing country	2	1
Developed country	4	8

Source: Sloman, 1999, pp.659–60.

surface of the globe. It must nestle everywhere, settle everywhere, establish connections everywhere ... All old-established national industries ... are dislodged by new industries, ... that no longer work up indigenous raw material, but raw material drawn from the remotest zones; industries whose products are consumed not only at home, but in every corner of the globe ... In place of the old local and national seclusion and self-sufficiency, we have intercourse in every direction, universal interdependence of nations.

(Marx and Engels, [1848] (1967), pp.83–4)

It is this continuing search for new markets, raw materials and labour supplies that underpins recent processes of globalization, which can be viewed as the latest chapter in the on-going geographical expansion of capitalism.

Colonialism entailed a forcible transformation of pre-capitalist societies in Asia, Africa and Latin America in the eighteenth and nineteenth centuries. As a result, these societies 'were no longer locally orientated but had now to focus on the production of raw materials and foodstuffs for the "core" economies' (Knox *et al.*, 2003, p.250). The raw cotton for the mills of Lancashire was supplied by slave plantations in the West Indies and, from the 1790s, the southern US, ensuring that the 'most modern centre of production thus preserved and extended the most primitive form of exploitation' (Hobsbawm, 1999, p.36).

At the same time, the significance of the colonies as outlets for manufactured goods increased greatly. Cotton exports, for instance, multiplied by ten times between 1750 and 1770 (Hobsbawm, 1999, pp.35–6) with the vast majority exported to colonial markets, originally in Africa, before India and the Far East took over from the middle of the nineteenth century. The early development of the modern cotton industry relied on Britain's effective monopoly of colonial markets in this period, reflecting both its commercial superiority and naval supremacy. In many cases, the export of manufactured goods from the core countries resulted in active deindustrialization in the periphery as traditional local industries were undercut and destroyed by modern factory-based production. Thus, the Indian textile industry was decimated by the much cheaper products of the Lancashire mills. The resultant distress amongst the native handloom weavers prompted Marx's ([1867] 1976, p.555) comment that 'the bones of the cotton-weavers are bleaching the plains of India'.

The geographical expansion of capitalism was facilitated by new transport and communications technologies during the nineteenth century. On the transport side, the development of railways and steamships resulted in a dramatic **'shrinking' of space**, transporting a growing volume of goods and people over long distances. In 1870, for instance, 336.5 million journeys by rail were made in Britain (Leyshon, 1995, p.23). The rapid construction of the English railway network in the 1840s had far-reaching consequences:

> In every respect this was a revolutionary transformation ... it reached into some of the remotest

Figure 4.1 A freight train passing through Laramie, Wyoming
Source: D. MacKinnon.

areas of the countryside and the centres of the greatest cities. It transformed the speed of movement – indeed of human life – from one measured in single miles per hour to one measured in scores of miles per hour, and introduced the notion of a gigantic, nation-wide, complex and exact interlocking routine symbolised by the railway timetable ... it revealed the possibilities of technical progress as nothing else had done ...

(Hobsbawm, 1999, p.88)

Railways were subsequently built across much of continental Europe and North America, as well as in European colonies overseas (Figure 4.1), opening up these territories to large-scale investment and trade and creating strong demand for coal, iron and steel. In

North America, the railways were crucial to the development of the economy during the nineteenth century, linking the Great Plains to the ports of the Great Lakes and the markets of the east coast and Europe (Leyshon, 1995, p.28). Huge cities such as Chicago grew as transportation hubs and agricultural markets and processing centres, linking the resources of the American interior to the wider world economy (Figure 4.2) (Cronon, 1991). The steamship also played an important role in both transforming the speed of movement between continents and providing a market for the heavy engineering and shipbuilding industries.

On the communications side, the late nineteenth century saw the invention of the telegram (the 1850s), the telephone (1870s) and the radio (1890s). In terms of facilitating communication over distance and

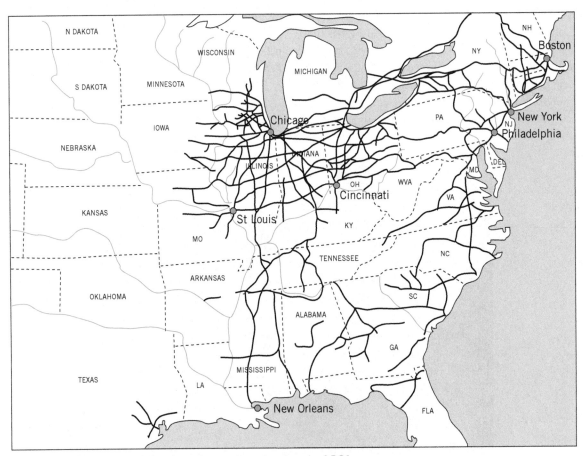

Figure 4.2 Chicago and the American railroad network, 1861

Source: 'Map: Nature's Metropolis with American railroads, 1861', from *Nature's Metropolis: Chicago and the Great West* by William Cronon. Used by permission of W.W. Norton & Company, Inc.

connecting up the realm of everyday life with a geographically dispersed 'out there', these developments were of great significance, paving the way for later developments such as television, the internet and satellite communications. As the German philosopher Martin Heidegger noted in 1916:

> I live in a dull, drab colliery village ... a bus ride from third rate environments and a considerable journey from any educational, musical or social advantages of as first class sort. In such an atmosphere life becomes rusty and apathetic. Into this monotony comes a good radio set and my little world is transformed.

(Quoted in Urry, 2000, p.125)

A telegraph cable was laid across the Atlantic Ocean in 1858, enabling rapid communication between North America and Europe for the first time (Leyshon, 1995, p.24). By 1900, the global telegraph system was complete, creating a communications system that dramatically reduced the time and costs of sending information between places (Figure 4.3). The telegraph

was particularly important, supporting the growth of trade and the growing integration of financial markets in particular. For example, foreign exchange could now be easily and rapidly exchanged between markets in London and New York, as the telegraph enabled traders to inform one another of the rate at which they would sell pounds for dollars and vice-versa (leading to the sterling–dollar exchange rate becoming known as 'cable') (ibid., p.25).

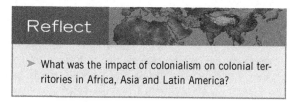

Reflect

➤ What was the impact of colonialism on colonial territories in Africa, Asia and Latin America?

4.3 The rise and fall of industrial regions

Industrialization was a regional phenomenon during the nineteenth century, originating in areas of northern and central England before spreading to parts of continental

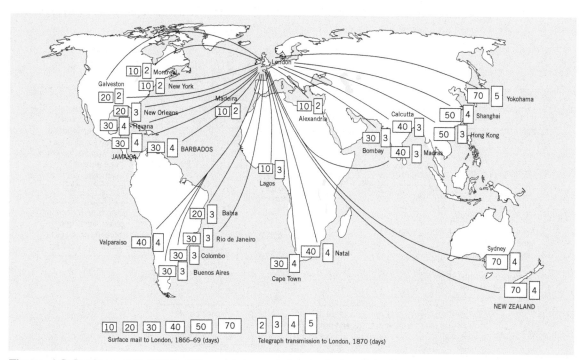

Figure 4.3 Surface mail (1866–69) and telegraph transmission (1870), times in days
Source: 'Annihilating space? The speed-up of communications', in *A Shrinking World? Global Unevenness and Inequality*, edited by Allen, J. and Hammett, C. (Leyshon, A., 1995). By permission of Oxford University Press.

Box 4.2

Myrdal's model of cumulative causation

A useful model of the process of uneven regional development is that of **cumulative causation**, derived from the work of the Swedish economist Gunnar Myrdal (1957). This explains the spatial concentration of industry in terms of a spiral of self-reinforcing advantages that build up in a particular area (Figure 4.4) and the adverse effect this has on other regions, creating a core–periphery pattern. Once an industry starts to grow in a region, for whatever reason, it will tend to attract ancillary (supporting) industry made up of firms supplying it with various inputs and services. At the same time, the expansion of employment and population created by the growth of industry creates a large market which draws in further capital and enterprise. The expansion of industry and the growth in population, moreover, creates increased revenues

for local government, resulting in the provision of an improved infrastructure for industrial development (Figure 4.4).

The process of cumulative causation in a growing region is linked to the fate of surrounding ones through flows of capital and labour. Myrdal identified two contrasting types of effects. The first he termed **'backwash' effects**, when investment and people are sucked out of surrounding regions into the growth region, which offers higher profits and wages. In this situation, the virtuous circle of growth in the latter is matched by a vicious circle of decline in the former, leaving such regions suffering from classic symptoms of underdevelopment such as a lack of capital and depopulation. The prevalence of 'backwash' effects, then, means that industry becomes concentred in growth regions,

creating an entrenched pattern of core-periphery differentials.

The second set of effects, however, are **'spread' effects**, where surrounding regions benefit from increased growth in the core region. One important mechanism here is increased demand in the core region for food, consumer goods and other products, creating opportunities for firms in peripheral regions to supply this growing market, (Knox *et al.*, 2003, p.243). At the same time, rising costs of land, labour and capital in the core region, together with associated problems such as congestion, can push investment out into surrounding regions. Rising costs reflect increased demand and the tendency for growth to out-strip the capacity of the underlying infrastructure to support it. This problem, which has periodically affected the economies of major core

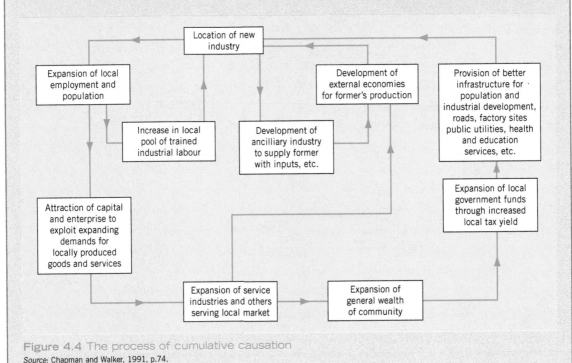

Figure 4.4 The process of cumulative causation
Source: Chapman and Walker, 1991, p.74.

Box 4.2 (continued)

regions such as south-east England, is often referred to as 'over-heating'. As a result, capital flows out into lower-cost regions, followed by labour. This is a process of **spatial dispersal**, where industry moves out of existing centres of production into new regions. It will create a geographically balanced economy if it is the dominant process over a long period of time.

Europe and North America (Pollard, 1981). It gave rise to a distinct pattern of **regional sectoral specialization**, involving particular regions becoming specialized in certain sectors of industry. Characteristically, all the main stages of production, from resource extraction to final manufacture, were carried out within the same region. This pattern of **spatial agglomeration** or concentration (Box 4.2) (section 11.2.2) contrasted with the system of proto-industrialization which had emerged in seventeenth- and eighteenth-century England where local merchants distributed or 'put out' raw materials to be manufactured by smallholders and artisans in their cottages and workshops. Such domestic industry was small scale and widely dispersed across the countryside, often based on the exploitation of local mineral resources and water power.

4.3.1 Nineteenth-century industrialization

As we emphasized in section 3.2.4, industrialization occurred in distinct waves, known as Kondratiev cycles. These cycles are not only historical phenomena; they also gave rise to distinct geographies as certain countries and regions assumed technological leadership, leaving others behind. When we look at patterns across different cycles, a more complicated and dynamic picture emerges. Many formerly leading regions (for example, nineteenth-century industrial areas in Europe and North America) have been challenged and eclipsed by 'rising' regions such as the US 'sunbelt' or East Asia since the 1970s. At the same time, a select number has been able to maintain their position – particularly metropolitan core regions around cites such as New York and London – while much of the developing world outside East Asia has remained peripheral to the world economy. In broad terms, the economic geography of

the world can be explained in terms of the interaction between the patterns of investment associated with successive Kondratiev cycles (Massey, 1984).

The first wave of industrialization, based on cotton, iron smelting and coal, took off in certain regions of Britain, principally Lancashire, the West Riding of

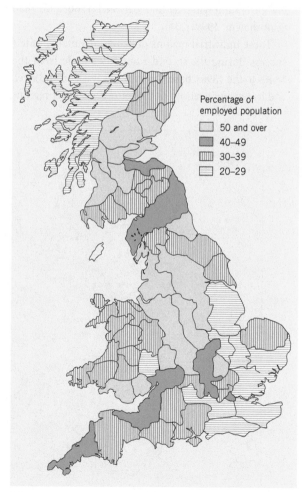

Figure 4.5 UK manufacturing employment, 1851
Source: Lee, 1986, p.32.

Yorkshire, the West Midlands, the north-east of England and west-central Scotland (Figure 4.5). Possession of substantial coal reserves gave these regions important advantages, as industry became increasingly dependent on coal for energy from the 1820s. The development and improvement of canal systems was crucial in enabling raw materials and finished goods to be transported economically.

The textiles industry – at the heart of the industrial revolution – was concentrated in two main regions: Lancashire, which specialized in cotton products, and the West Riding of Yorkshire, which focused on woollens. The city of Manchester, for example, experienced explosive growth as the industrial metropolis of the early nineteenth century ('cottonopolis'), with its population multiplying tenfold between 1760 and 1830 (Hobsbawm, 1999, p.34).

These industrial regions of Britain maintained their success during the second Kondratiev cycle from the 1840s to the 1890s, based on the railways, iron and steel, and heavy engineering. At the same time, other regions

of continental Europe, such as southern Belgium, the German Ruhr, parts of northern, eastern and southern France, together with the north-east of the US, experienced rapid industrialization (Figure 4.6). The third wave saw industrialization spread to 'intermediate Europe', parts of Britain, France, Germany and Belgium not directly affected by the first two waves, as well as northern Italy, the Netherlands, southern Scandinavia, eastern Austria and Catalonia (Pollard, 1981). The position of the US manufacturing belt in the north-east and Midwest was reinforced by new rounds of cumulative causation (Figure 4.7), while Japan also began to experience industrial growth.

Peripheral Europe, on the other hand – encompassing most of Spain and Portugal, northern Scandinavia, Ireland, southern Italy, east-central Europe and the Balkans – was left behind, becoming specialized in a subordinate role, supplying agricultural products and labour to the core regions (Knox et al., 2003 p.148). The relationship between these industrial cores and the surrounding territories was defined by 'backwash'

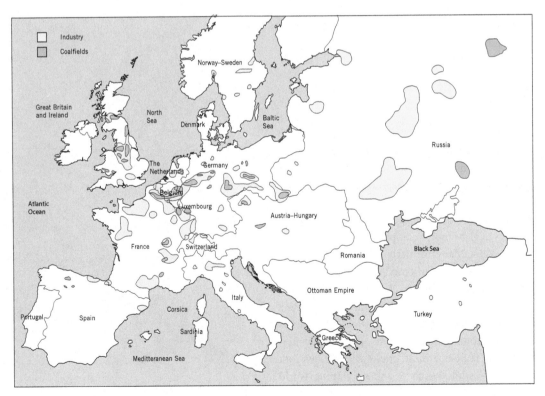

Figure 4.6 Europe in 1875
Source: *Peaceful Conquest: The Industrialization of Europe 1760–1870* (Pollard, 1981). By permission of Oxford University Press.

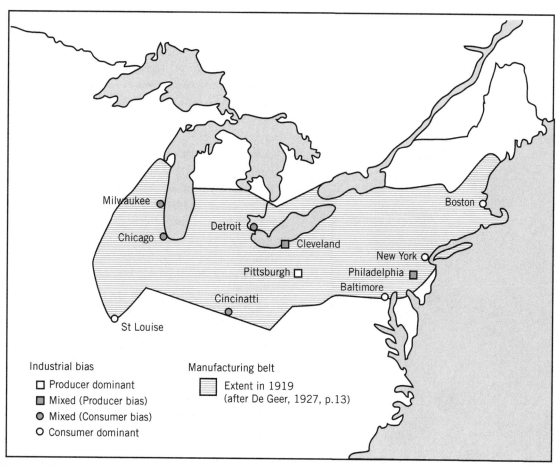

Figure 4.7 The US manufacturing belt in 1919
Source: Knox *et al.*, 2003, p.156.

rather than 'spread' effects, as industrialization sucked in capital and labour, creating an uneven economic landscape of urban-industrial cores surrounded by extensive rural peripheries (Box 4.2).

By the early twentieth century, the spatial concentration of industry in a small number of specialized industrial regions in Britain was readily apparent. These regions were built on a coalfield base, supplying the energy for the heavy industries that emerged from the 1840s. In addition to textiles, the iron and steel and shipbuilding industries became highly concentrated. In shipbuilding, for instance, north-east England and west-central Scotland accounted for 94 per cent of employment in Britain in 1911 (Figure 4.8). South Wales was another leading centre of heavy industry by this stage, focused around coal, iron and steel, while

the West Midlands became specialized in engineering and metal industries. These industrial regions developed a certain '**structured coherence**' as centres of heavy industry (Harvey, 1982), becoming working-class regions with strong Labour Party and trade union traditions.

4.3.2 Fordism and mass production

Under the fourth cycle, based on the mass production of consumer durables, certain established manufacturing regions such as the US manufacturing belt and the Midlands of England further consolidated their position. Proximity to the main centres of population became important for these market-oriented Fordist

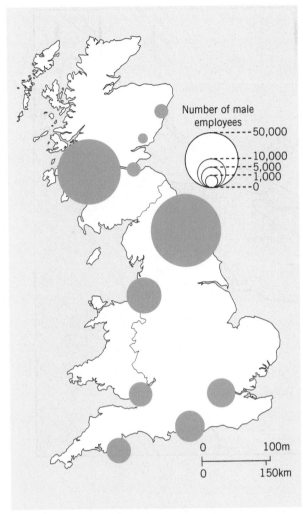

Figure 4.8 Shipbuilding employment in Britain, 1911
Source: Slaven, 1986, p.133.

driven by automobiles and electronics as well as more traditional heavy industries such as shipbuilding and iron and steel.

The pattern of regional sectoral specialization created by successive waves of industrialization in Europe and North America began to break down from the 1920s. As the economic difficulties of the 1920s crystallized into a major depression in the 1930s, the output of industries such as coal, cotton and ship-building slumped. The effects were devastating in terms of unemployment, dereliction and poverty. At the peak of the slump in 1931–32, 34.5 per cent of coalminers, 43.2 per cent of cotton operatives, 43.8 per cent of pig iron workers, 47.9 per cent of steelworkers and 62 per cent of shipbuilders and ship repairers in Britain were out of work (Hobsbawm, 1999, p.187).

Such unemployment was concentrated in the traditional industrial regions of northern England, Scotland and Wales, prompting the official recognition of the 'regional problem' by government in 1928 as central Scotland, north-east England, Lancashire and South Wales were designated as 'depressed regions' requiring special assistance (Hudson, 2003). As such, a growing north–south divide in economic and social conditions in Britain is apparent from the 1930s, although its roots stretch back further (Box 4.2). Rearmament brought a gradual economic recovery from the late 1930s, reinforced by war and subsequent reconstruction. By the 1960s, however, the underlying problems of nineteenth-century industrial regions in northern Britain, the US manufacturing belt, north-eastern France, southern Belgium and the German Ruhr were becoming increasingly severe.

By the 1960s and 1970s, a new phase of '**neo-Fordism**' was apparent – broadly corresponding to the decline phase of this Kondratiev wave – as mass production technologies became increasingly routine and standardized. This was creating a new pattern based on the geographical dispersal of industry to peripheral regions. As Massey (1984) demonstrated, a new '**spatial division of labour**' was emerging, in which different parts of the production process were carried out in different regions, reflecting underlying geographical variations in the cost and qualities of labour.

Increasingly, corporations were separating the higher-level jobs in areas such as senior management

industries. In Britain, for example, the new mass production industries were drawn to the Midlands and south-east England. In the late 1920s and 1930s, particularly, the Soviet Union experienced very rapid state-led industrialization, although this was very much focused on heavy capital goods such as iron and steel and heavy engineering rather than consumer goods. This was focused on a core manufacturing belt stretching south and east from St Petersburg and the eastern Ukraine through the Moscow and Volga regions to the Urals (Knox *et al.*, 2003, p.164) (Figure 4.9). The Japanese economy experienced very rapid growth, averaging around 10 per cent a year, in the 1950s and 1960s,

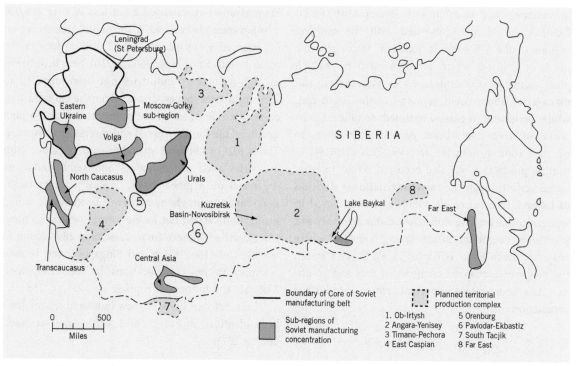

Figure 4.9 The manufacturing belt in the former Soviet Union
Source: Knox et al., 2003, p.164.

Table 4.2 The spatial division of labour in manufacturing

Functions	Characteristics of the division of labour	Location
Research and development	Conceptualization/mental labour; high level of job control	South-east England
Complex manufacturing and engineering	Mixed; some control over own labour process	Established manufacturing reigns like West Midlands
Assembly	Execution, repetition, manual labour, no job control	Peripheral regions such as Cornwall or north-east England

Source: Adapted from Massey, D., Spatial Divisions of Labour: Social Structures and the Geography of Production, 1994 Macmillan, reproduced with permission of Palgrave Macmillan.

and research and development from the lower-level and more routine jobs such as the processing or final assembly of products (Table 4.2). Through the emergence of large, multi-plant corporations, this division of labour takes on an explicitly spatial form, with companies locating the higher-order functions in cities and regions where there are large pools of highly educated and well-qualified workers, with lower-order functions locating increasingly in those regions and places where costs (especially wage rates) are lowest. As a result, the

organization of particular industries becomes spatially 'stretched' through corporate hierarchies, with different stages of production carried out in different regions (Massey, 1988). This contrasts with the nineteenth-century pattern of regional sectoral specialization where all the main stages of production were concentrated within the same region.

When applied to the UK, Massey's analysis reveals a sharp divide between London and the south-east of England where a disproportionate amount of company headquarters and research and development (R&D) facilities are located, compared with the outlying regions of the UK such as Scotland, Wales and the north of England, which are dominated by 'branch plant' activities. Concentrations of professional and managerial labour could be found in the South-east, while peripheral regions contained significant surpluses of lower-cost labour, particularly women, to perform routine work in factories. This dispersal of routine production has also occurred on an international scale through the '**new international division of labour**' (Froebel *et al.*, 1980), as MNCs based in western countries have shifted low-status assembly and processing operations to developing countries where costs are much lower (Chapter 6). Partly as a result, the newly industrialized countries of East and Southeast Asia have become important centres of industrial production.

4.3.3 Deindustrialization and 'new industrial spaces'

The geography of the 'fifth Kondratiev cycle' is based on the rise of new 'sunrise' industries such as advanced electronics, computers, financial and business services, and biotechnology, employing **flexible production** methods. At the same time, many traditional manufacturing regions have experienced serious deindustrialization since the late 1960s, as manufacturing industry has declined in the face of competition, over-production and reduced demand. The legacy for old industrial regions has been one of high unemployment, poverty, industrial dereliction and decay. The West Midlands of England, for example, part of the industrial heartland of the UK in the 1950s and 1960s, lost over half a million manufacturing jobs between 1971 and 1993: 50 per cent of its total manufacturing employment (Bryson and Henry, 2005, p.358).

The process of 'creative destruction', associated with the rise of a new technology system based upon information technology has led to some dramatic geographical shifts, in terms of, for example, the US 'rustbelt' (the north-east and Midwest) and 'sunbelt' (the south and west) and north-western and southern Germany (Box 4.3). For example, while the mid-Atlantic region of the US (New Jersey, New York and

Pennsylvania) experienced a net loss of over 175,000 jobs between 1969 and 1976, the South Atlantic region experienced a net gain of over 2 million jobs in the same period (Knox *et al.*, 2003, p.230). New investment in high-technology industries has been attracted to '**new industrial spaces**', distinct from the old industrial cores that offered attractive environments and a high quality of life for managerial and professional workers. This is part of a broader spatial division of labour, with industries such as computers and semi-conductors organized on a global basis. Typically, for instance, R&D functions might be based in Silicon Valley, skilled production carried out in the central belt of Scotland (the so-called 'Silicon Glen'), assembly and testing in the likes of Hong Kong and Singapore, and routine assembly in low-cost locations in the Philippines, Malaysia and Indonesia (Knox *et al.*, 2003, pp.235–6).

Three different kinds of 'new industrial spaces' have been identified in Europe and North America (ibid., 2003, p.237):

➤ Craft-based industrial districts containing clusters of small and medium-sized firms producing products such as textiles, jewellery, shoes, ceramics, machinery, machine tools and furniture (Box 1.7). Examples include the districts of central and north-eastern Italy, Jura in Switzerland, parts of southern Germany and Jutland in Denmark. This production system is based on high levels of sub-contracting and out-sourcing, often relying on family labour and artisan skills.

➤ Centres of high-technology industries such as advanced electronics, computer design and manufacturing, pharmaceuticals and biotechnology. Examples include Silicon Valley in California (see Box 4.4), Route 128 around Boston in the US; the M4 corridor and Cambridge region in the UK; and Grenoble and Sophia-Antipolis in France. Such areas are characterized by rapid growth and high levels of innovation, often based around small firm networks, although they have spawned some important MNCs. They are often close to major cities but offer a high quality of life for workers with an integrated local labour market that enables workers to switch jobs without leaving the area. Links with universities are often significant in terms of providing

Box 4.3

Britain's north–south divide

Concerns about a north–south divide in levels of wealth and prosperity in Britain have been periodically expressed since the 1930s (Massey, 2001). This became a topic of intense political debate in the 1980s and again in the late 1990s, reflecting concerns about the social and spatial impacts of the policies of the Conservative and New Labour governments respectively. In the 1980s, the combined impact of the neoliberal reforms of the Thatcher government and wider processes of deindustri-alization seemed to have resulted in the emergence of 'two nations': a prosperous and dynamic south in which most of the growth industries were located, and a stagnant and impoverished north, scarred by industrial dereliction, poverty and unemployment (Martin, 1988). The late 1990s saw evidence emerge that the divide had widened under New Labour (Massey, 2001), creating political problems for a government that drew many of its Ministers and Members of Parliament (MPs) from the north.

As Martin (1988) argues, the roots of the north–south divide stretch beyond the 1930s to the nineteenth century. While the image of a dynamic, industrial north and sleepy agricultural south in this period remains powerful, the real situation was more complex. For one thing, the south-east was actually the most affluent region throughout the nineteenth century, reflecting the diversified nature of its economy, and the gap with other regions widened from the 1850s. At the same time, the economies of many northern regions remained somewhat precarious, exhibiting a heavy dependence on a narrow range of export-oriented staple industries. This was particularly the case for what Martin calls the 'industrial periphery' of Scotland, the northern region and Wales (Figure 4.10). These industries col-lapsed in the difficult economic climate of the 1920s, plunging the north into an economic and social crisis. As we have indicated, moreover, the new consumer industries of the interwar years were drawn to the large market of the South-east and Midlands.

The post-war period saw the economic dominance of the South-east consolidated through the expansion of modern, lighter industries and the service sector, com-pared with the sluggish growth of much of the north. The deindustrialization of the 1970s and 1980s decimated the northern regions, whereas the financial and business service and high-tech manufacturing sectors, which grew rapidly in the 1980s and 1990s, were predominantly

located in the south and east of the country. While the recession of the early 1990s was most severe in the South-east, leading to a temporary narrowing of the divide, this was soon overtaken by the geography of economic recovery after 1993, with the South-east gaining the most new jobs between 1993 and 1997 (Peck and Tickell, 2000, p.157). Since 1997, over a million manufacturing jobs have been lost, many of them in the north (Office for National Statistics [ONS], 2005a). More recently, there are indications that northern regions such as Yorkshire, the West Midlands, north-west England and South Wales have suffered most from rising unemployment in the after-math of the 2008–09 recession, which had a particularly adverse impact on manufacturing output and employment (Industrial Communities Alliance, 2009), despite the focus on financial services and the City of London.

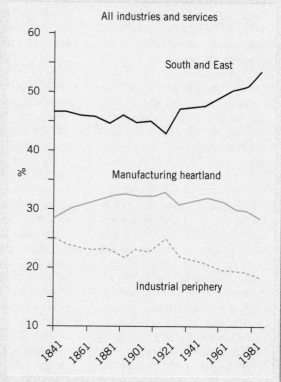

Figure 4.10 The regional distribution of employment in Britain, 1841–1986
Source: Martin, 1988, p.392.

the research and development infrastructure that supports innovation and learning.

➤ Clusters of advanced financial and producer services, often in central districts of large world cities such as London, New York and Tokyo. The City of London is a good example. The concentration of corporate headquarters functions and major financial institutions found in such metropolitan regions is supported by specialized networks of firms in activities such as accountancy, legal services, management consultancy and advertising. These regions are characterized by high levels of specialization and

---

Box 4.4

The development of 'Silicon Valley'

Over the last half century or so, Silicon Valley in California has become probably the most renowned centre of high-tech industry in the world. As such, it has attracted attention from a range of government and development agencies seeking to emulate its success. While the development of Silicon Valley reflects some specific historical circumstances, its experience does highlight some of the factors which explain why particular types of high-tech industry tend to become concentrated in particular places, including links with universities, the availability of skilled labour and the crucial role of social networks (Saxenian, 1994).

Silicon Valley is a strip of land in Santa Clara County to the south of San Francisco, stretching from Palo Alto to San Jose (Figure 4.11). In the 1940s and 1950s, this was a sparsely populated agricultural area focused on fruit production. Writing in the early 1990s, Castells and Hall (1994) summarize its subsequent development and transformation in terms of five main stages:

➤ The historical roots of technological innovation, dating back to the early twentieth century.

➤ The growth of high-tech industry in the 1950s around the Stanford Industrial Park. Here, the links with Stanford University were crucial. An engineering professor, Frederick Terman, was instrumental in the creation of the Stanford Industrial Park in the early 1950s, which attracted a growing number of innovative firms (Saxenian, 1994).

➤ The growth of innovative electronics companies in the 1960s through spin-offs from the original firms. The establishment of Fairchild Semiconductors in 1957 was particularly significant as a base for the growth of a number of spin-off firms, including Intel and National Semiconductors. This phase of growth was underpinned by a strong demand for electronics devices from the military.

➤ The consolidation of semi-conductor firms and the launching of the personal computer era in the 1970s, spawning firms such as Apple and Sun Microsystems as the Valley became increasingly specialized in advanced microelectronics and computers.

➤ The growing dominance of the computer industry, the internationalization of production and a new round of innovative spin-offs in the 1980s and early 1990s.

➤ We can add another stage to this, which has been based upon a new generation of internet technology from the early mid- to late 1990s with a phase of rapid growth (the 'dot.com boom') followed by a sharp recession triggered by the 'dot.com crash' in 2000 with Silicon Valley losing a fifth of its jobs in three years (*The Economist*, 2004). After slowly recovering from this nadir, Silicon Valley is being affected by the current economic downturn, although not as severely as in the early 2000s.

In explaining the success of Silicon Valley, Saxenian (1994) has drawn attention to its distinctive **social networks**. These have facilitated high levels of informal communication and cooperation between individual engineers and entrepreneurs, often working in different firms, allowing technical information and ideas to be rapidly circulated and shared. High levels of labour mobility represent a key mechanism for the diffusion of technology as people move between firms or leave to establish their own spin-off ventures. The success of the early pioneers provided a ready source of venture capital and created a culture of entrepreneurialism and individualism that has spurred subsequent generations of innovators. The great advantage of this industrial system is that it enabled the region to maintain competitive advantage through continuous innovation, something which, somewhat paradoxically, required cooperation between firms (Saxenian, 1994, p.46). In particular, close geographical proximity fostered face-to-face communication and trust between key individuals, providing the region with the adaptability and responsiveness that has allowed it to maintain its competitive advantage.

Box 4.4 (continued)

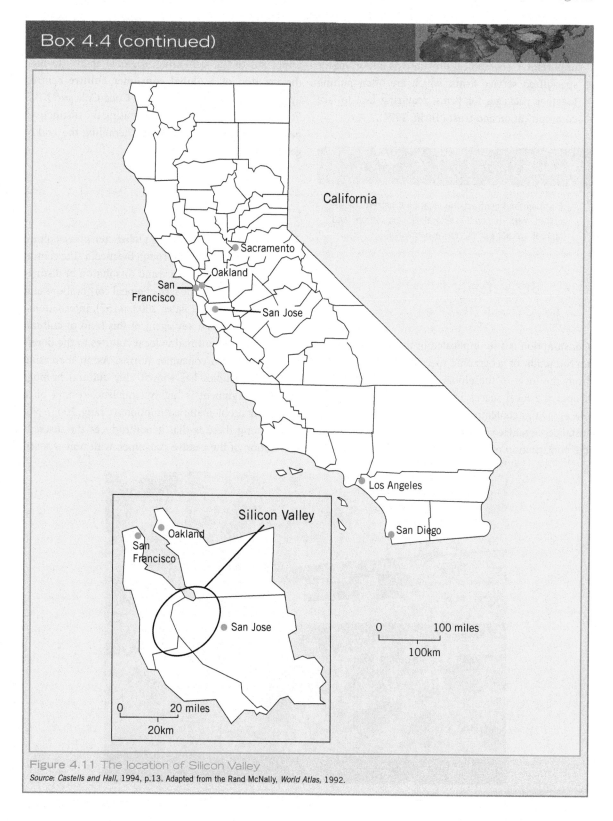

Figure 4.11 The location of Silicon Valley

Source: Castells and Hall, 1994, p.13. Adapted from the Rand McNally, *World Atlas*, 1992.

large pools of skilled, white-collar labour, although low-wage female and ethnic minority labour is also important. Geographical proximity is important for specialized service firms, which are often putting together packages for firms, requiring face-to-face communication and trust (Thrift, 1994).

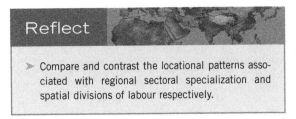

Reflect

➤ Compare and contrast the locational patterns associated with regional sectoral specialization and spatial divisions of labour respectively.

4.4 Spaces of consumption

Consumption is also implicated in the production and reproduction of geographic space at a range of scales, from the local to the global (Crang, 2005, p.360). It shapes the local spaces in which everyday life takes place, with contemporary urban landscapes, for instance, organized to facilitate consumption through the construction of shopping centres, retail parks and

the like, often in out-of-town locations accessible by car. Consumption has been viewed as increasingly important to the fashioning of global space, through the creation of a **global consumer culture** centred upon brands such as McDonald's, Coca-Cola and Nike. For many commentators, this is erasing the distinctiveness of local places and cultures, heralding the 'end of geography' (Ritzer, 2004).

4.4.1 A global consumer culture?

The notion of a homogenous global consumer culture has become popularized through the media. The central image here is of the erasure and dissolution of distinctive local cultures in the face of global corporations and brands (Figure 4.12) (Slater, 2003, p.157). International tourism is seen as a key agent of this kind of cultural imperialism, subordinating local cultures to the dominance of western consumer norms. As an increasing number of studies has shown, this cultural homogenization argument is highly simplistic, resting on a number of problematic assumptions (Crang, 2005). Not least among these is that it reintroduces the discredited notion of the passive consumer, with non-western

Figure 4.12 McDonald's in Beijing
Source: D. MacKinnon.

populations powerless to resist western consumer norms propagated by powerful corporate interests (Slater, 2003, p.158). At the same time, the authenticity of non-western cultures is seen as dependent on being pure and uncontaminated by external forces and influences. In reality, however, cultures are a product of the relationships and connections between places, blending elements from different sources (Massey, 1994). Think, for example, of the importance of drinking tea or eating a curry to contemporary British culture.

A number of recent studies has shown that global consumer cultures assume locally specific forms, blending with pre-existing local cultures in particular ways. Research on how McDonald's is consumed in East Asian countries, for example, found that the chain has been localized and incorporated into local practices, with restaurants functioning also as 'leisure centres, where people can retreat from the stresses of urban life' (Watson, 1997, quoted in Crang, 2005,

p.368). The consumption of such non-local products can actually be seen as part of the production of identifiable local cultures, with people appropriating such goods for their own ends, as shown by Miller's research on the consumption of Coca-Cola in Trinidad (Box 4.5).

There is a clear sense in which modern consumer culture actively embraces cultural and geographical difference with the globalization of food, for instance, presenting consumers in western countries such as the UK and US with the 'world on a plate', through a choice of ethnic cuisine from different cultural regions (Cook and Crang, 1996). In any large British city, for example, consumers are presented with a wide choice of ethnic restaurants, stretching beyond the popular Indian, Chinese, Mexican and Italian to include specialties such as Lebanese, Thai and Vietnamese, and an array of goods of diverse geographical origins in supermarkets. Rather than eradicating geographical difference,

Box 4.5

'Coca-Cola: A Black Sweet Drink from Trinidad'

Coca-Cola is usually regarded as one of the pre-eminent global brands, central to creation of a global consumer culture that is actively marginalizing and subordinating more authentic local cultures. As several commentators have observed, however, this view is highly simplistic (Crang, 2005). One study that demonstrates this is the economic anthropologist Danny Miller's work on the consumption of Coca-Cola in the Caribbean island of Trinidad. Rather than representing the dominance of Western consumer culture, Miller shows that Coca-Cola is consumed in locally specific ways in Trinidad, having become absorbed into local cultures and traditions. This leads him to term it 'a black sweet drink from Trinidad'.

Instead of being viewed as an imported western drink, Coke is seen as authentically Trinidadian. It is

regarded as a basic necessity and the common person's drink (Miller, 1998a, pp.177–8). For local consumers, the categories that frame and inform their choice of product are not those used by the producer and advertisers, but the distinctly local concepts of the 'black' sweet drink and the 'red' sweet drink. The latter is a traditional category particularly associated with the Indian population. The former is summarized in the centrality of a 'rum and coke' as the most popular alcoholic drink on the island, although the consumption of 'black' sweet drinks without alcohol is equally common. Coke is probably the most popular of these 'black drinks', becoming associated with the black African population. These links are historical associations rather than describing actual patterns of consumption with possibly a higher proportion of Indians than blacks

drinking Coke, identifying with its modern image, while many Africans consume red drinks, associating it with an image of Indian-ness that is an essential part of their Trinidadian identity.

These local cultural specificities and complexities impose real limits on the marketing strategies of the producers, indicating that consumption is shaped by a wide range of locally specific factors, beyond their control. More broadly, the case study shows how the consumption of a prominent global brand is dependent on locally specific cultural practices and traditions. In this way, mass consumption practices effectively transform global products into locally specific forms, suggesting that capitalism should be viewed as a diverse collection of practices rather than as a set of overarching economic imperatives.

consumption produces new geographies, presenting us with particular representations of the global and the local, the foreign and the domestic, etc. (Crang, 2005, pp.371, 372). Often, however, the geographies of production and distribution that actually generate the goods on sale in western shops are typically obscured by the emphasis we attach to the physical appearance and price of goods. This process of **commodity fetishism** is inherent to capitalism.

While much work on consumption has been cultural in orientation, and sometimes rather celebratory in tone, some political economists have sought to overcome commodity fetishism by reasserting the relations of production. This involves uncovering the working conditions and regimes through which particular commodities are produced and distributed before being sold to consumers (Harvey, 1996). The downside of this is the tendency to reduce consumption to production. One framework that helps to overcome the opposite extremes of commodity fetishism and productionism is the commodity chain approach (section 10.2). This traces how commodities link together diverse actors performing various roles such as farmers, labourers, haulage companies, shipping companies, retailers and consumers. The increasing globalization of the economy means that such actors are often located in separate countries and continents, emphasizing how commodity flows create a range of linkages between people, things and places (Mansvelt, 2005).

4.4.2 Places of consumption

Another major focus of attention has been particular **places of consumption**. Key sites include the department store, the mall, the street, the market and the home, as well as a host of more inconspicuous sites of consumption (for example, charity shops and car boot sales) (Mansvelt, 2005). Tourist regions and heritage parks are also sites of consumption, albeit of landscapes and experiences rather than material goods and services. The department store is seen as the classic consumption site of the late nineteenth and early twentieth century, particularly in the major cities of North America and western Europe such as New York, London and Paris (section 3.4.3). More recently, such

stores have been reinvented in the form of the 'flagship store', which incorporates its own labels, building on concepts introduced by chains such as Habitat. Good examples of such chains are Harrods and Harvey Nichols, which claim to be selling lifestyles rather than simply goods, combining 'designer interiors, rituals of display and leisure, sexuality and food' (Lowe and Wrigley, 1996, p.25).

The mall or shopping centre has attracted a lot of interest from geographers and other consumption researchers, representing perhaps the most visible and spectacular kind of retail environment. It consists of a range of shops and entertainment facilities within an enclosed space that is usually privately owned and managed (Mansvelt, 2005, p.61). Malls are widely viewed as the iconic space of contemporary retailing, representing the 'urban cathedrals' of contemporary capitalism (Goss, 1993). The world's first fully enclosed mall was opened in Southdale, Minneapolis in 1956, becoming the prototype for thousands of others over the succeeding decades.

Shopping centres are designed in order to maximize the exposure of consumers to goods, with those of the 1960s and 1970s designed as 'machines for shopping'. More recently, planners and developers have sought to provide spectacular places that people want to spend time in, thus maximizing spend, incorporating entertainment facilities, food courts and visual features as well as shops (Crang, 2005, p.373). A number of regional malls was opened in Britain in the 1980s such as the Metro Centre in Gateshead or Meadowhall in Sheffield. Further developments occurred through the development of mega malls in North America in the 1980s and 1990s such as the West Edmonton Mall and the Mall of America (see Box 4.6). Again, people's use of such spaces is not wholly determined by the intentions of developers and chains. They provide certain groups with a place to socialize or 'hang out' instead of shop, with teenagers, for instance, often coming into conflict with centre management (Crang, 2005, pp.374–5).

More recently, the focus of attention has moved away from malls as representing the grand and spectacular towards more mundane and everyday sites of consumption such as the street, the home and the likes of car boot sales. Streets provide a great variety of

Box 4.6

The Mall of America

The Mall of America (MoA) was opened in 1992 in Bloomington, Minnesota. It is probably the most spectacular example of the modern mall in the world today, representing 'the largest fully enclosed retail and family entertainment complex in America' (MoA, 1997, quoted in Goss, 1993, p.45). It receives between 35 and 40 million visitors a year, more than the Grand Canyon, Disneyland and Graceland combined. The Mall contains over 520 stores, over 50 restaurants, 14 movie screens, the largest indoor family theme park in the US, a 1.2-million-gallon aquarium and a range of other attractions (MoA, 2005). According to the owners, 'first in the industry to mix retail and entertainment, Mall of America has become the model for combining signature attractions with retail to create an outstanding entertainment venue' (MoA, 2005, History of Mall of America).

The account presented here is based largely on Goss's study (Goss, 1999). In contrast with other studies that focus on how consumers attach meaning and significance to consumption – utilizing methods such as interviews and focus groups – Goss's approach emphasizes the symbolic construction of the retail environment, 'reading' it in terms of the images and themes that the developers and owners seek to convey. It is important to appreciate that Goss does not consider how consumers actually receive and view the Mall since other work on consumption suggests that there are likely to attach their meanings to it which may be quite distinct from the intentions of the developers (Crang, 2005). While elements of his analysis may be disputed, Goss's research is indicative of how geographers have sought to understand the mall as a key site of consumption.

A key narrative (story) is that of authenticity, stressing the Mall's rootedness in the local environment and culture. An important source of this is the site itself, previously the Metropolitan Stadium; home of local sports teams the Minnesota Twins (baseball) and the Minnesota Vikings (American football). The connections with nearby Southdale are also stressed. As such, the site is a strong symbol of local identity, representing a natural place of congregation, and conveying a local 'sense of place' as rooted and authentic. Notions of travel and tourism are incorporated into the fabric of the mall itself, divided into districts based on imagined tourist-retail destinations (West Market, North Garden, etc.).

A close inspection of certain displays of goods provided some glaring examples of the process of commodity fetishism in terms of how the descriptions of goods served to obscure their actual origins. Most notably, perhaps, a stuffed bear in the shop 'Love From Minnesota' had a tag in its right ear saying 'Minnesotans who live deep in the northwoods among the loons, wolves and scented pine trees, listen to the gentle lapping of the waves of the shoreline while they handicraft unique memories of our homeland, like this one, to share with you'; a tag on the other ear said, 'Bear made by the Mary Meyer Corp., Townsend, Vermont ... Made in Indonesia' (Goss, 1999, pp.54–5).

In terms of time, the developers of the Mall sought to mobilize meaning and a sense of magic through the four key themes of nature, primitivism, childhood and heritage. In this way, a strong sense of nostalgia is evoked, building a collective dream of authenticity, which becomes attached to the products sold with the Mall. According to Goss (1999, p.72), '... it [the Mall] must promote the spontaneity of crowds in order to evoke [the] natural commerce of the marketplace'. The obvious point underpinning all of this is that elaborate narratives of authenticity are constructed to sell products, creating an aura of mystery in order to overcome the perceived meaningless and superficiality of modern life.

environments for consumption, with particular types of shop often tending to congregate in particular districts. Inconspicuous consumption spaces such as car boot sales, charity shops and retro-vintage clothes shops involve the valuing and purchase of second-hand commodities (Crewe and Gregson, 1998). Domestic space has also been re-examined as a site of consumption with research examining how a range of consumer goods is utilized within the home. The role of home-based shopping has also been examined in terms of catalogues, classified adverts and Tupperware, for example. Food and cooking represent another strand of research, with studies focusing on this as an expression of social relations, particularly those of gender, within the household. The body can also be seen as an important site of consumption, being central to the creation of identity through appearance and image, underpinning the consumption of items such as clothes and cosmetics.

The growth of ICTs has had a profound impact on retailing and consumption, opening up a new sphere of **electronic commerce**. Major retailers were routinely using electronic point-of-sale data to automate and control flows of commodities between stores, warehouse and distribution networks by the late 1980s (Hudson, 2005, p.160). More recently, ICT has facilitated new and expanded forms of home-based consumption through the telephone, internet and cable networks. There has been significant growth in internet auction sites, for example, with the best known of these, eBay, established in the US in 1995, overtaking Amazon.com, the online bookstore, as the world's most popular shopping site (Mansvelt, 2005, p.3). More generally, the internet is often used in conjunction with more conventional sites such as shops, providing consumers with the ability to scan websites, identify items of interest and gather information before visiting shops to purchase them. 'Thus the internet becomes a new search technology rather than an electronic space of sale *per se*' (Hudson, 2005, p.162).

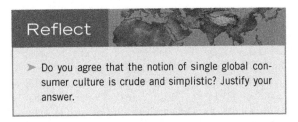

Reflect

➤ Do you agree that the notion of single global consumer culture is crude and simplistic? Justify your answer.

4.5 Summary

In geographical terms, the capitalist system has expanded progressively over time from its roots in western Europe to encompass much of the globe by the early twenty-first century. While facilitated by successive revolutions in the 'space shrinking' technologies of transport and communications, this process has been fundamentally driven by a quest for new markets, fresh sources of raw materials and unexploited reservoirs of labour. The geography of economic activity has shifted from the pattern of regional sectoral specialization that characterized the nineteenth century, supported by colonial markets and raw material supplies, to the spatial divisions of labour and expertise of today. The former pattern involves all the main stages

of production being concentrated within a particular region, while spatial divisions of labour and expertise are based on the geographical 'stretching' of production as different parts of the overall process are carried out in different countries. The 1970s and 1980s saw the dispersal of routine manufacturing activities to lower-cost locations in the developing world (Froebel *et al.*, 1980), paralleled by the second 'global 'shift' of selected consumer services in recent years (Bryson, 2008). At the same time, higher-value activities such as research and development and financial services have become increasingly concentrated in metropolitan regions and world cities.

Spaces of consumption were assessed at both global and local scales, highlighting the links between economy and culture (Slater, 2003). The notion of a single global consumer culture defined by major brands such as McDonald's, Coca-Cola and Nike is hugely influential, but it rests on a limited and simplistic view of cultures as separate and pure. In reality, even highly symbolic 'global' products are consumed in locally specific ways, as Miller's research on Coca-Cola in Trinidad demonstrates. At the same time, different cultures have become increasingly mixed and entangled, with different elements defined as global and local, foreign and domestic (Crang, 2005). Geographers have also focused on local sites of consumption, with research on large malls and shopping centres giving way to work on mundane and everyday spaces such as car boot sales, urban markets and charity shops. The broader point to take from this chapter is that of the continuing importance of place and locality within a globalized economy, focusing attention on the interaction between local conditions and global processes.

Exercise

Concentrate on the city, region or state in which you live. Investigate the restructuring of its economy since the early nineteenth century. Prepare an essay that addresses the questions identified below.

What have been the main industries located there? What markets did these industries serve and where were raw materials and components supplied from? In what ways has this area been affected by successive waves of industrialization? How have the economic fortunes of the place changed over time. What key institutions have shaped the process of economic development? Can this be understood in relation to changing spatial divisions of labour and expertise? Has the area been affected by deindustrialization since the 1970s? Have any new industries or new forms of investment grown in the same period? What are the economic prospects of the area in the early twenty-first century?

Key reading

Bryson, J.R. (2008) 'Service economies, spatial divisions of expertise and the second global shift', in Daniels, P.W., Bradshaw, M., Shaw, D. and Sidaway, J. (eds) *Human Geography: Issues for the 21st Century*, 3rd edn, Harlow: Pearson, pp.339–57.
A useful introductory summary of the changing nature of service economies, introducing the concept of spatial divisions of expertise and assessing how it is shaping the international relocation of services.

Crang, P. (2008) 'Consumption and its geographies', in Daniels, P.W., Bradshaw, M., Shaw, D. and Sidaway, J. (eds) *Human Geography: Issues for the 21st Century*, 3rd edn, Harlow: Pearson, pp.376–94.
A good introduction to contemporary debates on consumption. Focuses particularly on the notion of a global consumer culture and studies of local places of consumption.

Leyshon, A. (1995) 'Annihilating space? The speed-up of communications', in Allen, J. and Hamnett, C. (eds) *A Shrinking World? Global Unevenness and Inequality*, Oxford: Oxford University Press, pp.11–54.

Worth revisiting (Chapter 1 reading) for the historical perspective on processes of time–space compression during the nineteenth century, driven by developments such as railways and telegraph.

Knox, P. and Agnew, J. (2003) *The Geography of the World Economy*, 4th edn, London: Arnold, pp.143–56.
An informative summary of the changing geography of production in developed countries in terms of successive waves of industrialization and the development and restructuring of particular core regions.

Massey, D. (1988) 'Uneven development: social change and spatial divisions of labour', in Allen, J. and Massey, D. (eds) *Uneven Re-development: Cities and Regions in Transition*, London: Hodder and Stoughton, pp.250–76.
An introduction to the concept of spatial divisions of labour, illustrated with reference to the changing geography of the UK.

Useful websites

http://www.oecd.org/
Organisation for Economic Development and Cooperation which represents mainly the developed countries. Offers access to a wealth of statistics, documents and surveys on trends and effects of economic restructuring.

http://en.wikipedia.org/wiki/Consumption.
A brief definition of consumption and summary of contemporary understandings of the term which links to related concepts.

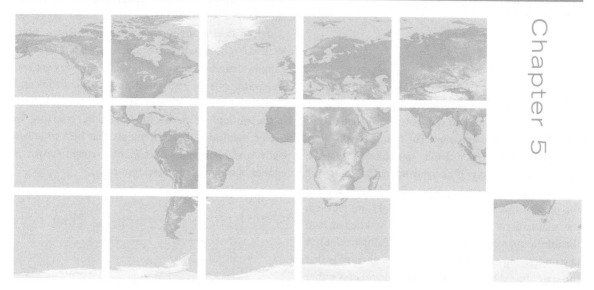

Chapter 5

The state and the economy

Chapter map

In the introduction to the chapter, we define the state and examine its role in the economy. Section 5.2 focuses on the general concept of the state, emphasizing that states are the outcome of historical processes rather than existing as natural entities and highlighting the key notion of the 'qualitative state' (O'Neill, 1997). Section 5.3 reviews the development of the Keynesian welfare state, dominant in developed countries from the 1930s to the 1970s. This is followed by an examination of the notion of a 'developmental state', focusing particularly on the experience of the East Asian 'tiger' economies. We then turn to consider contemporary changes in state–economy relations, assessing how the state has been reformed since the late 1970s in response to processes of globalization and the spread of neoliberal or free market policy. Key aspects of change such as the internationalization of the state through the expansion of bodies such as the European Union (EU), the shift from welfare to workfare and the introduction of new forms of regional development policy are discussed. The main points of the chapter are summarized in the conclusion

5.1 Introduction

The role of the state in the economy is wide-ranging, but often invisible to individual consumers, shaping the provision of goods and services in ways that are not always immediately apparent. For example, whenever you go to the pub for a drink, you will find that the state decides how long it can stay open; the size of the measures of beer, wine or spirits offered; how much of the price is taken in tax; how the drinks are labelled; the standards of hygiene governing the kitchen; and the minimum wages paid to the staff (Painter, 2006). A key underlying argument made in this chapter is that the **state** should be viewed as a dynamic process rather than a fixed 'thing' or object (Peck, 2001). Instead of focusing solely on the size of the state, expressed in terms of levels of taxation or expenditure, for example, we should examine how states intervene in economic life, the forms of economic policies that the state pursues and the effects of these on different social groups and regions (O'Neill, 1997).

Our analysis of the geography of state intervention is informed by our political economy approach, incorporating a regulationist position that rejects the notion of an autonomous, self-regulating economy. The idea of a self-regulating economy is central to mainstream, neo-classical economics, emphasizing the role of the market in ensuring that supply and demand are balanced through the price mechanism, consigning the state to a limited role of upholding property rights and enforcing business contracts. Instead, we believe that the economy is regulated through a wide range of political, social and cultural mechanisms in addition to market forces (Aglietta, 1979). The state plays a key role in harnessing and coordinating these different mechanisms, formulating a wide range of rules and laws covering matters such as business taxation, trade policies, employment standards and financial markets (Table 5.2). The role of the state in the economy is generally directed towards the promotion of economic growth, attempting to create the conditions that allow businesses to make profits and workers to find employment, thereby generating revenue through various forms of taxation. From a geographical perspective, states play a key role in regulating wider processes of uneven development, sometimes introducing policies that are focused on particular types of place (for example, depressed regions). Beyond these very general dimensions of state regulation, the specific forms and functions of the state change over time, as highlighted by the notion of modes of regulation, referring to specific institutional arrangements that help to create relatively stable periods of economic growth known as regimes of accumulation (section 2.4.3).

5.2 Understanding the 'qualitative state'

5.2.1 Defining the state

The state is the basic organizing unit of political life (Figure 5.1). Before proceeding further, it is important to distinguish the state from the **nation**. By the state, we are referring to a set of institutions for the protection and maintenance of society (Dear, 2000, p.789). These institutions include parliament, the civil service, the judiciary, the police, the armed forces, the security services, local authorities, etc. As this suggests, the state is a complex entity stretching beyond what is normally referred to as government (the national executive of ministers and civil servants) (section 3.5). States exercise legal authority over a particular territory, holding a monopoly of legitimate force and law-making ability (Mann, 1984). The nation, by contrast, refers to a group of people who feel themselves to be distinctive, on the basis of a shared historical experience and cultural identity, which may be expressed in terms of ethnicity, language or religion. The two come together to form nation-states in cases where the state territory contains a single nation. This is often presumed to be the norm, but there are many examples of multi-national states that contain different national groups (the UK for one is made up of English, Scots, Welsh and Northern Irish). The corollary of this is that not all national groups have their own states (for example, the Kurds or the Basques).

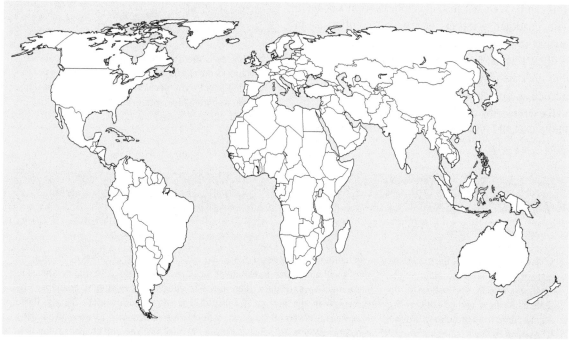

Figure 5.1 A world of states

5.2.2 The growth of the state

The modern state is often regarded as a natural phenomenon that has always existed, in much the same way as many economists have naturalized the modern capitalist economy. In reality, however, states are a product of historical processes operating in the modern era (Painter, 1995, p.31), gradually taking shape in western Europe following the Treaty of Westphalia in 1648. The growth of the state was closely associated with the rise of absolute monarchies with centralized bureaucracies, taxation systems and large armies. The state has played a crucial military role, concentrating resources to attack rival states or to defend its territory against potential threats from outside its borders. It has also accumulated administrative power, gathering information about its population and territory through various forms of surveillance and information storage, from the invention of writing to the advanced computer systems of today (Giddens, 1985). Such technologies have underpinned the development of information-gathering devices such as registration

systems, population censuses and surveys. Through such devices, information is collected, recorded within the state bureaucracy and used to govern the population – a process that is crucial to the consolidation and perpetuation of state power.

5.2.3 The changing nature of the state

As emphasized in the introduction to this chapter, it is useful to view the state as a dynamic process rather than regarding it as a fixed object or thing. This is consistent with our regulationist perspective that emphasizes the role of the state in stabilizing and sustaining capitalist forms of development in terms of both its 'accumulation' and 'legitimation' functions (section 3.5), focusing attention on specific institutional arrangements and processes. Rather than focusing on the size of the state and the extent of its intervention, measured in terms of its share of national wealth, the level of taxation and the volume of welfare payments, recent work has emphasized the nature of such intervention and the social and economic goals that it is directed towards.

This can be seen as a shift of emphasis from a concern with quantitative aspects of state intervention to an interest in its qualitative characteristics (Painter, 2000, p. 363). This new approach examines the specific roles that the state plays in the economy, the key policies it introduces and the geographical effects of these. It is well expressed by the Australian economic geographer Philip O'Neill (1997) in his concept of the '**qualitative state**' (Box 5.1).

Reflect

> How does the qualitative concept of the state as a process differ from traditional understandings of the state as a fixed object with a widely accepted function and a highly centralized structure?

Box 5.1

The 'qualitative state'

O'Neill's conception of the qualitative state is based on three main points. First, he rejects the notion of the state as a single, unified entity with a highly centralized structure. Instead, it is structured by a continual process of interaction between state agencies such as the UK Treasury and non-state actors and forces, for example business organizations such as the Confederation of British Industry (CBI). Second, states always play a crucial role in the construction and operation of markets, including those operating at the international scale. Two key ways in which states have actively constructed markets in recent years are through the privatization of formerly state-owned industries such as electricity or telecommunications and the fostering of globalization through policies that have sought to lower trade barriers and promote competition with organizations such as the WTO playing a key role. An appreciation that the state is always involved in the operation of markets focuses attention on 'the nature, purpose and consequences of state action' (O'Neill, 1997, p.290). Third, a qualitative view of the state overcomes the politically disabling argument that the powers of the nation-state are being eroded by globalization and neoliberalism, emphasizing, instead, that they are being transformed in particular ways. Rather than being powerless in the face of globalizing processes, governments can still regulate markets to achieve broader social goals such as full employment or universal healthcare.

This emphasis on the importance of state action in supporting the operation of market echoes the writings of Karl Polanyi in the 1940s and 1950s. As we saw, Polanyi argued that the state played a crucial role in matching the supply and demand of the 'fictitious commodities' of land, labour and capital during the great transformation wrought by industrialization in the nineteenth century (Box 1.7). The underlying point of Polanyi's work is that markets do not operate in a social and political vacuum. The state intervenes in the economy in a wide variety of ways, implementing a number of measures and programmes to support its goal of promoting economic development (Table 5.2). This involves it engaging with a wider range of non-state interests such as MNCs, financial institutions, business organizations, trade unions and consumer groups. Rather than being a single, fixed entity, the state is made up of a large number of different institutions and agencies, with the relationship between these often characterized by tension and conflict. Getting these diverse agencies to work in harmony is always a real challenge for governments and politicians. Successive attempts to reform the US healthcare system, for example, highlight the difficulties of achieving reform in face of opposition from various interests, with President Obama eventually succeeding in getting his reform bill passed by Congress in early 2010.

The notion of the 'qualitative state' can be used to inform analyses of contemporary processes of state restructuring. It helps to focus attention on the changing forms and functions of the state in relation to the interrelated processes of globalization and neoliberal reform. The qualitative perspective highlights the importance of the state itself as an actor in processes of globalization and economic restructuring, rather than assuming that the state is always acted on by other more powerful forces, such as MNCs and financial markets. States have contributed to the process of globalization through measures such as the reduction of trade barriers and the abolition of controls on the movement of capital.

5.3 The Keynesian welfare state

5.3.1 Origins and development

Comprehensive welfare states were established in the developed countries of Europe and North America in the 1940s. The Keynesian element of the **welfare state** refers to the methods of economic management and planning developed by the state in the post-war period, guided by the theories of the British economist John Maynard Keynes. While **Keynesian economic policies** and the welfare state are distinct entities, they emerged together and became closely inter-linked. The former ensured full employment to fund welfare payments and the latter supported mass consumption and helped to reproduce the labour force through the provision of social services such as health, education and housing (Martin and Sunley, 1997, p.287). The growth of the welfare state in the post-war period is evident from Table 5.1 which shows the coverage of unemployment schemes among a group of industrialized countries in 1925 and 1975 (Painter, 2002, p.163).

5.3.2 Key features

The main features of the Keynesian welfare state are summarized in Table 5.2. It was closely bound up with the consolidation of Fordism as a regime of accumulation or mode of growth based on mass production and consumption, providing the system of state regulation to support and underpin it (Table 2.2). The primary geographical scale at which the Keynesian welfare state operated was that of the nation-state, regarded as the key unit of economic organization (Table 5.2). Keynesianism assumed a high degree of closure of national economies to international trade and financial flows, enabling increased demand for goods to be met by increased supply from within the domestic economy.

Keynesian economic theory emphasized the need for the state to take an active role in managing the national economy through fiscal policies, referring to taxation and government expenditure. The key focus here was on managing the aggregate level of demand for goods and services in the economy as a whole or the total spending of firms and households on purchasing goods and services. Maintaining high levels of spending and consumption is crucial. This involved stimulating demand in downturns by increased government expenditure – supporting public works and employment schemes, for example road building projects – or reducing taxes (Box 5.2). Such policies would

Table 5.1 Unemployment insurance: members as a percentage of the labour force, selected European countries

Country	1925	1975
Austria	34	65
Belgium	18	67
Denmark	18	41
France	0	65
Ireland	20	71
Italy	19	52
Norway	4	82
Switzerland	8	29
United Kingdom	57	73

Source: Painter, 2002, p.163.

Table 5.2 The Keynesian welfare state

Primary scale	Economic policies	Social policies	Spatial policies
National, centralized regulation and management of national economy. Some local delivery of welfare services.	Full employment, demand management. Provision of infrastructure to support mass production and consumption.	Collective bargaining and state help to expand mass consumption. Expansion of welfare and redistributionist policies.	Goal of spatial integration and regional balance. Redistribute resources between rich and poor regions.
National	Keynesian	Welfare	Spatial Keynesianism

Sources: Adapted from Jessop, 2002, p.59; Martin and Sunley, 1997, pp.279–80.

create additional demand in the economy, leading to increased employment and income. In periods of economic growth, the state should dampen down demand by increasing taxes or reducing expenditure, thus preventing inflation. Interestingly, interest in Keynesianism was re-kindled in response to the current economic downturn, focusing attention on the need for fiscal stimuli through increased government expenditure in 2008–09, although this seemed to be giving way to austerity packages involving substantial expenditure cuts from early 2010 (Elliott, 2010).

In the context of Fordism, Keynesian policies were relatively successful, delivering sustained growth between the late 1940s and early 1970s (Table 5.2). Long-term growth in output, exports and the stock of fixed capital between 1950–73 was much greater than in earlier periods (Table 5.3). Increases in output outstripped the growth in employment, reflecting rises in productivity associated with the introduction of new machinery and technology (Armstrong *et al.*, 1991, pp.118–19).

Box 5.2

The Keynesian multiplier

In the 1920s and 1930s, Keynes, based at the University of Cambridge in England, began to criticize key elements of the prevailing economic wisdom. The cherished assumptions of neo-classical theory were clearly at odds with economic reality during the depression of the 1930s, and Keynes sought to reconstruct economic theory and practice accordingly. The depression was being intensified, Keynes argued, by orthodox economic policy, focused upon balancing the budget and limiting the role of the state, which merely depressed the level of demand still further by restricting expenditure. His aim was to save the capitalist system from collapse through the development of better

policies to manage its fluctuations, staving off the threat of prolonged slumps in output and employment (Skidelsky, 2003). Keynes's economic theories were set out in *General Theory of Employment, Interest and Money*, published in 1936.

The concept of the economic multiplier is a crucial part of Keynesian theory. It refers to the snowball or knock-on effects of additional expenditure or investment within the economy. A new factory funded by state expenditure or private investment will buy inputs and services from other firms, creating employment and income, which is subsequently spent on other goods and services. At the same time, the

workers employed in the factory will spend part of their incomes on goods and services, supporting other sectors of the economy such as retail and entertainment. The multiplier effect works in both directions, of course, meaning that relatively small downturns in investment and expenditure can also have significant effects on the overall level of activity in the economy. This makes it all the more important for governments to intervene to counter the fluctuations of the economic cycle.

Aware that his ideas would be opposed by many other economists and officials wedded to traditional doctrines of limited intervention and the balanced budget, the concept of

Box 5.2 (continued)

the multiplier allowed Keynes and his disciples to demonstrate that additional expenditure would be self-financing, generating increased income and receipts. This helped to overcome some initial resistance and Keynes' theories gained broad acceptance in the late 1930s and 1940s (Skidelsky, 2003, pp.546–51). An emerging generation of economists were won over as rearmament and programmes such as the New Deal in the US led to a process of economic recovery. In this way, the Keynesian revolution redefined economics and provided the intellectual basis for a new form of state intervention in the economy. By the early 1970s, the influence of Keynesianism was so pervasive that the American President, Richard Nixon, remarked that 'we are all Keynesians now'. Ironically, however, it was just about this time that belief in Keynesian policies faltered as the post-war boom turned to recession, precipitating the monetarist revolution of the 1970s and 1980s (section 5.5).

Table 5.3 Long-term growth rates, 1820–1970

	Output	Output per head of population	Stock of fixed capital	Exports
1820–70	2.2	1.0	n.a.	4.0
1870–1913	2.5	1.4	2.9	3.9
1913–50	1.9	1.2	1.7	1.0
1950–73	4.9	3.8	5.5	8.6

Source: Armstrong *et al.*, 1991, p.118. Figures refer to the average of the 'advanced capitalist countries' of the US, UK, Germany, Italy, France, Japan and Canada.

The Keynesian welfare state was associated with **national collective bargaining** processes, where representatives of employers, trade unions and government got together to agree pay rates and awards, often on an annual basis. The basic deal, under Fordism, was that labour would gain higher wages while business gained higher productivity. The expansion of welfare supported mass consumption by establishing a minimum income level, helping to underpin demand. Rather than being regarded as merely a cost of production, workers were now seen as sources of demand, vital to the economic health of the nation. Welfare was designed to be broadly redistributionist, supported by a system of progressive taxation (where the wealthy pay more than the poor), although research has shown that most of the welfare benefits in the UK actually went to the middle classes, having only limited success in alleviating poverty among the working class (Goodin and Le Grand, 1987).

5.3.3 Regional policy

The Keynesian welfare state also sought to foster spatial integration, creating a regionally balanced economy. The aim of such '**spatial Keynesianism**' (Martin, 1989) was to close the gap in income and wealth between rich and poor regions. The establishment of national systems of progressive taxation and welfare expenditure contributed to this aim, transferring resources from core to peripheral areas. This was accompanied by **regional policy**, through which government induced companies to locate factories and offices in depressed regions. It was a highly top-down approach, offering grants and financial incentives to companies to locate factories or offices in depressed regions (Box 5.3). At the same time, development in core regions such as southeast England and Paris was restricted. Firms seeking to expand in such regions had to gain official approval from the government, which was often not forthcoming if the new factory or office could be located in the depressed regions. Classical regional policy reached

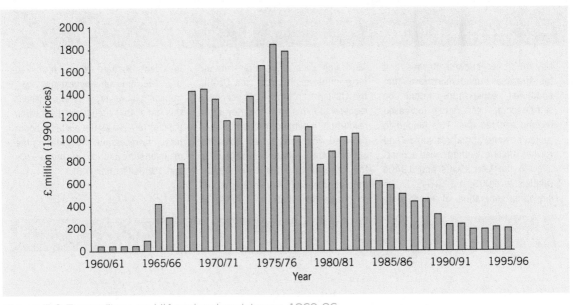

Figure 5.2 Expenditure on UK regional assistance, 1960–96

Source: 'UK regional policy: an evaluation', in *Regional Studies*, 31, Taylor & Francis Ltd (Taylor, J. and Wren, C., 1997), http://www.tandf.co.uk/journals.

its peak in the 1960s and 1970s (Figure 5.2), helping to reduce the income gap between rich and poor regions in Europe (Dunford and Perrons, 1994).

5.3.4 The stagflation crisis of the 1970s

The Keynesian welfare state experienced growing problems from the late 1960s, reflecting the fact that the post-war Fordist growth dynamic, based upon the link between increased productivity and rising wages, was losing momentum. Mass markets for consumer durables such as cars and washing machines were becoming increasingly saturated in Europe and North America. At the same time, full employment and the increased strength of trade unions meant that wages rose rapidly, outstripping productivity growth by the late 1960s. International trade and investment grew marketedly in the 1960s as capital became increasingly internationalized. Increased international flows of goods and capital undermined Keynesian economic policies, based on the assumption of relatively closed national economies. The position of the US in world manufacturing and trade weakened in the 1960s, in face of domestic problems and rising competition from Europe and Japan, in particular. This meant that the

Box 5.3

French regional policy

French regional policy in the post-war era has been organized in a highly top-down fashion, reflecting the post-revolution tradition of a strongly centralized state. Regional policy, known as *aménagement du territoire*, was established in the 1950s, addressing the problem of uneven development, particularly the gap between Pairs and the provinces (Faludi, 2004). It emphasized national unity during the *Trente Glorieuses* period of sustained economic growth (roughly 1945–75) and the need to redistribute the fruits of growth equally between regions (Ancien, 2005). Regional planning was the responsibility of a specialist state agency, DATAR (*Délégation à l' Aménagement do Territorie at à l'Action Régionale*), charged with formulating and implanting development programmes for the 22 administrative regions. DATAR offered grants

Box 5.3 (continued)

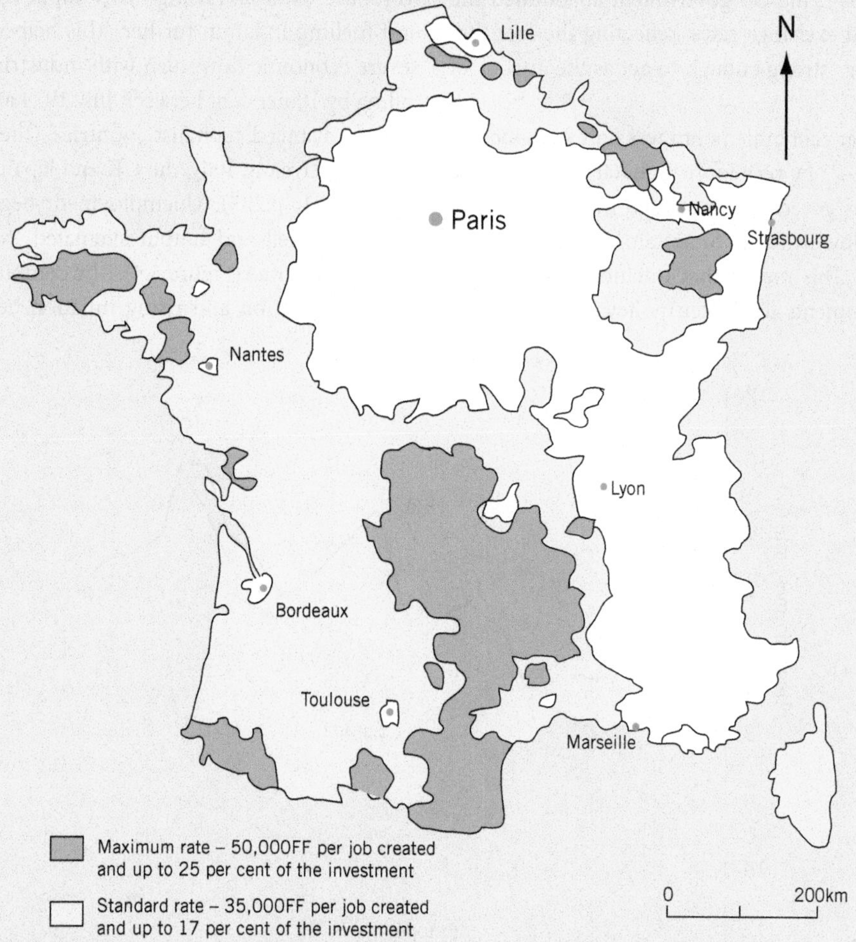

Figure 5.3 Areas eligible for regional development grants in France
Source: Tuppen and Thompson, 1994, p.126.

Map legend:
- Maximum rate – 50,000FF per job created and up to 25 per cent of the investment
- Standard rate – 35,000FF per job created and up to 17 per cent of the investment

Map locations: Lille, Paris, Nancy, Strasbourg, Nantes, Lyon, Bordeaux, Toulouse, Marseille

of up to 25 per cent for projects that created jobs in the '*zones critiques*' or assisted areas which covered much of the south and west as well as selected parts of the north-east affected by deindustrialization (Figure 5.3). Key investment projects located in the periphery included the Citroën automobile factory in Rennes in western France, employing 12,000 workers.

At the same time, the decentralization of government facilities and departments such as new research centres at province level was another important strand of regional policy, along with the designation of eight counterweight 'metropoles' – Lille, Nancy, Strasbourg, Lyon, Marseille, Toulouse, Bordeaux and Nantes – to provide a focus for growth outside Paris (Figure 5.3). Toulouse, for example, became the capital of the aircraft industry, also attracting electronics investment and research laboratories. Restrictions on growth in the Paris region were also introduced, through the *agrément*, a form of official certification similar to the industrial development certificate used in Britain (Tuppen and Thompson, 1994). Factories and offices were required to look for alternative locations outside the capital, although the majority of the decentralization occurred within a zone of some 200 kilometres around Paris, which contained several new towns designated for this purpose.

Bretton Woods system of fixed exchange rates (created in 1945) – whereby other currencies were pegged to the US dollar at fixed rates – came under increasing pressure. In 1971, the US government abandoned the system of fixed exchange rates, reflecting the fact that it was no longer strong enough to act as the anchor of the system.

The post-war economic boom was only sustained in the years 1969–73 by very loose monetary policy in the US and UK, as governments printed money and kept interest rates low in order to sustain growth (Harvey, 1989a, p.145). This meant that inflation rose rapidly, leading governments to tighten policy again in 1973,

resulting in reduced growth and rising unemployment. Oil prices quadrupled in the winter of 1973–74, following the Middle Eastern oil producers' decision to reduce exports, raising firms' input costs massively and fuelling inflation further. This helped to trigger a severe economic downturn with industrial production falling by 10 per cent between July 1974 and April 1975 in the 'advanced capitalist countries' (the US, Canada, France, Germany, Italy, the UK and Japan) (Armstrong *et al.*, 1991, p.225). Unemployment began to rise as investment fell and output stagnated, while inflation continued to rise (Figure 5.4). The combination of economic stagnation and rising inflation became known

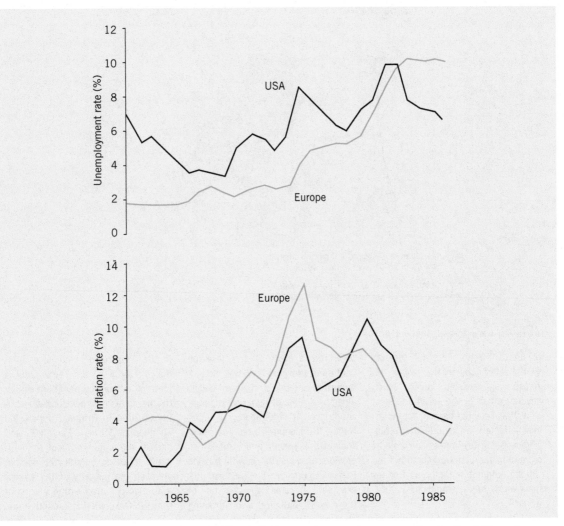

Figure 5.4 The economic crisis of the 1970s: inflation and unemployment rates in the US and Europe, 1960–87
Source: Harvey, 2005, p.14.

as 'stagflation', a term that has come to symbolize the economic crisis of the 1970s.

One of the consequences was to plunge governments into financial difficulties as tax receipts fell in the recession of the mid-1970s while welfare expenditure continued to grow, fuelled by rising unemployment. This so-called 'fiscal crisis of the state' (O'Connor, 1973) saw New York City become technically bankrupt in 1975, for example, while the UK was forced to seek an emergency loan from the International Monetary Fund (IMF) in 1976 in response to its chronic budgetary problems. The conditions attached to the loan involved the cutting back of government expenditure to reduce the budget deficit and control inflation, prompting the Labour Prime Minster, James Callaghan, to announce that 'the party is over', referring to the post-war experience of rising wages and increased welfare expenditure. Keynesian economic policies were seen as implicated in the crisis, prompting the development of an alternative neoliberal agenda (section 5.5.1).

Reflect

> In what ways did Keynesian economic policies contribute to the crisis of the 1970s? Was the abandonment of such policies in the 1980s and 1990s inevitable?

5.4 The developmental state

5.4.1 Industrialization strategies

Developing countries faced a choice of two types of industrialization strategy in the 1950s. **Import-substitution industrialization** (ISI) involves a country attempting to produce for itself goods that were formerly imported. Newly created 'infant industries' are protected from outside competition through the erection of high tariff barriers, allowing the country's economy to be diversified and dependence on foreign technology and capital reduced (Dicken, 2003a, p.176). By contrast, the second type of strategy is known as **export-oriented industrialization** (EOI), based on producing goods for external markets. EOI is compatible with traditional notions of free trade and comparative advantage (Box 4.1) in contrast with ISI which involves high levels of protection and state intervention.

Many developing countries followed ISI strategies in the 1950s and 1960s, with only limited success. Outside the larger Latin American countries such as Brazil and Mexico, domestic markets proved too small to stimulate growth while high tariff barriers against imports generated inefficiency and higher prices for consumers. In response, a number of developing countries began to turn towards EOI from the late 1960s. Aggressive export-oriented strategies have subsequently become particularly associated with the Asian newly industrialized countries (NICs) of South Korea, Taiwan, Hong Kong and Singapore.

Crucially, export-orientation should not be associated with a reduction of government intervention in the economy; instead, the activities of the state have been crucial in fostering export-led expansion. This has given rise to the notion of a distinctively Asian **developmental state** (Douglass, 1994). The first step was often the devaluation of currencies to make exports cheaper in world markets, followed by a raft of export promotion measures, such as the setting of targets and incentives for firms, creation of export promotion agencies and establishment of export processing zones (EPZs). Restrictions on imports remained in place as the state sought to protect strategic '**infant industries**' from outside competition until they were strong enough to compete in global markets (Brohmann, 1996, pp.115–16). South Korea, for example, protected strategic infant industries such as chemicals, steel, shipbuilding and electronics during the 1970s and early 1980s (Dicken, 2003a, p.182). Such an approach, adopted from Japan and the experience of European countries such as Germany, is informed by the interventionist doctrines of the nineteenth-century German economist Fredrich List rather than the classic liberal theory of comparative advantage (Brohmann, 1996).

5.4.2 Characteristics of the developmental state

Weiss's (2000, p.23) definition of the developmental state emphasizes three key criteria:

➤ The priorities of the state, which are focused upon enhancing the productive powers of the nation, raising a surplus for investment and ultimately closing the economic gap with the industrialized countries

➤ Its organizational arrangements focused particularly on the establishment of a strong government department to coordinate and promote industrial development. Such ministries are staffed by an elite bureaucracy and are typically relatively insulated from both short-term political pressures and the demands of particular social groups. The prototype of the powerful coordinating agency is the Japanese Ministry of International Trade and Industry (MITI), created to promote economic growth and to protect 'infant industries' (Brohmann, 1996, p.117). South Korea's Economic Planning Board and Singapore's Economic Development Board have operated in a similar fashion.

➤ Its links with business, which stress the value of close cooperative ties rather than arm's length relations with firms and sectors, allowing the development of new technologies and methods through joint projects. In South Korea, for instance, the state facilitated the development of a small number of large and highly diversified firms – the *chaebol* – large family-based corporations such as Samsung, Hyundai, Daewoo and Lucky-Goldstar – that have dominated the economy (Dicken, 2003a, p.181).

The first two criteria above reflect the autonomy of developmental states from key interests and groups in society (e.g. the landlords, local merchants and the military), helping them to mobilize society towards the goal of economic development. This partly reflects the absence of a strong indigenous capitalist class and the elimination of landed elites in South Korea and Taiwan by land reform programmes in the 1950s, creating an agricultural sector dominated by large numbers of smallholders. This new pattern of landowning leads to increased productivity, helping to generate a surplus for the state to invest in industry and generating domestic demand for manufactured goods (Gwynne, 1990, pp.185, 187). Cultural factors also seem to have underpinned the success of developmental states in East Asia through common 'Confucian' values emphasizing duty,

Figure 5.5 Hong Kong skyline
Source: D. MacKinnon.

loyalty and responsibility, providing acceptance for the role of an authoritarian state bureaucracy.

In geographical terms, the NICs' proximity to Japan allowed them to benefit from its powerful growth dynamic in the 1960s and 1970s. Complementarities between Japan, the NICs and a third group of developing countries such as China, Indonesia, Malaysia and Thailand have fostered the development of an elaborate **regional division of labour** (Brohmann, 1996, p.120). Japan and, increasingly, the NICs (Hong Kong and Singapore in particular (Box 5.4)) have provided capital and focused on high-level managerial and professional activities, promoting themselves as centres for foreign MNCs to locate their regional headquarters while production is carried out in neighbouring lower-wage countries (Figure 5.5).

While the role of the developmental state has clearly been central to the Asian miracle, lifting millions of people out of poverty, there are also negative aspects to this process. These include the perpetuation of authoritarian regimes, the repression of labour and pro-democracy groups, the exploitation of young female workers in export-oriented industries and widespread environmental degradation (Brohmann, 1996). The repression of large sections of the labour force has been one of the main ways in which the NICs have responded to the challenge of maintaining competitiveness through low labour costs, while increasing the productivity of labour. South Korea, for example, was governed by a succession of authoritarian, military backed and strongly nationalist governments between 1948 and 1988, when some liberalization occurred (Dicken, 2003a, p.181). Under the leadership of Park Chung Hee (1961–79), economic policy was directed by the state through a series of five-year plans. In the 1970s, as the popularity of the regime declined, elections were dispensed with, the press silenced, non-state organizations tightly controlled and labour movements subject to direct repression (Douglass, 1994, p.561).

5.4.3 Differences between NIC strategies

At the same time, it is important not to gloss over the considerable differences that exist between the specific strategies adopted by individual NICs (see Douglass, 1994). In Taiwan, industrialization involved the direct state ownership of key sectors while the Korean state implemented policy through its close relations with the *chaebol*. In Singapore, by contrast, the attraction of foreign direct investment (FDI) was the central plank of industrialization from the late 1960s (Box 5.4). Hong Kong pursued a less state-directed approach, reflecting its higher level of development in the 1950s, with colonialism having bequeathed a legacy of entrepreneurialism and a skilled labour force (Douglass, 1994, p.550). In both South Korea and Singapore, the state has directly repressed labour and trade unions, while Hong Kong and Taiwan have relied upon more traditional norms of paternalism and loyalty stressing the mutual obligations between worker and employers within locally-owned factories (ibid., p.55).

Box 5.4

Development states in action: the case of Singapore

By far the smallest of the East Asian NICs, Singapore has been described as the world's most successful economy (Lim, quoted in Huff, 1995, p.1421). In common with the other NICs, the role of the state has been central in driving and shaping development. Although it is a parliamentary democracy, Singapore has been governed by one party, the People's Action Party (PAP) since the early 1960s. Indeed, one powerful politician, Lee Kuan Yew, was in charge until his retirement in 1990, becoming the longest serving Prime Minister in the world. Singapore occupied an important position within the British Empire, becoming a major trading centre and port, reflecting its strategic geographical position astride the major east–west global trade route and its fine natural harbour.

The break from Malaysia in 1965 greatly reduced the size of the domestic market, militating against the pursuit of an ISI-based strategy. In response, the PAP regime pursued an aggressive policy of export-oriented, labour-intensive

Box 5.4 (continued)

manufacturing development based on attracting FDI. This was seen as a question of national survival for a newly independent state without any natural resources. The policy was highly successful, with manufacturing employment increasing fourfold between 1967 and 1979, and manufacturing exports growing from 13.6 to 47.1 per cent of GDP over the same period (Huff, 1995, p.1423). Investment was focused in sectors such as electronics, petroleum and shipbuilding. Control of labour was central to the export-based strategy, with the state achieving a rare degree of acquiescence and cooperation from the labour movement. This was based on the incorporation of labour into corporatist bodies such as the National Wages Council and the redistribution of some of the proceeds of growth in the form of employment, housing, education and healthcare programmes (Coe and Kelly, 2002).

In response to fears about the erosion of its labour costs advantage by low-wage competitors, the government changed policy direction in the 1980s, focusing on the expansion of the financial and business services sector. This approach included an Operational Headquarters Scheme to encourage foreign MNCs to locate high-level management, administrative, and R&D functions there (Dicken, 2003a, p.187). This is closely linked to a strategy of 'regionalizing' the economy by establishing Singapore as the control centre of a regional division of labour, hosting high-level functions such as corporate headquarters, business services and R&D, while labour-intensive manufacturing and assembly is carried out in lower-wage neighbours. The Singapore–Johor Bahru (Malaysia)–Batam/Bintan (Indonesia) growth triangle is the best-known manifestation of this policy (Figure 5.6).

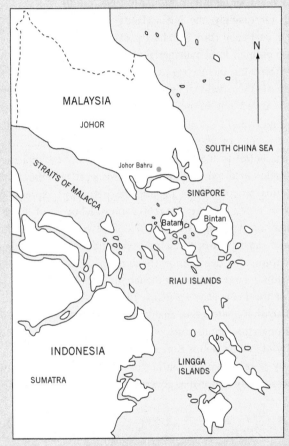

Figure 5.6 The location of the Singapore–Johor Bahru–Batam/Bintan growth triangle of South-east Asia
Source: Sparke et al., 2004, p.487.

Reflect

➤ To what extent does the NIC experience of economic development provide a distinctive model for other developing countries to follow?

5.5 Reinventing the state: neoliberalism, globalization and state restructuring since 1980

Established state structures have undergone considerable restructuring and change since the late 1970s, driven by **neoliberal** reform programmes and the globalization of the economy. In Europe and North America, these structures were based on strong interventionist states, focused on managing their economies according to Keynesian theories and dispensing universal welfare services. Most attention has focused on the abandonment of Keynesianism and the reform of welfare states in the developed world, but states in developing countries have faced similar pressures, not least through the activities of international organizations such as the IMF and World Bank. Neoliberalism can be seen as providing the basis of the new mode of regulation that has emerged since the early 1980s, informing the introduction of a range of institutional experiments and reforms. Whether it offers the stability and order required for the consolidation of a coherent 'post-Fordist' regime of accumulation is, however, highly questionable (Peck and Tickell, 2002).

In considering the impact of neoliberalism and globalization, it is important to recall our earlier discussion of the qualitative state, emphasizing the nature and purpose of its interactions with the economy rather than their magnitude. We are not witnessing the demise of the state, but a set of on-going and multi-faceted changes to its organizational forms and policy functions. While states may have lost powers in some areas, such as trade and the regulation of financial markets, they have maintained or even gained them in other spheres, for example foreign policy, crime or immigration (Peck, 2001). In this section, we examine four key dimensions of contemporary state restructuring: the rise and spread of neoliberalism as a set of ideas shaping state policy; globalization and increased transnational links between states; moves towards a 'competition state' and the introduction of workfare schemes; and the increased prominence of the local and regional scales in terms of economic development policy.

5.5.1 The evolution of neoliberalism

The late 1970s and early 1980s represent a key turning point in the recent history of capitalism (Harvey, 2005, p.1). Following the 1973 military coup, Chile introduced neoliberal economic policies, while the Chinese communist leader, Deng Xiaoping, launched a far-reaching economic reform programme in 1978. A year later, Margaret Thatcher came to power in the UK, espousing a radical new brand of free market liberalism. This was followed by the election of a former Hollywood actor named Ronald Reagan as US President in 1980. 'From these several epicentres ... revolutionary impulses spread out and reverberated to remake the world around us in a totally different image' as the doctrine of neoliberalism was 'plucked from the shadows of relative obscurity ... and transformed ... into the central guiding principle of economic thought and management' (Harvey, 2005, pp.1–2).

As a political and economic ideology, neoliberalism is based on a belief in the virtues of individual liberty, markets and private enterprise. Neoliberals are hostile towards the state, believing that its role in the economy should be minimized to that of enforcing private property rights, free markets and free trade (ibid., p.2). Furthermore, in areas where markets do not exist because of excessive state intervention and regulation, the state should create them through policies of **privatization** (transferring state-owned enterprises into private ownership), **liberalization** (opening up protected sectors to competition) and **deregulation** (relaxing the rules and laws under which business operates). Beyond this, however, the state should not venture, since it cannot possess enough information to second-guess market signals (prices) based on the preferences of millions of individuals (ibid., p.2).

Since the 1970s, neoliberalism has evolved considerably. Three distinct phases can be identified (Peck and Tickell, 2002). The first, proto-liberalism, refers to its early development in the 1970s, when ideas that were deeply unfashionable for most of the twentieth century were developed and promoted by a New Right group of important thinkers and politicians in the UK and US (including the economists Milton Friedman and Friedrich von Hayek) in think-tanks, universities and the media. Their views became increasingly influential, appearing to offer radical solutions to the economic crisis of the 1970s in terms of reducing inflation, cutting welfare spending and restricting trade union power, while restoring individual liberties through the promotion of free markets. The New Right essentially sought to reassert traditional nineteenth-century liberal principles in the circumstances of the 1970s (hence the term neoliberalism).

After the election victories of Thatcher and Reagan, a second phase of 'roll back' neoliberalism ensued. Neoliberal ideas involving the reduction of state intervention in the economy and the curbing of trade union rights were put into practice. Inflation was tackled by applying the monetarist theory of Friedman, based on reducing the supply of money in the economy. This form of 'shock therapy' succeeded in lowering inflation, in the short term, but at the expense of deepening the recession of the early 1980s and increasing unemployment. State intervention in the economy was reduced through policies of privatization, liberalization and deregulation. Many conservative politicians and commentators criticized the welfare state in the US and UK during the 1980s and 1990s, attacking it for encouraging individuals to become dependent on the state, undermining work incentives and imposing a high tax burden. While successive 'reform' programmes have been launched amid considerable fanfare, it has proved more difficult to achieve significant reductions in welfare expenditure (section 5.5.3).

Since the early 1990s, a new form of 'roll-out' neoliberalism has emerged. By this stage, neoliberalism had become normal, regarded as simple economic 'common sense'. As such, it could be implemented in a more technocratic and low-key fashion by governments and agencies such as the World Bank and IMF. Following the conversion of key figures in the early 1980s, the latter two agencies in particular played a crucial role in spreading neoliberal doctrines across the globe, acting as 'the new missionary institutions through which these ideas were pushed on the reluctant poor countries that often badly needed their grants and loans' (Stiglitz, 2002, p.13). In the early 1990s, neoliberal 'shock therapy' in the form of privatization and liberalization was rapidly implemented in the former communist countries of central and eastern Europe.

In the third phase, neoliberalism became softer and more mainstream, informing the **'third way'** policies of centre-left leaders such as Blair, Clinton and Schroder which aim to find a new path between the conflicting extremes of free market capitalism and state socialism. Marrying the efficiencies of markets to a revived social democratic notion of social justice is the key notion. At the same time, neoliberal principles have been incorporated into the design and operation of particular institutions and agencies. Examples would include the establishment of the WTO to implement further trade liberalization, the creation of the euro and European Central Bank (governed by anti-inflationary goals) and the decision to grant the Bank of England independence (insulating it from alternative political agendas). The implementation of neoliberal policies has been a highly uneven process, however, as individual states have tended to adopt particular aspects of the neoliberal package while ignoring others. Elements of neoliberalism have, moreover, interacted with pre-existing institutional arrangements and practices in complex ways, generating a wide range of distinctive local outcomes (Box 5.5) (Peck and Tickell, 2002).

As a political and economic practice, rather than a 'pure' theory, neoliberalism has generated a number of contradictions and tensions (Harvey, 2005, pp.79–81). In terms of gaining and retaining power, neoliberals have had to foster the loyalty and support of citizens, often prompting an appeal to nationalism that has no place in the theory. At the same time, the political authoritarianism that has often been associated with the implementation and enforcement of neoliberal reforms, sometimes against popular opposition, sits uneasily with the emphasis on individual freedoms. Furthermore, the focus on market freedoms and the commodification of a wide range of cultural activities, social services and environmental resources threaten to lead to social fragmentation and anarchy as traditional social bonds and ties are dissolved in the face of an aggressive individualism.

In recent years, the effects of the 2007–08 financial crisis and subsequent global recession have prompted discussions of what could be seen as a fourth phase of 'post-neoliberalism' (Sader, 2009). This reflects a sense in which the crisis has exposed many of the failures

Box 5.5

Neoliberalism 'with Chinese characteristics' (Harvey, 2005, p.120)

Following the momentous decision of the Communist leadership to open up to foreign trade and investment in 1979, China has experienced a rate of economic growth almost unsurpassed in recent history, averaging 9.5 per cent a year between 1980 and 2003 (Wolf, 2005) and over 10 per cent between 2003 and 2008 in GDP terms (Kujis, 2010). Reform was initially couched in terms of the 'four modernizations' (referring to agriculture, industry, education and science and defence), before a period of retrenchment after the Tiananmen Square massacre of 1989. The pace of reform accelerated again in the early 1990s, after an ageing Deng Xiaoping declared, in a tour of the southern regions in 1992, that 'to get rich is glorious' and 'it does not matter if it is a ginger cat or a black cat as long as it catches mice' (quoted in Harvey, 2005, p.125). The implementation of neoliberal reforms in China has created a curious hybrid of communism and capitalism, creating real tensions between economic liberalization and political authoritarianism, in the form of the continuing control of the Communist Party.

The Party leadership initially sanctioned the establishment of four economic zones to attract foreign investment as local experiments which would have little effect on the rest of the economy. Three of these zones were located in the southern Guangdong province, adjacent to Hong Kong, and the other was situated in Fuijan province, across the straits from Taiwan (Figure 5.7). Such experiments proved hugely successful, with the Guangdong province, in particular, acting as a magnet for foreign capital. Two-thirds of foreign investment in China was

being channelled through Hong Kong in the mid-1990s (Harvey, 2005, p.136). Other externally oriented areas such as coastal cites and export processing zones were created in the 1990s with the growth of Shanghai, China's largest city, proving particularly explosive (Dicken, 2003a, p.190), while China's main comparative advantage lay in labour-intensive goods such as textiles, footwear and toys, with hourly wages in textiles in the late 1990s standing at 30 cents compared with $2.75 in Mexico and South Korea, $5 in Hong Kong and Taiwan, and over $10 in the US (ibid., p.332).

Rapid economic growth has created a number of social and political contradictions and tensions in

China. One of the major difficulties that China has faced is how to absorb its huge labour surplus, fuelled by massive migration from the rural areas to the coastal cities, officially estimated at 114 million workers in the reform period (Harvey, 2005, p.127). The principal response in recent years has been the development of massive infrastructure projects, financed by borrowing, such as the Three Gorges Project to divert water from the Yangtze to the Yellow River. Social inequality has increased hugely since the 1980s, with the income divide between the urban rich and rural poor comparable with some of the poorest nations in Africa. Regional inequalities have deepened as the southern and eastern coastal zones have surged

Figure 5.7 The geography of China's 'open door' trading policy
Source: Dicken, 2003a, p.190.

ahead of the interior and north-eastern 'rustbelt' region.

The rapid pace of growth in recent years has made China heavily dependent on imported raw material and energy, with China consuming 30 per cent of the world's coal production, 36 per cent of the world's steel and 55 per cent of the world's cement (ibid., p.139). Environmental degradation and pollution has become a huge problem, with China containing some of the world's most polluted cities. The government has recognized the problems caused by rapid growth, expressing concerns about the pace of development and quality of growth, although the extent to which it can now control the process is questionable (Olesen, 2006). While China has maintained controls on capital and exchange rates, protecting it against the dangers of financial speculation, the conditions of WTO membership (achieved in 2001) require these to be phased out over the next few years (Harvey, 2005, p.141). Rapid economic growth and the spread of a modern consumer culture have helped to contain demand for political liberalization, but an upsurge of riots and protests in recent years suggests that the Communist Party may face problems in retaining control.

of neoliberalism, relating particularly to the deregulation and liberalization of financial markets in the 1980s and 1990s and the encouragement of personal indebtedness to fuel consumption (Blackburn, 2008). In this context, the term 'post-neoliberalism' implies a movement beyond neoliberalism, involving a transition towards a new type of state regime ordered by different principles. This can be seen to be somewhat premature, with Peck *et al.* (2010) remaining sceptical about whether the current crisis does really herald the final collapse of neoliberalism, in view of its dynamic and uneven nature.

There is little real indication that official responses to the financial crisis (section 9.5), comprising large-scale state support for the financial sector and modest forms of re-regulation as represented by the Obama administration's financial reform package in the US (Noyes, 2010), represent a decisive break with neoliberalism (Wade, 2010). Indeed, in Harvey's terms, measures such as the 'bail outs' of insolvent banks in the US and UK can be seen as reinforcing the class project of neoliberalism, as state power is used to support financial interests (Harvey, 2005). The subsequent introduction of austerity packages to reduce public deficits by European governments and international regulators in response to the sovereign debt crisis in the Eurozone of April–May 2010 (Elliott, 2010) would also seem to represent a reinforcement of neoliberalism through the entrenched structural power of markets and international regulatory agencies such as the IMF and European Commission.

5.5.2 Globalization and the state

The dominant view of the relationship between globalization and the state has been that of the hyper-globalists (section 1.2.1). Their theory holds that globalization has eroded the capacity of nation-states to regulate their economies, leaving them unable to intervene meaningfully in markets in order to protect jobs or social conditions. As several commentators have argued, however, this simplistic 'endist' account must be rejected (Weiss, 2000). Rather than offering an objective analysis of state restructuring, it is informed by neoliberal prescriptions, representing a form of wishful thinking that views the reduction of state powers as a 'good thing'. Instead, the 'qualitative state' is experiencing a multi-faceted and on-going process of reorganization, which can partly be viewed as a response to globalization pressures.

The growth of MNCs and global financial markets is certainly associated with changes in state policy since the 1970s, with states exercising political choice in responding to globalization in certain ways. In particular, the huge rise in the volume of mobile capital and the increased openness of national economies led to the widespread abandonment of Keynesian polices in the 1980s. The goals of full employment and modest income redistribution through progressive taxation and high levels of welfare expenditure were dropped. Instead, governments across the world have focused on providing a 'good business climate' for firms, defined in terms of low inflation, reductions in taxation and flexible labour markets. Gaining the confidence of financial markets and investors requires states to ensure low inflation, which is crucial in maintaining the value of financial assets and wealth over the long term (Leyshon, 2000, p.439).

Interest rates were raised by governments such as the US and UK in the early 1980s, in order to reduce inflation and attract mobile capital, deepening the recession and leading to increased unemployment. Taxes on higher earnings, in particular, were reduced in order to increase work incentives and stimulate entrepreneurship. In the US, President Reagan's tax cuts, coupled with increased military expenditure associated with the intensification of the Cold War, resulted in a record budget deficit. More **flexible labour markets** were created by curbing trade union powers, relaxing controls on wages and conditions, and encouraging local company-level bargaining instead of the integrated national bargaining that occurred under Keynesianism.

As outlined in section 5.5.1, a more technocratic phase of 'roll-out' neoliberalism has ensued since the early 1990s. Neoliberal policy prescriptions became consolidated into the so-called **Washington Consensus**, reflecting how this agenda has been embraced and enforced by the US Treasury, World Bank and IMF, all headquartered in the city. The Washington Consensus consists of the following key elements (Peet and Hardwick, 1999, p.52):

➤ Fiscal discipline: minimizing government budget deficits.

➤ Public expenditure priorities: promoting economic competitiveness not the provision of welfare or redistribution of income.

➤ Tax reform: lowering of tax rates and strengthening of incentives.

➤ Financial liberalization: determination of interest rates and capital flows by the market.

➤ Trade liberalization: eliminating restrictions on imports.

➤ Foreign direct investment: removing barriers to the entry of foreign firms.

➤ Privatization: selling off of state enterprises.

➤ Deregulation: abolishing rules that restrict competition.

In adopting the policies outlined above, states have effectively acted as key agents of globalization. Discipline is exercised by the prospect of capital flight (investors withdrawing their money) if alternative policies stressing full employment or the redistribution of wealth are chosen, threatening higher inflation, and by the power of the IMF and World Bank to refuse debt rescheduling for developing countries and declare them uncreditworthy.

At the same time, states also shape processes of globalization through their relations with a range of organizations such as MNCs, financial institutions, supranational authorities and international agencies. The emergence of a more prominent **supranational tier of government**, involving bodies such as the WTO and EU, is largely the result of state action in that states have come together to create such bodies. Membership provides a forum for states to assert and extend their power, requiring regular contact and discussion with other states. It is state representatives who negotiate over trade in the WTO, while states have sought to exert some supervision over international financial transactions through bodies such as the Bank for International Settlements (O'Neill, 1997, p.297). Regional economic integration offers states access to larger markets and protection against competition from outside the regional bloc. The best example is the EU, which has evolved from a customs union between six countries into a monetary union containing 27 states (Box 5.6). Other regional trade blocs include the North American Free Trade Association (NAFTA), incorporating Canada, Mexico and the US, and the ASEAN Free Trade Agreement (AFTA) in Southeast Asia (Table 5.4). States have certainly lost some powers to the EU, particularly in the economic sphere, such as those over trade, competition and monetary policy (for those within the Eurozone), as well as in sectors such as agriculture and fisheries. Even in these areas, state representatives develop policy and national government continues to exercise some authority. In other spheres there is increased cooperation between national states in face of threats such as terrorism, drug trafficking, organized crime, etc. The operation of such

Table 5.4 Major regional economic blocs

Regional group	Membership	Date	Type
EU (European Union)	Austria, Belgium, Czech Republic, Cyprus, Denmark, Estonia, France, Finland, Germany, Greece, Hungary, Ireland, Italy, Latvia, Lithunian, Luxembourg, Malta, Netherlands, Poland, Portugal, Slovakia, Slovenia, Spain, Sweden, United Kingdom	1957 1967 1992	European Economic Union Communities European Communities European Union
NAFTA (North American Free Trade Agreement)	Canada, Mexico, United States	1994	Free trade area
EFTA (European Free Trade Association	Iceland, Norway, Liechtenstein, Switzerland	1960	Free trade area
MERCOSUR (Southern Common Market)	Argentina, Brazil, Paraguay, Uruguay	1991	Common market
AFTA (ASEAN Free Trade Agreement)	Brunei Darussalam, Cambodia, Indonesia, Laos, Malaysia, Myanmar, Philippines, Singapore, Thailand, Vietnam	1967 (ASEAN), 1992 (AFTA)	Free trade area

Source: Adapted from Dicken, 2003a, p.149.

organizations, then, seems to reinforce and reconstruct state power, rather than usurping it (ibid.).

States in the developing world have also been affected by neoliberalism and globalization. In return for providing grants and loans to poor countries that are short of capital, the World Bank and IMF require governments to meet strict conditions, referred to as **'conditionality'** – a highly controversial issue in development circles. In response to the debt crisis facing many developing countries in the 1980s and 1990s, the agencies made the provision of further support dependent on governments agreeing to undertake a

Box 5.6

The EU and state reorganization

The institutional structure of the EU is unique. It is neither a federation nor a simple inter-governmental organization. Decision-making is centred upon the 'institutional triangle' formed by the European Commission, Council and the Parliament. The Council of Ministers is the main decision-making body, representing the member-states. The heads of government and foreign ministers meet at least twice a year in the European Council – high-profile summits that attract widespread media coverage. The main roles of the Commission (27 members appointed by the Council) are to initiate legislation and proposals, implement European legislation, budgets and programmes, and represent the Union on the international stage. The Parliament, consisting of directly elected Members of the European Parliament (MEPs), has powers over legislation and the budget that are shared with the Council. It also exercises democratic supervision over other EU institutions, particularly the Commission.

The EU operates on the basis of agreement and negotiation between independent states, although the member states have delegated some of their functions to the central institutions. Member states continue to shape how the powers of the EU are exercised and implemented, often seeking to further their national interests within the supra-national space it provides. This is evident even in areas such as trade, where the EU is the decision-making body. While the EU is represented as one body in trade negotiations, different member states have adopted different stances. Britain's liberal free trade views, for example, are countered by the more cautious position of countries such as France, making the job of the Commission officials particularly difficult.

set of economic reforms, generally known as structural adjustment programmes (SAPs) (section 8.3.3). These are based on the Washington Consensus specified above, requiring countries to devalue their currencies, reduce public spending and inflation, liberalize trade and privatize state enterprises. The effects of these policies are often damaging, reflecting how they have been imposed from above, irrespective of economic conditions in developing countries. According to Joseph Stiglitz, the former Chief Economist of the World Bank:

> Forcing a developing country to open itself up to imported products that would compete with certain of its industries, industries that were dangerously vulnerable to competition from much stronger counterpart industries in other countries can have disastrous consequences. Jobs have systematically been destroyed … even worse, the IMF's insistence on developing countries retaining tight monetary policies has led to interest rates that would make job creation impossible even in the best of circumstances. And because trade liberalisation occurred before safety nets were put in place, those who lost their jobs were forced into poverty.
>
> (Stiglitz, 2002, p.17)

In this way, the burden of adjustment has been placed on the poor countries themselves rather than the northern institutions and banks that lent the money in the first place. Over the past decade or so, the emergence and consolidation of a new discourse of 'good' governance and poverty reduction has prompted talk of a post-Washington Consensus, although these elements have been added to the principles underpinning the original Consensus, resulting in its augmentation rather than replacement (Sheppard and Leitner, 2010).

5.5.3 Workfare and the 'competition state'

Since the early 1980s, the overriding purpose of the state has been redefined by neoliberals as that of promoting national economic development rather than the provision of welfare services to its citizens. National economic prospects have come to be viewed in terms of ensuring and promoting **competitiveness**. This refers to the underlying strength of the economy in terms of

its capacity to compete with other countries, generating increased wealth and improving the standard of living (Bristow, 2010). It is based on the assumption that nations and regions compete for global market share in a similar fashion to firms. Key aspects of competitiveness include the levels of innovation, enterprise and workforce skills and the task of the 'competition state' is to foster these capacities (Jessop, 1994). Labour market flexibility is an important aspect of competitiveness, requiring that workers be prepared to expand their skills, and modify their wage claims and working practices in response to prevailing economic conditions. The emphasis on the promotion of innovation and competitiveness and the subordination of social policy to these primary objectives are central features of the 'competition state' (Jessop, 1994).

Concerns about the welfare state making benefit claimants dependent on the state and eroding their incentives to participate in paid work have prompted the introduction of a number of **workfare** initiatives, particularly in the US and UK. Workfare can be defined as a system that requires people to work in exchange for welfare benefits and payments. Its introduction over the last two decades has involved a number of schemes that encourage people off benefits and into work, education or training, relying on incentives where possible but resorting to compulsion if necessary, exercised by cutting the benefits of those who have failed to comply.

US states have pioneered the introduction of workfare schemes, beginning under the Reagan administration in the 1980s. Experiments in states such as Massachusetts, California and Wisconsin generated wide international interest from politicians and officials looking for ways to address the problem of welfare dependency. Under 'roll-out' neoliberalism, workfare has become part of the mainstream centre-left consensus with Clinton and Blair stressing its central role in the politics of the 'third way' (section 5.5.1) (Peck, 2001). Elements of **international policy transfer** are evident in terms of the introduction of workfare to Britain through national programmes such as the Jobseekers' Allowance introduced by the Conservatives in the mid-1990s and Labour's New Deal programme for unemployed young people. Polices such as workfare are increasingly regarded as geographically mobile by social scientists, reflecting how specific ideas, programmes and initiatives are circulated and 'churned' rapidly between policy actors located in different territories (McCann and Ward, 2010; Peck and Theodore, 2010).

Workfare programmes reflect the neoliberal conception of unemployment as stemming from the inadequacies of the individual in terms of a lack of suitable skills or unwillingness to work. By contrast, in the 1960s and 1970s unemployment was seen as the product of wider economic and social forces such as a lack of sufficient demand in the economy. The introduction of workfare schemes signals a crucial move away from the welfarist notion of universal entitlement to benefits and voluntary programme participation in favour of compulsion, selectivity and active inclusion in the labour market (Peck, 2001, p.424). For critics of workfare, it rarely leads to permanent employment, and governments would be better advised to create real jobs by expanding the economy. According to Jamie Peck – an economic geographer who has conducted detailed research on the subject – 'workfare is not about creating jobs for people that don't have them; it is about creating workers for jobs that nobody wants' (quoted in Painter, 2002, p.169).

The notion of a workfare-oriented 'competition state' captures some of the underlying shifts that seem to be redefining the role of the state, particularly in North America and Europe. Its importance should not be exaggerated, however. In spite of the rhetoric of reform and the many experiments that have been introduced, the welfare state remains surprisingly resilient, as is evident, for instance, from successive UK governments' preoccupation with the issue of welfare reform. Levels of public expenditure in the UK have not dropped significantly since the late 1960s, fluctuating in line with the economic cycle (reflecting levels of unemployment) (Figure 5.8). More broadly, the public sector still accounts for a large share of the economy in developed countries, ranging from 34.3 per cent in Australia to 54.8 per cent in France (Table 5.5) The social-democratic tradition remains particularly strong in continental Europe with welfare reform proposals often running up against public pressure to protect existing rights and services (Painter, 2002, p.170). As such, the introduction of substantial cuts in welfare as part of austerity packages designed to reduce budget deficits may spark protest and social unrest (Elliott, 2010).

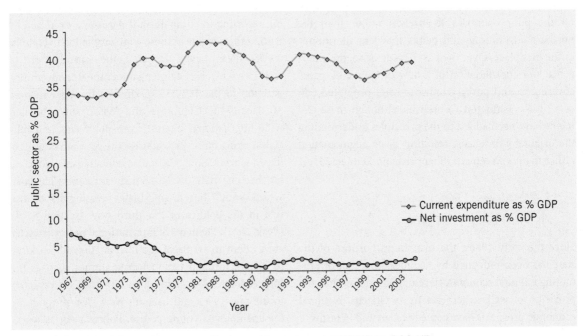

Figure 5.8 UK public sector expenditure and investment 1967/8 to 2004/5
Source: HM Treasury, 2005, *Public Expenditure Statistical Analyses 2005*. Reproduced under the terms of the Click-Use Licence.

Table 5.5 Size of state sector (per cent of GDP)

	1998	2003	2008
Australia	35.2	34.6	34.3
Canada	44.8	41.2	39.7
Czech Republic	43.1	47.1	45.4
France	52.7	53.2	54.8
Germany	48.1	48.4	43.8
Greece	44.4	44.7	48.3
Italy	49.3	48.3	48.7
Japan	42.5	38.4	37.1
South Korea	23.5	28.9	30.0
Sweden	60.2	56.0	51.8
United Kingdom	39.5	42.4	47.5
United States	34.6	36.3	38.8
Euro area	48.6	48.0	46.9
Total OECD	40.6	40.8	41.4

Source: OECD Economic Outlook 86 database.

5.5.4 New forms of local and regional development

One of the contradictory effects of globalization is to have increased rather than diminished the significance of geographical differentiation and place (Harvey, 1989b). A key element of this has been the **resurgence of local and regional levels of government**, particularly as a result of the devolution of powers from the central state. This is closely bound with the growing emphasis on economic competitiveness and growth that has seen local and regional organizations become increasingly active in seeking to attract investment and support innovation and entrepreneurship within their regions. As the Keynesian mode of regulation associated with Fordism has been dismantled in the face of globalization and neoliberalism, regions have become increasingly exposed to the effects of global competition. In particular, the deindustrialization of many traditional industrial regions has forced local authorities and development agencies to focus on the problem of economic regeneration, requiring the attraction of new investment to generate growth and employment.

As part of the move way from Keynesian policies towards a 'competition' state, traditional regional policy (section 5.3) has been downgraded. The objective of spatial redistribution and the interventionist measures associated with it were relaxed in the face of uncertain economic conditions and neoliberal ideology. Regional policy was viewed as an expensive luxury as governments sought to reduce expenditure and adopt less interventionist, market-friendly policies (Knox et al., 2003, p.384). In the UK, the Thatcher government's diagnosis of the 'regional problem' pointed to rigidities in the operation of labour markets, with workers and unions demanding excessive wages (Mohan, 1999, p.185), deterring investment and creating unemployment. This view was encapsulated by the Employment Minister, Norman Tebbit's, advice that the unemployed should 'get on their bikes' in search of work (quoted in Jenkins, 1987, p.326). The Conservatives cut regional assistance from £842 million in 1979 to £560 million in 1985 (Mohan, 1999, p.185) and reduced the areas eligible for assistance in both 1983 and 1993 (Figure 5.9). In France, restrictions on expansion in the Paris region were dropped in the 1980s as the government realized the need to encourage its development as a

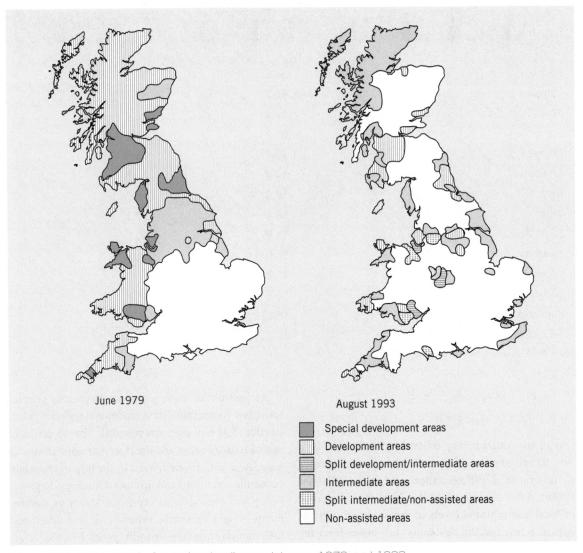

Figure 5.9 Areas eligible for regional policy assistance, 1979 and 1993
Source: Mohan, 1999, p.182.

world city capable of attracting corporate headquarters, and advanced financial and business services.

A range of agencies and organizations has become involved in local and regional economic development since the early 1980s. As well as traditional local authorities, special purpose agencies were created by central government to address specific problems. For example, Urban Development Corporations were created in certain parts of England and Wales in the 1980s in order to regenerate derelict inner-city areas (Imrie and Thomas, 1999), while Regional Development Agencies were established in 1999 to cover each of the standard regions of England. Other organizations involved in local economic development include business organizations such as chambers of commerce and more specialist trade associations, educational institutions and training agencies. In attempting to address local economic problems, local agencies often work together in '**partnership**' where they develop common objectives and share resources. This organizational complexity has prompted the widespread use of the term 'local **governance**' in place of the traditional notion of local government, incorporating the role of special purpose agencies, business interests

Box 5.7

Regional development planning and innovation in Ontario

As an industrial region, which has experienced prolonged economic restructuring since the 1980s, prompting a succession of government initiatives, the Canadian province of Ontario (Figure 5.10) highlights some of the key shifts in regional development policy. As the country's most populous region, Ontario has been the industrial heartland of Canada for decades, accounting for over 50 per cent of manufacturing GDP in recent years (Wolfe and Gertler, 2001). A key trend since the 1980s has been increasing economic integration within North America, culminating in the signing of the North American

Free Trade Agreement (NAFTA) with the US and Mexico in 1994. As a result, according to Pike *et al.*, (2006, p.210), Ontario 'has reoriented itself from a provincial economic heartland and focal point for the trans-Canadian economy to a North American region-state, building upon its close geographical proximity to major US markets in the Great Lakes'.

The recession of the early 1990s was the most severe in Ontario since the Great Depression of the 1930s. Between 1989 and 1991, real provincial GDP declined by over 5 per cent and nearly 200,000 jobs were lost, 150,000 of them in the

manufacturing sector (Wolfe and Gertler, 2001: 579). Subsequently, the economy recovered slowly, punctuated by further economic shifts emanating from the telecommunications bust of 2001–02 and the global financial crisis of 2007–08 (Wolfe, 2010). During this recovery, high-tech sectors experienced high output and productivity growth, while the overall number of plants and jobs fell (Wolfe and Gertler, 2001). The automotive industry continued to dominate manufacturing as it recovered strongly from the recession, with Ontario representing the second largest producer in North America

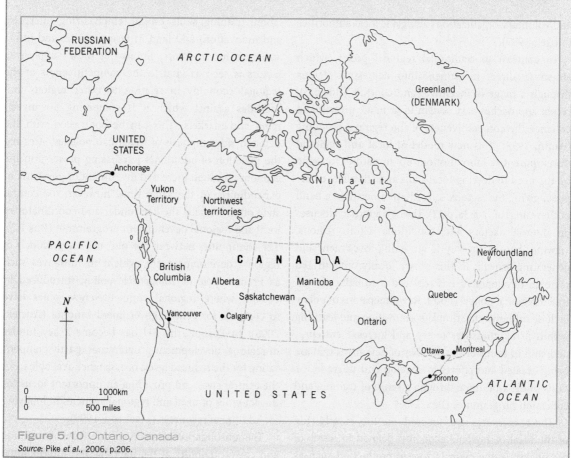

Figure 5.10 Ontario, Canada
Source: Pike *et al.*, 2006, p.206.

Box 5.7 (continued)

after the US state of Michigan (Pike *et al.*, 2006, p.210), although it was severely affected by the fall-out from the global recession of the late 2000s (Wolfe, 2010). Other important industries include electrical and electronic products, food and chemicals.

Both the provincial and federal government have played key roles in supporting processes of industrial restructuring and technological upgrading in Ontario. Broadly speaking, the strategy has to foster the emergence of a modern, knowledge-based economy (Wolfe and Gertler, 2001; Wolfe, 2010). In response to high levels of foreign ownership, associated with a 'branch plant' economy that lacked high-status and highly-skilled jobs, successive governments have sought to strengthen the research base, particularly by encouraging close links between universities and the private sector (Wolfe and Gertler, 2001). Second, policy-makers have supported the growth of knowledge-intensive sectors, prompted by the success of home-grown firms such as Nortel and Newbridge in electronics and Bombadier in aerospace, in addition to a number of smaller firms in the computer and digital media industries. In contrast to traditional 'top-down' regional policy, this approach has generally emphasized the importance of supporting entrepreneurialism and civic leadership at the local level (Wolfe, 2010). Third, governments have invested heavily in post-secondary education since the 1960s. These efforts have met with considerable success, evident in the growth of high-tech clusters such as 'Silicon Valley North' in the Ottawa–Carleton region and the technology sector in Kitchener–Waterloo–Cambridge. Nonetheless, the economic difficulties resulting from the recession of the late 2000s are likely to test the flexibility and resilience of Ontario's decentralized institutional networks to the limit.

and voluntary organizations alongside local authorities (Stoker, 1999).

In contrast to traditional regional policy, which aimed to direct investment into depressed regions through a range of incentives and controls (Box 5.3), recent approaches have sought to facilitate growth and enhance the competitiveness of the regional economy (Amin, 1999). This **new model of local and regional development** is more 'bottom-up' in nature, focusing on the need to develop local skills and stimulate enterprise, giving the regions 'a hand up rather than a hand out' (Amin *et al.*, 2003a, p.22). The new approach focuses on internal factors and conditions within regions, viewing these as the key to attracting investment and generating growth. In this sense, 'locally-orchestrated regional development has replaced nationally-orchestrated regional policy' (ibid.). Key dimensions of policy include initiatives to stimulate innovation and learning within firms, measures to try and increase entrepreneurship in terms of the number of new firms that are being created and efforts to develop and upgrade the skills of the workforce through a range of training and education programmes (Box 5.7).

These measures are focused on the **supply-side** of the local or regional economy, defined in terms of the quality of the main factors of production such as labour training, skills, capital (emphasizing enterprise and innovation) and land in terms of sites and infrastructure for investors. Improving these supply-side factors is seen as vital to the competitiveness of the regional economy in relation to other regions and localities against which it is competing for investment and markets. This can be contrasted with the demand-side emphasis of Keynesian policies, stressing the injection of additional purchasing power into the economy through increased investment or government expenditure. At the same time, however, the central state often remains the key funder and coordinator of local and regional development programmes (Box 5.7). The continuities between the 'old' and 'new' models of regional development are evident in initiatives such as France's *pôle de compétitivé* policy, introduced in 2004/5, where regional competitiveness poles have to compete for central government funding (Ancien, 2005). In Europe, the EU has become a key funder of regional development programmes, partly compensating for the reduced levels of assistance available from the central state and providing an important focus for the activities of local and regional agencies in lobbying for additional resources (MacLeod, 1999).

The consolidation of neoliberal doctrines of free trade and capital mobility at the national and international

Table 5.6 Regional inequalities. Variance of the log of regional GDP per capita

Country	Percentage change		
	1980–90	1990–00	1980–00
Developing countries			
China	−16.31	20.21	0.61
India	7.11	16.96	25.27
Mexico	−1.29	13.57	12.11
Brazil	−17.16	1.33	−16.06
Developed countries			
USA	11.75	−2.69	8.74
Germany	2.18	−0.96	1.2
Italy	1.55	3.01	4.6
Spain	−3.92	10.47	6.14
France	8.63	−0.31	8.3
Greece	1.22	0.13	1.35
Portugal		1.82	
European Union		11.25	

Source: Rodríguez-Pose and Gill, 2004, p.2098.

scales by national states and bodies such as the EU, WTO and World Bank has established a framework of inter-regional competition within which local and regional development organizations operate. Regions have effectively become 'hostile brothers' directly competing for investment, markets and resources (Peck and Tickell, 1994). The unequal relationship between mobile capital on the one hand and communities requiring employment and income on the other has allowed MNCs to play regions off against one another, on the basis of labour costs and the financial assistance offered by government agencies, sparking fears of a 'race to the bottom', as wage rates and living standards are progressively undermined. Despite this, the prevailing supply-side approach presents regional development as a 'race to the top' where all regions can be winners if they follow the right policies (Bristow, 2010).

A key issue for economic geographers is the relationship between local development strategies and **regional inequalities**. Following a period of regional convergence under 'spatial Keynesianism' in the post-war era, regional disparities have widened since 1975 (Dunford and Perrons, 1994; Rodríguez-Pose and Gill, 2004). Table 5.6 shows trend in regional inequalities for selected countries with ten of the 11 experiencing increased disparities between 1980s and 2000. This outcome is hardly surprising, since current policy treats 'unequal regions equally' (Morgan, 2006, p.189), assuming that the same broad strategy should be adopted in all regions, in contrast to the spatial Keynesianism of the 1960s and 1970s which sought to direct investment to depressed regions and restrict growth in core areas. The magnitude of regional inequalities is generally modest for developed countries, but greater for developing countries, with the exception of Brazil where regional inequalities fell.

Reflect

➤ Do neoliberal polices provide the basis for a distinctive post-Keynesian mode of regulation, ensuring prolonged growth and stability (Table 2.2)? Justify your answer.

5.6 Summary

The state refers to a set of institutions that holds sovereignty over a designated territory, exercising a monopoly of legitimate force and law-making ability. The key concept of the qualitative state has been used to frame and inform this chapter, emphasizing that we should focus on the 'nature, purpose and consequences' of state intervention in the economy rather than its extent or magnitude (O'Neill, 1997, p.290). The Keynesian welfare state was the key expression of the state in developed countries over the middle decades of the twentieth century. Shackled to the Fordist system of economic organization, it was associated with over two decades of sustained economic growth between 1945 and the late 1960s. The long post-war boom ended in the 1970s, when rising inflation and high unemployment led to a crisis of the state. At the same time, the NICs of East Asia were experiencing rapid industrialization, shaped by the actions of 'developmental states', which co-ordinated the process of economic development through powerful government agencies, focusing on the protection of strategic 'infant industries' until they were strong enough to compete internationally, the channelling of investment into these industries and the promotion of exports.

The main political response to the crisis of the Keynesian welfare state was that of the New Right. Its neoliberal agenda has spread across the globe, underpinning the Washington Consensus, implemented through bodies such as the WTO, World Bank and IMF. Neoliberalism has sought not only to reduce the role of the state in the economy, but also to reshape the 'internal' structures of the state. Key trends include a move towards a 'competition' state, the introduction of workfare initiatives and a focus on enhancing the economic competitiveness of regions through a range of supply-side measures. In geographical terms, the supranational and regional scales of governance have become increasingly important (Jessop, 1994). It would be a mistake to conclude from this, however, that the role of the national state has been reduced. As suggested by the notion of the qualitative state, the increased prominence of certain scales of state action should not be regarded as resulting in the erosion or decline of others. The overriding conclusion of this chapter is that the national state remains a crucial actor in the regulation of the economy and its geography, not least through its role in coordinating activities carried out across different scales. While the nature, purposes and consequences of state action have changed since the 1970s, the overall economic significance of the state has not been reduced.

Exercise

Select a particular region, referring to basic economic statistics (GDP, income, growth, employment and unemployment) available from the appropriate government publications or website, to get a basic sense of economic conditions within it. Examine and review the current economic strategy for that region, identifying its strengths and weaknesses. What are the key agencies and organizations involved in the formulation and implementation of the strategy (in England, for example, the RDAs are responsible for regional economic strategies, which have to be approved by the national government)? What are the key elements of the strategy? How realistic or appropriate is it in relation to regional economic conditions and needs? What assumptions is the strategy based on? To what extent is it framed by the new model of local and regional economic development identified in section 5.5.4? Are there any major omissions from the strategy? Are there any potential tensions or conflicts between different objectives? What alternative objectives or priorities would you like to see included?

Based on your analysis, sketch your own economic strategy for the region. How would you characterize this strategy (e.g. Keynesian, neoliberal, alternative)?

Key reading

Douglass, M. (1994) 'The "developmental state" and the newly industrialised economies of Asia'. *Environment and Planning A*, 26: 453–66.

An account of the notion of the 'developmental state', focused on the four NICs of East Asia. The paper offers a good introduction to the concept, examining the key relationships between the state, capital and labour. It highlights important differences between the strategies adopted by individual countries.

Harvey, D. (2005) *A Brief History of Neoliberalism*, Oxford: Oxford University Press.

A stimulating account of the impact of neoliberalism across the world from the leading Marxist geographer. Harvey assesses the growth of neoliberal theories, their impact on state polices and the effects on economic growth and development. He views neoliberalism as a project to restore upper-class power and wealth, which was developed in response to the economic crisis of the 1970s.

O'Neill, P. (1997) 'Bringing the qualitative state into economic geography', in Lee, R. and Wills, J. (eds) *Geographies of Economies*, London: Arnold, pp.290–301.

The key article in which the concept of the qualitative state is developed. O'Neill criticizes the focus on the extent of state intervention, focusing attention on the 'nature, purpose and consequences' of state action (p.290). Instead of heralding a decline in the importance of the state, the period since the 1970s has seen the nature and purpose of its intervention in the economy change significantly.

Painter, J. (2002) 'The rise of the workfare state', in Johnston, R.J., Taylor, P. and Watts, M. (eds) *Geographies of Global Change: Remapping The World*, 2nd edn, Oxford: Blackwell, pp.158–73.

A concise and accessible introduction to recent debates on the changing nature of the state. Reviews the rise and fall of the Keynesian welfare state and explores the alleged shift towards a workfare state since the 1990s. Less sceptical about the significance of these changes than Martin and Sunley.

Peck, J., Theodore, N. and Brenner, N. (2010) 'Postneoliberalism and its malcontents'. *Antipode*, 42: 94–116.

An important contribution to recent debates around 'postneoliberalism' – the idea that political and economic relations are moving beyond neoliberalism in the wake of the financial crisis. This paper takes a cautious approach, arguing that neoliberalism has become deeply entrenched as a set of institutional rules that structure the relations between different scales and spheres of economic activity.

Sheppard, E. and Leitner, H. (2010) '*Quo vadis* neoliberalism? The remaking of global capitalist governance after the Washington Consensus'. *Geoforum*, 45: 185–94.

An incisive assessment of recent changes to the system of global capitalist governance, focusing on the emergence of a 'post-Washington Consensus' discourse around poverty reduction and good governance and the growth of a new development economics which advocates Keynesian principles. Ultimately, however, Sheppard and Leitner view these as supplements to the original Washington Consensus which reassert the underlying belief among policy-makers that capitalism is the solution to the problem of global poverty.

Useful websites

http://europa.eu/index_en.htm
The official site of the European Union. Contains a wealth of information on the EU's activities, divided into specific topic areas. See especially the sections on 'economic and monetary affairs', 'enterprise', 'external trade' and 'regional policy'.

http://www.ontariocanada.com/ontcan/1medt/en/home_en.jsp
The official site of the Ontario Ministry of Economic Development and Trade. Provides a useful insight into current regional development policy, containing details of range of programmes and initiatives.

http://web.inter.nl.net/users/Paul.Treanor/neoliberalism.html
Offers a useful introduction to neoliberalism, covering its origins, theoretical background and definition.

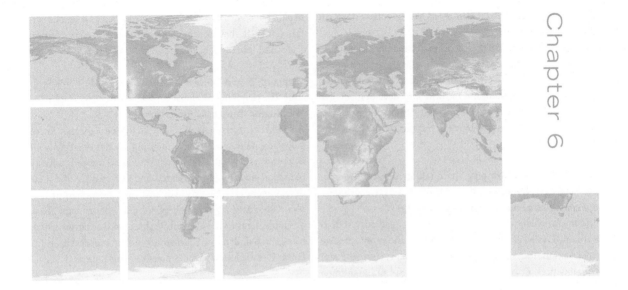

Chapter 6
The changing geography of multinational corporations

Topics covered in this chapter

- ➤ The changing geography of foreign direct investment (FDI).
- ➤ The emergence of the multinational corporation and the 'New International Division of Labour' (NIDL).
- ➤ The 'globalness' of multinational corporations and their geographical embeddedness in their home countries.
- ➤ The internationalization of services through FDI.
- ➤ The implications of FDI for host regions.

of its emergence and growing importance in the global economy. Section 6.2 looks at the changing geography of FDI by corporations since the 1960s, while section 6.3 outlines the reasons for the growth of the MNC and examines recent developments in its evolution. Section 6.4 explores the extent to which MNCs have become global entities, questioning the 'myth of the placeless corporation' (Dicken, 2007) and highlighting their continuing embeddedness in their home regions. In section 6.5 we consider the internationalization of services through FDI, which has grown considerably in recent years. This is followed by an examination of the relationships between MNCs and the 'host' regions in which investments are located, assessing the regional development effects of FDI.

Chapter map

We begin the chapter with a definition of the multi-national corporation (MNC), followed by an analysis

6.1 Introduction

The **multinational corporation (MNC)** has become one of the dominant actors in globalization and advanced capitalism, reflecting the large number of these firms and the geographical reach of their operations. MNCs are fundamental to the operation of the global economy, representing one of the principal mechanisms through which global economic integration takes place. Through their geographically dispersed production networks, MNCs connect up different places in increasingly complex international divisions of labour. In contrast with the hyper-reality and virtual spaces of global financial networks, MNCs' organizational networks seem to provide a more tangible and material economic geography of globalization. MNCs are increasingly at the heart of 'webs of enterprise' (Dicken, 2003a) that connect headquarters, assembly plants, research and development facilities and increasingly complex supply chains to produce individual goods and services.

Despite having this critical role in the functioning of the global economy, it is important to dispel the image of all-powerful, 'placeless' organizations that characterizes some of the business and media literatures (e.g. Ohmae, 1990). Instead, as we will demonstrate in this chapter, to understand the workings of MNCs we need to appreciate their diverse organizational geographies. While they may operate trans-nationally, individual MNCs create their own organizational geographies, influenced both by the national and regional business cultures from which they originate, and by the

characteristics of the regions in which they locate. For this reason, we prefer to use the term **multinational** here rather than **trans-national**. This is to signify corporations that operate in more than one country but that are geographically embedded or rooted in particular places, rather than the popular notion that corporations are now footloose and not tied to anywhere, hence 'trans-national' (section 6.4).

6.2 The changing geography of FDI

Foreign direct investment (FDI) can be seen as a key mechanism of international economic integration or globalization, alongside trade and financial investment. It refers to direct investment across national boundaries, where a firm based in one country buys a controlling investment in a firm in another country or sets up a branch or subsidiary in another country (Dicken, 2007, p.36). As such, FDI can be distinguished from (indirect) portfolio investment whereby a firm or investor buys shares in another firm for financial reasons.

After the end of the Second World War, peace and political stability in Europe underpinned the internationalization of business activities, leading to a dramatic increase in FDI. By the 1960s, rates of growth

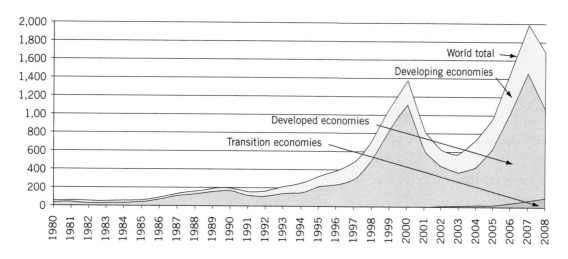

Figure 6.1 FDI inflows, global and by groups of economies, 1980–2008 (billions of dollars)
Sources: UNCTAD FDI/INC database (http://www.unctad.org/fdistatistics) and UNCTAD Secretariate estimates.

in FDI were growing 40 per cent faster than exports (Dicken, 2003a, p.52). After a brief slump in the 1970s, FDI levels experienced further growth from the early 1980s onwards. As Figure 6.1 indicates, the FDI growth has not been continuous, with periods of dramatic expansion followed by slumps in investment during recessions. Rapid growth ensued from the mid-1990s following the downturn of the early 1990s, peaking in 2000 prior to the downturn of 2001–03. Renewed growth saw global FDI flows reach a new peak in 2006 before declining as a result of the global financial and economic crisis (Figure 6.1).

The changing geography of FDI over the past century highlights important continuities and changes in patterns of global economic development. If we consider, first, the stock of outward FDI (i.e. from the country of origin), the period from 1914 up until the mid 1970s marks the decline of the UK as the dominant source and the growth of the US (Table 6.1a). The expansion overseas of US MNCs was particularly dramatic after the end of the Second World War, when the development of a stable trade regime under the Bretton Woods agreement, the recovery of the western European economy and the dominance of the US dollar

Table 6.1a Foreign direct investment outward stock by leading investing countries as % of world total, 1914–2008

	1914	1960	1978	2000	2008
Developed countries	100.0	99.0	96.8	85.4	84.1
USA	18.5	49.2	41.4	21.7	19.5
Europe	–	–	–	53.6	55.0
UK	45.5	16.2	12.9	14.8	9.3
Germany	10.5	0.8	7.3	8.9	9.0
France	12.2	4.1	3.8	7.3	8.6
Japan	0.1	0.7	6.8	4.6	4.2
Developing countries	–	1.0	3.2	14.2	14.5

Sources: Derived from Knox and Agnew, 1989, Tables 8.4, 8.5, pp.255–56 and UNCTAD, *World Investment Report 2009*, Annex Table; B.2, pp.251–4.

Table 6.1b Foreign direct investment inward stock by world region and selected countries as % of world total, 1914–2008

	1914	1960	1978	2000	2008
Developed countries	37.2	67.3	69.6	68.8	68.5
United States	10.3	13.9	11.7	21.8	15.3
Western Europe/Europe[1]	7.8	22.9	37.7	39.6	46.5
Japan	0.2	0.2	1.7	0.9	1.4
Developing countries	62.8	32.2	27.8	30.2	28.7
Latin America	32.7	15.6	14.5	8.7	7.9
Africa	6.4	5.5	3.1	2.7	3.4
Asia	20.9	7.5	7.0	18.7	17.3

[1] Includes the 10 EU accession countries of 2004, Bulgaria and Romania in 2000 and 2008 figures.
(*Sources*: Derived from Dicken 2003a, Table 3.8 p.57; Knox and Agnew 1989, Table 8.4, 8.5, pp.255–6; UNCTAD *World Investment Report 2009*, Annex Table Annex Table; B.2, pp.251–4.)

as a global exchange currency presented new market opportunities (Wright, 2002). The 1970s and 1980s were marked by the growing importance of overseas investments by Japanese firms, in particular, alongside the growth of European countries such as France and Germany. More recently, the share of developing countries in outward FDI has grown, with a small number of **newly-industrialized countries** (NICs) in East and Southeast Asia playing a key role, with Hong Kong particularly prominent. The US remains the single largest national source of outward FDI, although its share has decreased markedly since 1960. While Europe has become more important as a bloc through economic integration, its highly uneven economic geography continues to be reflected in outward FDI, with Germany, France and the UK accounting for just under half of the EU total.

In terms of inward FDI (Table 6.1b) – host countries for investment by MNCs – in the early part of the twentieth century, developing countries were the dominant destination, largely linked to resource and raw material extraction to supply the industries of the developed world, a set of relations that we can term the '**old international division of labour**'. But after the First World War and especially since the end of the Second World War in 1945, there has been a shift towards greater investment flows between different regions of the developed world, with FDI linked more to the setting up of manufacturing plants to gain access to affluent consumer markets. Western Europe in particular became the major destination for FDI, initially from US MNCs in the 1960s, followed by Japanese and East Asian corporations in the 1980s and 1990s. The US itself experienced an unprecedented surge of inward investment from Japanese and European MNCs in the 1980s and 1990s. Another key trend was the marked growth in FDI to Asia, largely East and South-east Asia, attracted by access to growing markets and the availability of low-cost labour and land, in contrast with the declining and stagnant performance of Latin America and Africa respectively.

Reflect

> To what extent is the global map of FDI in 2000 different to that in 1914?

6.3 Understanding the emergence and development of MNCs

There are two main ways in which firms internationalize: through the setting up of new plant or facilities; and by the acquisition of existing factories or firms (Dicken 2007). The choice of strategy in turn will be dependent on the motivations behind the initial decision to internationalize. For example, if a firm is attracted to a country because labour is cheap, it might set up a new production plant there to serve global export markets, based on the assembly of largely imported materials. Conversely, a firm may have identified a new growth market in the destination country which can provide a better return than some of its existing activities at home, but for reasons of cost and uncertainty, decided that the acquisition of an existing plant is a less risky option than developing a new plant from scratch. Acquisition might also be driven by a desire to reduce the competition in its market by taking over a foreign rival.

6.3.1 Why do firms internationalize?

In addressing the issue of why firms internationalize, we can return to some of the insights about the functioning of the economy in Chapter 3. Firms operating in a capitalist system have the ultimate aim of realizing surplus value or profit. The rate of profit must increase over time for a firm to reinvest and stay ahead of the competition, as we saw through the operation of the circuit of capital (Figure 3.1). In successful firms this is usually accompanied by expansion both at the level of the organization itself, with economies of scale providing increasing returns to capital, and growth in geographical terms as firms seek to overcome the initial restrictions of place.

Increases in profits are achieved by increasing the productivity of existing operations, by increasing the exploitation of labour or by developing new technology or new forms of organization (as we noted in Chapter 3). This can continue successfully for some time, but at some point the returns to capital from remaining in its original location begin to fall, either through the growing organization and resistance of labour, or increased competition from other firms, or the achievement of productivity limits in existing technologies and methods of production. Addressing these falling profit rates requires firms to seek out cheaper locations than existing arrangements. The need to resolve this dilemma and develop a new geography of production has been termed the **'spatial fix'** (Harvey, 1982), whereby capitalists seek to set up operations elsewhere. This will often involve relocation within the existing national economy in the first instance, creating a new spatial division of labour (Massey, 1984), but over time, as new limits to profitability are reached at the national level, firms will start to internationalize their operations.

Investment overseas is usually driven by the desire for access to either markets or new sources of labour and raw materials (section 4.2). In this respect, it is likely that market access accounts for the largest share of FDI, given that developed economies accounted for 68 per cent of FDI inflows in 2007 (UNCTAD, 2009, p.247). Certainly, the flow of US investment into Europe in the 1950s and 1960s was linked to taking advantage of the opportunities in the rapidly growing European economy that experienced an average growth rate over the period above that of the US (Williams, 1995). Equally, the wave of inward investment from Japan and East Asia into North America and western Europe from the 1980s onwards was driven by market access. In order to overcome barriers to market entry, principally involving local sourcing and production rules, Asian firms gradually shifted from export-based strategies to setting up overseas plants.

6.3.2 The new international division of labour

While market access has been the dominant factor, there has also been a growing level of FDI activity linked to the search for cheaper labour costs. This has led to a growing level of investment in the developing world, associated with a phenomenon known as the **'new international division of labour'** (NIDL) (e.g. Hymer, 1972; Froebel *et al.*, 1980), replacing or, more accurately, running alongside the OIDL. The simple model envisaged by economist Stephen Hymer, dating from the 1960s, was one in which MNCs developed a new geography of production by reorganizing the division of labour within the firm. Higher level decision-making and R&D activities would remain concentrated in the major metropolitan regions of the advanced world (e.g. London, New York, Paris, etc.) while the more routine production activities would be dispersed, depending upon labour skills and levels of technology to more peripheral locations at home (see Table 6.2). Over time, the more basic production functions will be relocated to cheaper locations overseas with a 'race to the bottom' encouraging the search for low-cost locations in the third world or global south. The developed countries would, however, retain the higher value added activities such as long-term strategic planning and R&D.

Three underlying factors shaped the NIDL, according to Wright (2002, p.73):

➤ an extended technical division of labour in production which reduced work to simple routine functions that needed minimal training, allowing them to be performed by semi-skilled workers (see section 3.3.2);

➤ the development of advanced transportation technologies such as containerization and air freight that allow materials and finished/semi-finished goods to be moved cheaply and efficiently over large distances, making production 'footloose';

➤ the release of agricultural labour as a low-wage resource onto urban labour markets in less developed economies through the modernization and intensification of agriculture under the 'Green Revolution'.

The archetypal example of production organized through the NIDL is the related industries of clothing and textiles. Wage rates vary enormously, globally, from over $20 per hour in some wealthy western European countries to less than $0.50 per hour in Bangladesh (Dicken, 2007, p. 257). The sector is still dominated by American and European manufacturers and retailers, who control the marketing, design and technologically more sophisticated production processes, but an

Table 6.2 Hymer's stereotype, in which the space–process relationship takes the form A–B–C

Level of corporate hierarchy	Major metropolis (e.g. New York)	Type of area Regional capitals (e.g. Brussels)	Periphery (e.g. South Korea, Ireland)
1. Long-term strategic planning	A		
2. Management of divisions	D	B	
3. Production, routine work	F	E	C

Source: Sayer, 1985, p.37.

increasing amount of the more routine production is located overseas. For European MNCs, a first wave of relocation in the 1960s and 1970s to low-cost production sites in geographically proximate areas (such as Portugal, Greece and North Africa) has been followed in the period since 1980 by a further wave to even cheaper locations in South-east and East Asia. As a consequence, employment in Europe in the clothing sector fell by 50% between 1980 and 1998, while growing in Asia by 35 per cent (ILO, 2000). At a global level, China has been the largest employer since 1980, with a clothing workforce of over 3.6 million, almost four times that of the next largest country, the US, in 1998 (ibid.).

For developing countries, these gains in employment were often associated with poor working conditions and low pay by global standards, although wages were often higher than for other types of work available in the host economy. Thus, a new form of exploitation replaced the old one, while, at the same time, the livelihoods of the working classes in the core economies were undermined. The NIDL now applies to routine work in some service sectors, as well as manufacturing activities, particularly in areas such as banking and telecommunications, where new information technology allows 'back-office' and customer-service functions to be relocated overseas to lower cost regions (see section 6.3.5).

6.3.3 Variations in internationalization

While market access and labour supply explain why some firms choose to internationalize production, this is clearly only part of the story and in practice many firms choose not to internationalize production at all. An alternative approach is the 'eclectic' paradigm of international business theorists such as John Dunning (1980), which emphasizes that there are no simple models of MNC development, but rather that a range of motives related to 'accidents of history' and the particular development of individual firms explain why and when internationalization occurs. It is important to remember, however, that the vast majority of firms do not internationalize – even in the contemporary economy. Overseas investment, particularly in a new and unfamiliar environment, involves many risks for the individual firm, such as concerns about the quality of the labour force and the abilities of local management. The lack of such business knowledge about foreign economies remains an important deterrent.

The extent of trans-nationalization also varies between different sectors of economic activity, reflecting differences in the nature of the production process and markets being served. According to Dicken (2003a), MNCs are particularly evident in three types of industry: high-technology industries such as pharmaceuticals and electronics; large volume, medium-technology industries such as motor vehicles; and mass consumer products such as jeans and t-shirts. These are industries that require both high levels of technology and resources (both monetary and labour) but for which demand is highly variable, therefore suiting firms that can organize at the transnational level and take advantage of variations in local conditions.

Within individual industrial sectors, firms can choose completely different geographical strategies when faced with the same market conditions, suggesting the importance of particular corporate cultures and even individual management decisions in shaping firm action (Schoenberger, 1997) (see Box 6.3).

6.3.4 A 'newer' international division of labour and changing organizational forms

By the early 1980s it was evident that a significant shift had occurred in the organization of production within MNCs. One pair of geographers at the time talked of a 'truly global system of production', remarking that:

> With a global network of subsidiaries set up, the question for most large multinationals had changed from what would be the most profitable area in which to expand next to which of the existing areas could be relied upon to produce the highest returns.

> (Taylor and Thrift, 1982, p.1)

The implication of this remark was that MNCs were becoming increasingly flexible and footloose in their organizational geographies. Underlying such developments, the rather simplistic set of bi-lateral relations underpinning the NIDL was beginning to give way to a much more complex set of organizational arrangement for MNCs.

Coffey (1996) has suggested that a new phase of MNC development has been underway since 1980 as part of what he terms the 'newer international division of labour'. MNCs are increasingly shifting strategy from setting up their own production facilities in developing countries to subcontracting work out to locally based firms. Such a change provides MNCs with greater locational flexibility to switch suppliers and avoid the **sunk costs** that are involved in committing investment to fixed assets. Sunk costs refer to the costs of investment that are not directly recoverable if a firm

were to pull out of a particular location (Clark, 1994) (see section 6.6.3). The clothing firm Nike is the oft quoted example of an MNC that maintains spatial flexibility through a vast web of 'independent' suppliers. Headquartered in Beaverton, Oregon, Nike styles itself as 'a firm of marketers and designers', with only 8,000 employees, and no in-house production (Coffey, 1996). As such, production is wholly out-sourced to subcontractors' factories, most of which are located in Asia. The geography of its supply chain has changed over time with continual relocation between countries in search of lower costs; for example, suppliers in Japan and Hong Kong were abandoned in the 1980s in favour of new producers in less-developed countries such as Thailand, Malaysia and Indonesia (Donaghu and Barff, 1990). More recently, China and Vietnam have become key locations for sub-contractors. Nike had contracts with over 700 factories throughout the world in 2005, including 124 factories in China employing over 200,000 workers and 34 in Vietnam employing another 84,000 people (Coe et al., 2007, p.239).

As Dicken (2007) has noted, an increasing amount of MNC activity takes place through external relationships with other firms. A growing trend of collaboration and partnership between MNCs themselves has become apparent in the form of **international strategic alliances** between firms who are competitors in the same markets. One study recorded a sixfold increase in strategic alliances from 1,000 at the start of the 1990s to over 7,000 by the end of the decade, with North American firms dominating the practice (Dicken, 2007, p.166). Such alliances tend to be more pronounced in sectors that require high levels of research and product development and therefore have high start-up costs for new products, such as computers, biotechnology, automotive production or aircraft manufacture. But they also reflect the vulnerability of MNCs to changes in global markets and the desire to minimize uncertainty (see Box 6.1).

6.3.5 Services and the 'second global shift'

Another element with Coffey's 'newer international division of labour' was the growing importance of FDI in services alongside manufacturing. Over the past decade or so this has attracted enough attention to be regarded as a key process in its own right. Indeed, Bryson (2008) goes so far as to identify a '**second global shift**' involving the international relocation of certain service activities to developing countries. This process of relocation or '**offshoring**' parallels the global shift of manufacturing in the 1970s and 1980s. While it is enabled by new ICTs and is a response to differentials in labour costs, it is not a simple one-dimensional process. Rather, Bryson argues that it is underpinned by 'spatial divisions of expertise' that are characteristic of expertise-intensive occupations such as finance, business services and engineering in which the nature of embodied expertise transfers significant power from employers to employees (Bryson and Rusten, 2008). Spatial division of expertise are evident in the emergence of educated and expert workers with English-language skills in developing countries. Here, India has emerged as the key 'offshore' destination, accounting for around 25 per cent of outsourced information technology enabled services globally (Russell and Tithe, 2008, pp.615–16), and other important sites include South Africa, the Philippines, Malaysia and China.

There is nothing particularly new about the spatial dispersal of services with '**back-office**' **functions** – routine clerical and administrative tasks like the maintenance of office records, payroll and billing, bank checks and insurance claims long having been subject to relocation out of city centres to surrounding suburbs where property (rent) and labour costs are typically much lower. What is new is that ICT opens up the possibility of relocation at the regional and, especially, global scales. One of the earliest examples of the international relocation of services was New York financial firms relocating administrative work to western Ireland, actively encouraged by the Irish government. These firms sent life insurance claims to a site located near Shannon airport for completion and checking before the processed data were sent back to New York via fibre-optic cable or satellite (Figure 6.2). Similarly, the Caribbean has become a favoured destination for US back-office functions. This reflects its combination of low wages and geographical proximity, giving it a similar relationship to the US in terms of back office functions as Mexico holds for manufacturing (albeit on a smaller scale). In the example shown in Figure 6.2, American Airlines sent its tickets for processing to a subsidiary in Barbados, a process that began in 1981. The subsidiary opened a second office in the Dominican Republic, where wages are one-half as high as Barbados, in 1987.

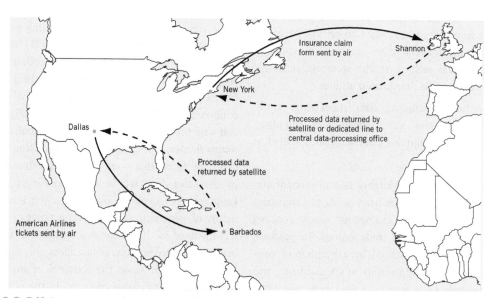

Figure 6.2 Offshore processing in the airline and insurance industries
Source: 'Telecommunications and the changing geographies of knowledge transmission in the late twentieth century', in *Urban Studies*, 32, Taylor & Francis Ltd (Warf, B., 1995). http://www.tandf.co.uk/journals.

Recent processes of service dispersal seem to reflect a combination of three main factors (Gordon *et al.*, 2005, p.19):

➤ Technical developments in telecommunications and internet provision, ensuring that 'many clerical tasks have become increasingly footloose and susceptible to spatial variations in production costs' Warf (1995, p.372).

➤ Increased pressures to reduce costs across a range of service industries, as the sustained growth of the 1990s gave way to a period of slower growth and increased competition from 2000.

➤ The discovery and, to some extent, the creation, of new pools of skilled labour within certain low-wage economies.

As a consequence, the spatial division of labour between regions and nations can be extended into the delivery of consumer services, as firms take advantage of geographical variations in the costs of labour.

In addition to various back-office or business support functions, two other types of service activity are subject to spatial dispersal or offshoring to low-wage locations:

➤ Call centres, focusing on marketing, routine customer enquiries and more sophisticated technical support. Recently rebranded as contact centres or customer service facilities, they have grown very rapidly since the early 1990s. Call centres were pioneered in the financial sector and have subsequently grown in areas such as travel, telecommunications, mail order and the deregulated utilities.

➤ IT functions, including data processing, code-checking, software development and modification, operations support, publishing and statistical analysis.

The second 'global shift' highlights the scope for companies to transfer service work from developed countries to lower-cost locations, prompting much concern among policy-makers and trade unions. In practice, however, relocation is limited by a number of constraints such as the impossibility of standardizing and digitizing some service activities (often those associated with innovation and creativity that typically require face-to-face communication), the need for proximity to existing customers, the nature of information which is personal, sensitive and confidential, and national standards and regulations (UNCTAD, 2004, p.152). Perhaps the best estimate is that around 20 per cent of total service employment has the potential to be relocated as a result of rapid developments in ICT (Bryson, 2008, p.354). Developing countries such as India are likely to make further inroads into more complex and high-value tasks, but the offshoring of services is not a zero-sum game and service employment is likely to continue growing in developed countries also.

Reflect

➤ Compare and contrast the 'first' and 'second' global shifts, referring to the relocation of manufacturing in the 1970s and 1980s and the offshoring of services over the last decade or so.

6.4 How global are they?

An important debate has been waged regarding the globalness of MNCs. From the late 1980s, business school theorists began to talk about the shift in corporate organization from the 'multinational', defined as 'an enterprise that engages in foreign direct investment and owns or controls value-adding activities in more than one country' (Dunning, 1993), to a 'transnational' form, viewed as an organization with the ability 'to strut across the global stage in a footloose manner [with] an attachment to any one particular country cast off' (Allen and Thompson, 1997). A classic text was the work of Bartlett and Ghoshal, *Managing Across Borders*, published in 1989, predicting the emergence of firms that were increasingly trans-national in character, reflected in the increasingly dispersed geography of corporate operations, the international nature of ownership and the diverse range of nationalities involved in the senior management of firms. The image portrayed was one of 'placeless' corporations, no longer dependent upon the activities of any one state or territory, but capable of roaming the global stage, playing off one location against another in search of competitive advantage.

6.4.1 Trans-national activity and foreign assets

In assessing the notion of the 'placeless' corporation, it is useful to draw upon the United Nations **trans-nationality index** (TNI), which compiles data on the international extent of MNC operations, by calculating the average of three ratios for individual corporations: foreign assets to total assets; foreign sales to total sales; and foreign employees to total employees. From this data, the economic geographer Peter Dicken (2003b) has undertaken an analysis of the extent to which firms are becoming more global in their operation. His findings reveal that between 1993 and 1999 there was no significant increase in the 'trans-nationality' of firms. He also noted that the majority of the world's leading MNCs continue to have over 50 per cent of their activities located in their home countries. He did find considerable variation by country; for example, Japan and the US had relatively low levels of internationalization among their leading firms, while the UK recorded one of the highest. Such variations reflect differences in the size of the domestic economy, the way activity is organized and the openness of the economy to foreign trade. Japan has traditionally operated a protectionist industrial policy that has encouraged the investment in large-scale industrial complexes at home, while the US, as one of the world's largest economies, records a low score because of its size rather than a lack of internationally oriented MNCs. The UK has a relatively open economy with few restrictions on outward or inward investment that reflects its long imperial and trading history.

While the TNI is a useful measure, it arguably does not necessarily reveal the most powerful and influential international actors in the world economy. For example, the top ten places in Table 6.3 include a number of firms that would not immediately spring to mind as 'global movers and shakers'. By contrast, the largest MNCs are, in most cases, far more embedded geographically in their own domestic economies. The largest automobile producers, in particular, while controlling extensive

Table 6.3 The world's most trans-national MNCs and selected other MNCs rankings, 2003

TNI Rank	Index	Corporation	Home economy	Industry	Foreign share of Assets (%)	Empl. (%)
1	95.4	Thomson Corp	Canada	Publishing & printing	98.6	92.5
2	95.2	Nestle	Switzerland	Food/bevs	88.9	97.2
3	94.1	ABB	Switzerland	Food/bevs	88.2	96.3
4	93.2	Electrolux	Sweden	Electrical	92.9	90.4
5	91.8	Holcim	Switzerland	Constr materials	91.9	93.4
6	91.5	Roche Group	Switzerland	Pharmaceuticals	90.4	85.6
7	90.7	BAT	UK	Food/tobacco	84.0	96.8
8	89.3	Unilever	UK/Neths	Food/bevs	90.4	90.5
9	88.6	Seagram Co.	Canada	Bevs/media	73.1	
10	82.6	Akzo Nobel	Neths	Chemicals	85.0	81.0
75	36.7	General Electric	US	Electrical	34.8	46.1
77	36.1	Ford	US	Motor vehicles		52.5
83	30.7	General Motors	US	Motor vehicles	24.9	40.8
18	73.7	BP	UK	Petroleum	74.7	77.3
22	68.0	Exxonmobil	US	Petroleum	68.8	63.6

Source: UNCTAD, *World Investment Report 2005*, Table 3.

foreign operations, are still heavily rooted in their countries of origin. Interestingly, recent data point to a slowdown in the rate of internationalization with the TNI of the 100 largest MNCs remaining the same in 2008 as 2007 (UNCTAD, 2009, pp.18–19). This is almost certainly a cyclical effect of the global economic crisis; long-run trends indicate that internationalization is likely to increase again when the global economy recovers (Figure 6.1).

A problem with the TNI is that it is based on the proportion of foreign activities relative to total firm operations rather than absolute size. A distortion that arises as a result is that firms based in smaller national economies are likely to be over-represented because of the necessity to internationalize at an earlier stage in their development, as they quickly reach the capacity limits of the domestic market. Thus, firms from, Switzerland, Canada and Sweden appear in the top 10 (Table 6.3). A more useful indicator of the most powerful MNCs is the value of their foreign assets (Table 6.4), which reveals a more familiar list of

names, pertinent in demonstrating where power lies in the global economy. The world's major economies dominate the list, with petroleum, electricity, motor vehicles and telecommunications being the leading sectors. While most of the largest MNCs tend to be involved in manufacturing and resource extraction, the number of firms from the service sector in the top 100 has increased from 14 in 1991 to 24 in 1998 and 26 in 2007 (UNCTAD, 2009, p.19). Many of these service MNCs operate in telecommunications and utilities.

The UNCTAD figures represented in Tables 6.3 and 6.4 are a useful reminder of both the extent and limit of global integration of business activities. But they remain a rather crude measure of the geographical networks within which MNCs are embedded, particularly given the increased use of foreign suppliers over direct investment that was noted above. For instance, a firm can be predominantly nationally oriented in terms of its own operations, but have enormous global influence through its purchasing and sourcing strategies. The classic example used to illustrate this point is

Table 6.4 The world's top 15 non-financial MNCs, ranked by foreign assets, 2007 (millions of dollars)

Corporation	Home economy	Industry	Assets (US $ ml)	
			Foreign	Total
General Electric	US	Electrical	420,300	795,337
Vodafone	UK	Telecomms	230,600	254,948
Royal Dutch / Shell	Nthlds	Petroleum	196,828	269,470
British Petroleum	UK	Petroleum	185,323	236,076
Exxonmobil	US	Petroleum	174,726	242,082
Toyota	Japan	Motor vehicles	153,406	284,722
Total	France	Petroleum	143,814	167,144
Electricite de France	France	Electricity	128,971	274,031
Ford	US	Motor vehicles	126,854	276,459
E.ON AG	Germany	Electricity	123,443	202,111
ArcelorMittal	Luxembourg	Metals	119,491	133,625
Telefonica SA	Spain	Telecomms	107,603	155,856
Volkswagen	Germany	Motor vehicles	104,382	213,981
ConocoPhillips	US	Petroleum	103,457	177,757
Siemens	Germany	Electronics	103,055	134,778

Source: UNCTAD, World Investment Report 2009, Annex Table A.1.9, p.225.

the US retailer Wal-Mart (see Dicken, 2003b), which employs around one-and-a-half million workers, three-quarters of whom are in the US (UNCTAD, 2004). Taken overall, however, the data illustrate the continued importance of geography in shaping corporate strategy. While the global economy is becoming more integrated – or perhaps interconnected is a better term – MNCs are still heavily embedded in and reliant upon their domestic economies and institutions (see Box 6.1).

Box 6.1

The limits to global corporate power: MNC restructuring in the global oil industry

The oil industry is perhaps the most globally integrated of industries and indeed has been organized on a global basis, through the activities of dominant multinationals, for a long period. As early as 1870, the US oil industry was selling two-thirds of its output to foreign markets (Vernon, 1971, p.28) and by the mid-1950s the leading seven multinationals – known as the 'Seven Sisters' because of their perceived market collusion – controlled over 90 per cent of the world's oil reserves and 70 per cent of refining capacity (Jacoby, 1974). In today's global economy, it comes as no surprise that the leading oil MNCs figure prominently among the top non-financial MNCs, ranked by size of foreign assets, with five of the six so-called 'supermajors' in the top 15 (Table 6.4).

From 1986 onwards, the industry underwent an intensive period of restructuring, driven initially by low oil prices and the long-run decline in reserves. Typically, firms have attempted to shed their less profitable non-core assets in established regions such as North America and western Europe, particularly in the 'downstream' activities of marketing, refining and associated petroleum products, to free-up resources for 'upstream' exploration in new and growing oil producing regions such as the former Soviet Union and West Africa. This resulted in a streamlining of operations, with the ten leading firms shedding over 140,000 jobs between 1993 and 1998 (Cumbers and Martin, 2001, p.38), and a wave of merger and acquisition activity in the late 1990 and early 2000s (Table 6.5).

Rather than being a sign of the industry's strength and increased global dexterity, the merger wave represented recognition of the **territorial embeddedness** and vulnerability of many MNCs. Indeed, despite their global aspirations, the reality confronting most MNCs was a continuing dependence upon their home countries or particular world regions for oil supplies and markets. Apart from Exxon, other US MNCs in particular were still heavily dependent upon producing oil from ageing and increasingly costly US fields. The sudden spate of mergers in the late 1990s reflected both the imperative to reduce costs and the desire for a greater global reach. The BP–Amoco merger is a case in point, fusing BP's Eurasian orientation with Amoco's predominantly US interests (the latter was still dependent on US-based resources for 60 per cent of its total revenue prior to merger).

BP can now be considered to be one of the most globalized of firms,

Table 6.5 Key oil and gas mergers and acquisitions since 1998

Date of deal	Companies involved		Comments
August 2002	Phillips	Conoco	Merged firm worth $53.5 billion
September 2001	Chevron	Texaco	$45 billion takeover by Chevron
February 2000	Total	Elf	'Friendly merger'
February 1999	BP Amoco	Atlantic Richfield	$26.8 takeover by BP
December 1998	Total	PetroFina	$7.0 billion takeover of PetroFina
December 1998	Exxon	Mobil	$75.3 billion takeover
August 1998	BP	Amoco	$48.2 billion merger

Sources: BP Amoco website http://www.bpamoco.com/; *Financial Times*, 21 December 1998, p.22.

Box 6.1 (continued)

with a TNI figure of 71.3 and activities on every continent, although it remains embedded in important ways in its home base. Forty per cent of company shares are still held by British institutions and individual shareholders (Table 6.6). Its ability to successfully 'globalize' was also underwritten by its own government from the 1980s, which, after intensive lobbying from BP and the other British oil company MNC Shell, reduced tax rates on oil extracted from the North Sea in March 1993, allowing the corporation to finance new exploration and investment overseas (Woolfson *et al.*, 1997, p.310). The attempt to develop a greater global reach has on the whole been achieved through joint venture and partnership activities. It operates in a partnership with Norwegian company Statoil in Azerbaijan, Kazakhstan, Vietnam, Angola and Nigeria, and in a joint venture with Russian company TNK in Siberia (Company Annual Reports). The desire to work increasingly in partnership or in alliance, both with competitors and suppliers, reflects the continued vulnerability of MNCs to geopolitical uncertainties, given that many of the world's major oil producing regions are characterized by political instability.

Table 6.6 Ownership of BP plc as at 31 December 2009

	Institutions	Individuals	Total
UK	33	7	40
US	25	14	39
Rest of Europe	19	–	10
Rest of World	7	–	7
Miscellaneous	4	–	4
	79	21	100

Source: BP website, 'Ownership Statistics',
http://www.bp.com/extendedsectiongenericarticle.do?categoryId=9010453&contentId=7019612.

6.4.2 The continuing embeddedness of MNCs in their home countries on MNC strategies

As the example of the oil industry shows, even the most trans-national of firms continue to be geographically embedded in their countries and regions of origin. This territorial embeddedness has a number of dimensions, including the nationality of the directors and their families, the legal frameworks binding company action, the financial systems governing firm decision-making and particular national cultures of 'doing business' (Dicken, 2007). For example, while there is an emerging elite of multinational corporate executives, most firms are still dominated by managers from their home countries at the highest levels. One international survey of the directors of MNCs, for example, found that only 10 per cent of board members were foreign nationals (Knox *et al.*, 2003). Among Japanese firms, foreign board members were almost non-existent, being described as 'as rare as British sumo wrestlers' (ibid., p.84).

At a very basic level, all MNCs develop from particular local contexts and, while the local scale itself has an influence upon the practices and strategies of the firm, the national scale continues to be critical in shaping MNC actions. Despite increased global economic integration, researchers continue to highlight diversity in forms of capitalism between places and between MNCs (Hutton, 1995; Whitley, 1999; Albert, 1993; Weiss, 1997). For example, a distinction is often made between US, German and Japanese MNCs (see Table 6.7). Whereas US MNCs are largely financed by stock markets that discipline firms to deliver increasing

Table 6.7 Differences between US, German and Japanese TNCs

	US TNCs	German TNCs	Japanese TNCs
Corporate governance and corporate financing	Constrained by volatile capital markets; short-termist perspectives. Finance-centred strategies	Relatively high degree of operational autonomy except during crises. Long-term perspectives. Conservative strategies	Bound by complex but reliable networks of domestic relationships. Long-term perspectives. Market-share-centred strategies
	High risk of takeover. 90% of firm shares held mainly by individuals, pension funds, mutual funds. Less than 1% held by banks	Low risk of takeover. Firm shares held mainly by non-financial institutions (40%). Significant role of regional bodies	Very low risk of takeover – mainly confined to within network
	Banks provide mainly secondary financing, cash management, selective advisory role	Banks play a lead role. Supervisory boards of companies are strongly bank-influenced	High degree of cross-shareholdings within group. Lead bank performs a steering function
	Ratio of bank loan/corporate financial liabilities 25–35%	Ratio of bank loan/corporate liabilities 60–70%	Ratio of bank loan/corporate liabilities 60–70%
Research and development	Corporate R&D expenditure peaked in 1985 at 2.1% of GDP. Declining	Corporate R&D expenditure declined steeply in late 1980s/early 1990s. At 1.7% of GDP lower than US and Japan	Corporate R&D grew very rapidly in 1980s. Overtook US in 1989. Peaked at 2.2% of GDP. Real cuts made only as last resort
	Diversified pattern; innovation-oriented	Narrow focus	High-tech and process orientation
	Some propensity to perform R&D abroad	Some propensity to perform R&D abroad	Very limited propensity to perform R&D abroad
Direct investment and intrafirm trade	Extensive outward investment. Substantial competition from inward investment	Selective outward investment. Moderate competition from inward investment	Extensive outward investment. Very limited competition from inward investment
	Moderate intrafirm trade; high propensity to outsource	High level of intrafirm trade	Very high level of intrafirm and intragroup trade

Source: Based primarily on Purdy and Reich, 1997.

returns and dividends to shareholders, thereby instilling a 'short-termist' approach, German and Japanese MNCs have greater freedom due to the greater role of banks in providing finance on a longer-term and more stable basis. A range of other motivations, such as increasing market share, safeguarding employment and a commitment to training and modernizing production, often dominate over narrow profit concerns. In locational terms, US MNCs have shown a greater propensity to invest overseas, both in basic production activities but also in some R&D. In contrast, Japanese MNCs still do very little R&D overseas and, up until the 1980s, were essentially domestic producers for export markets (see Box 6.2).

Researchers have also highlighted continuing diversity in the approach to labour relations by MNCs from different countries. While MNCs in continental Europe generally still accept trade unions as legitimate social partners, US and UK MNCs tend to display a strong anti-trade union bias, in many cases failing to

Box 6.2

Contrasting geographies of the global automobile manufacturers

While the automobile industry is undoubtedly one of the most 'global' of industries, with some of the largest MNCs, it continues to be characterized by considerable geographical variation. This reflects the distinct strategies employed by national groups of firms. The greatest contrast, until recently, was in the strategies of US compared with Japanese firms. While the two largest US car producers, GM and Ford, were already producing a high proportion of their cars overseas in the 1980s (41.9 per cent and 58.7 per cent in 1989) (Dicken, 2003a, p.374), the largest Japanese firm, Toyota, was still primarily a domestic producer, with only around 8 per cent of its operations overseas. Like other Japanese firms, it had been highly successful at competing overseas without having to relocate production from its home base:

> Before the early 1980s, Japanese automobile producers had shown themselves perfectly capable of serving the North American and European markets by exports from the Japan. Their production costs efficiencies more than offset any transport cost penalties

arising from Japan's geographical distance form major markets.
> (Dicken, 2007, p.299)

Japanese firms began to expand their production overseas in the 1980s in response to growing political opposition to imports in North America and western Europe. As result, they have become more globalized in their operations, although the majority of production continues to be based in Japan (see Figure 6.3). This pattern of majority home country production is echoed by General Motors, although Ford is more trans-national. Both GM and Ford have been in crisis in the mid-to-late 2000s, with the effect of decreasing shares of the US market and growing international competition greatly compounded by the slump in demand associated with the global economic downturn from 2007.

Japanese auto producers have also tended to subcontract out more of their production than US firms, amounting to 80 per cent of inputs in 1989 (Hayter, 1997, p.356). This meant that Toyota's workforce was 11.9 per cent of GM's in 1989 (ibid.), although its sales turnover was half of

its US rival's. The associated need for closer interaction with suppliers has direct geographical implications, as Japanese firms have tended to concentrate their production systems in localized clusters, whereas US firms have often favoured a more dispersed pattern, basing their locational decisions largely on cost.

Changes in the organization of the industry in the 1990s have partly diluted the national character of many MNCs. In a similar fashion to the oil industry, there has been a series of mergers and acquisitions, the most important being the takeover of US MNC Chrysler by the German firm Daimler-Benz, the purchase of Swedish firm Saab by GM and the buying of a 36.8 stake in Nissan by the French firm Renault. Additionally, the increase in cooperation between firms in recent years means that the industry at a global level has become what Dicken describes as 'a veritable spider's web of strategic alliances ... that stretches across the globe' (Dicken, 2003a, p.376). Nevertheless, national variation and embeddedness continue to play an important factor in shaping MNC strategies in the auto industry.

recognize the rights of workers to collective action, as well as demonstrating a greater preference to use strategies of employment flexibility (see Chapter 7).

Particular local or national cultures of production (section 2.5.3) are also 'exported', as corporations internationalize their operations, although this does not subsume existing cultures of production in the host countries. Indeed, Gertler's research into German machine manufacturers' supplying markets in North America found that they were unable to reproduce their home production practices in their own American plants, let alone those customers they were supplying.

Conversely, researchers have noted the frustration of US firms setting up in Germany, eventually having to accept German norms on employment relations and working practices such as the recognition of trade unions and more restrictive working hours than in the US (Gertler, 2004).

Reflect

> In what ways do national business cultures continue to shape MNC operations?

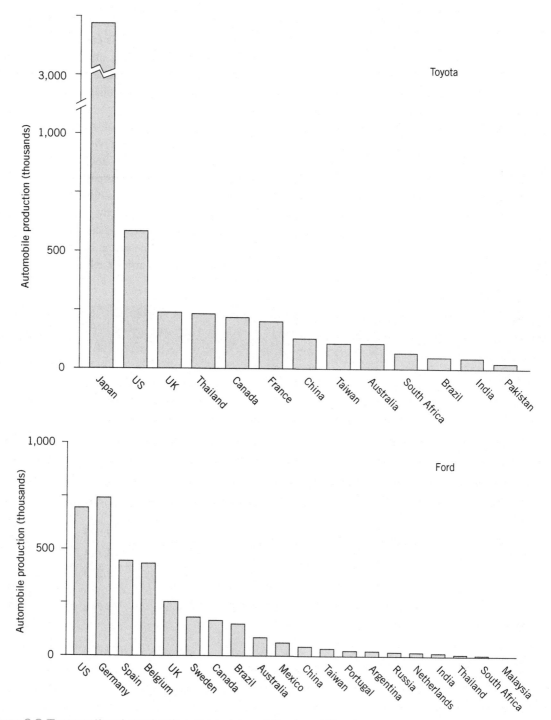

Figure 6.3 Trans-national production geography of Ford and Toyota
Source: calculated from SMMT, 2005, *World Automotive Statistics*.

6.5 The internationalization of services

While most of the academic and policy literature on FDI has focused on manufacturing industries such as clothing, electronics and automobiles, the internationalization of services has received increased attention in recent years (Bryson, 2008; Coe and Wrigley, 2007). The turnover of overseas MNC affiliates in services is generally higher, compared with the total services exports of these countries, than the same ratio in the manufacturing sector. This indicates that FDI is a more important mode of internationalization than exporting in service industries (OECD, 2005b). It reflects the fact that much service provision is embodied in the individuals supplying it, requiring either direct investment in new overseas offices to provide services to clients or indirect expansion through affiliations and agreements with foreign firms.

FDI in services has gown substantially since 1990 to eclipse manufacturing, with services accounting for 65 per cent of world outward FDI stock in 2007 (Table 6.8). Until 1990, FDI in services was particularly focused on finance and trade, but subsequently other services such as business services, electricity, telecommunications and water supply have experienced substantial growth (Table 6.8). In the airline sector, by contrast, FDI is of little importance, with airlines relying, instead, on the

Table 6.8 Estimated world outward FDI stock by sector and industry, 1990 and 2007

Sector/industry	1990			2007		
	Developed countries	Developing countries	World	Developed countries	Developing countries	World
Total	1,765,278	20,306	1,785,584	14,227,765	1,909,575	16,207,225
Primary	154,688	25,830	157,251	1,110,525	45,152	1,161,165
Manufacturing	769,479	7,217	776,696	4,051,964	163,876	4,217,443
Services	836,691	9,843	846,354	8,833,715	166,368	10,511,848
Electricity, gas and water	9,306	–	9,306	201,435	11,283	213,233
Construction	17,650	107	17,757	55,890	9,503	64,812
Trade	137,858	1,714	139,573	928,547	148,114	1,077,723
Hotels and restaurants	6,896	–	6,896	114,918	9,733	1,124,694
Transport, storage and communications	3,8471	455	38,925	652,586	75,763	728,402
Finance	416,522	6,114	422,636	3,248,047	2,747,889	3,524,674
Business activities	81,748	1,268	8,3016	2,776,980	1,115,725	3,901,514
Public admin and defence	–	–	–	7,982	4	8,009
Education	417	–	417	1,518	29	1,552
Community, social and personal	3,315	–	3,315	65,033	4,275	69,308
Other services	108,965	175	120,9140	233,149	10,327	243,476

Source: UNCTAD, *World Investment Report 2009*, Annex Table A.1.5, p.219.

development of alliances (for example, code sharing, frequent flyer programmes, cargo agreements) with other operators, the number of which increased from 20 worldwide in the early 1990s to 1,222 by 2001 (UNCTAD, 2004, p.108).

Many business service firms have internationalized by following their main MNC clients overseas, building on existing relationships to serve the needs of these clients in new locations. The accountancy industry, for example, has become highly global, mirroring the trans-nationalization of the various industries that use its services, through the establishment of networks and partnerships with local accounting firms under one brand name and through mergers (UNCTAD, 2004, p.110). The emphasis on partnerships reflects firms' efforts to circumvent nationally specific regulations about trade and commercial presence, accounting standards and the recruitment of staff. This involves

adding new units to networks of firms that are usually legally separate, locally owned and locally managed (ibid.).

Another service industry that has seen significant internationalization in recent years is retail. Although retailing has traditionally been seen as a very nationally specific economic activity, in common with many other services, the period since the late 1990s has witnessed what Coe and Wrigley (2007, p.342) term a 'deluge of retail FDI', focused largely on emerging markets in East Asia, central and eastern Europe and Latin America. This has involved a leading group of retail MNCs (Table 6.9) rapidly establishing stores in these markets, supported by extensive regional and global sourcing networks. This process of retail internationalization is clearly driven by market access with these emerging markets offering the prospect of sustained future growth, in contrast to saturated and

Table 6.9 The world's 15 largest food retail TNCs, ranked by foreign assets, 2007 (millions of dollars and numbers of employees)

| Rank | Corporation | Home economy | Assets | | Sales | | Employment |
			Foreign	Total	Foreign	Total	Total
1	Wal-Mart Stores	US	62,961	163,415	90,640	374,526	205,500
2	Metro AG	Germany	29,627	49,863	55,950	94,711	253,769
3	Carrefour	France	28,507	76,449	65,549	120,930	490,042
4	Tesco	UK	21,286	60,425	24,888	94,748	413,061
5	McDonald	US	178,855	29,392	13,970	22,787	390,000
6	Delhaize	Belgium	10,402	12,889	21,342	27,715	138,000
7	Ahold	Netherlands	9,158	19,845	22,423	41,158	118,715
8	Sodexo	France	8,101	11,671	11,985	18,247	342,380
9	Compass Group	UK	7,578	12,615	126,985	20,920	361,317
10	Seven and I Holdings	Japan	6,101	37,042	18,553	55,223	55,815
11	China Resources Enterprise	China	6,137	7,779	4,761	6,603	135,000
12	Yum! Brands	US	3,746	6,952	5,219	10,416	301,000
13	Autogrill	Italy	2,759	4,481	4,170	7,236	49,053
14	Alimentation Couche Tard	Canada	2,342	3,047	9,880	12,400	45,000
15	Safeway	US	2,197	17 651	6,015	42,286	201,000

Source: UNCTAD, *World Investment Report 2009*, Table A.III.7, p.242.

highly competitive domestic markets. It was facilitated by access to low-cost capital in the growth years of the late 1990s and early 2000 through debt and equity financing and the substantial liberalization of the retail sector in many emerging markets (ibid.). Retail MNCs' expectations of growth in emerging economies were derived not only from the prospect of rapid economic development, but also from the underdevelopment of the modern retail sector and the preponderance of traditional outlets.

Most of the leading retail MNCs have adopted a regional rather than fully global strategy over the past decade or so, focusing on developing and consolidating their activities in a limited number of markets in selected regions such as East Asia rather than seeking to 'collect countries' on a global basis (Box 6.3) (Dawson, 2007). In general, the retail sector is subject to high levels of territorial embeddedness, reflecting several underlying characteristics of retail as an industry (Coe and Lee, 2006, p.68). First, firms are closely connected to the property markets and planning systems of host countries, since they require an extensive network of stores. Second, retailers need to be highly attentive to local patterns of consumption, which reflect distinct cultural preferences, tastes and attitudes, raising a host of socio-cultural questions beyond the purely economic (section 4.4.). Third, retail firms typically source a wider range of products from local suppliers, alongside the growing emphasis on global and regional sourcing. Such embeddedness can be contrasted with the situation in high-volume manufacturing sectors in which MNC plants often have few links with the host economy, relying on a host of imported materials that are assembled for export (section 6.6.2).

Box 6.3

Strategic localization in practice: Samsung-Tesco in South Korea

Expansion into overseas markets has been a major component of Tesco's growth strategy since the mid 1990s, helping it to become the third largest food retail TNC in the world today (Box 3.5). The company has moved into 13 overseas markets since 1994, comprising five countries in eastern Europe and the Republic of Ireland, six in Asia and the US (Figure 6.4). While the UK remains dominant, accounting for 68 per cent of turnover in 2010, overseas store space and sales have grown markedly since 2000 (Table 6.10). In recent years, Tesco has moved into the potentially lucrative Chinese, Indian and US markets, following its earlier

Table 6.10 The increased importance of Tesco's international operations, 2000–10

	2000	2005	2010
Turnover			
UK	90	80	68
Rest of Europe	7	11	16
Asia	3	9	22
US			1
Total sales area			
UK	70	47	35
Rest of Europe	20	28	29.5
Asia	1025	33.8	
US			1.5

Sources: Coe and Lee, 2006, p.72; Tesco plc, *Annual Report and Financial Statements 2010*, p.6.

Box 6.3 (continued)

UK

	Revenue £m	Stores	Employees
Total	**38,558**	**2,482**	**287,669**

US

	Revenue £m	Stores	Employees
Total	**349**	**145**	**3,246**

Europe

Region	Revenue £m	Stores	Employees
Republic of Ireland	2,282	119	14,158
Poland	1,942	336	23,655
Hungary	1,698	176	20,079
Czech Republic	1,287	136	12,949
Slovakia	891	81	8,105
Turkey	595	105	7,630
Total	**8,695**	**953**	**86,576**

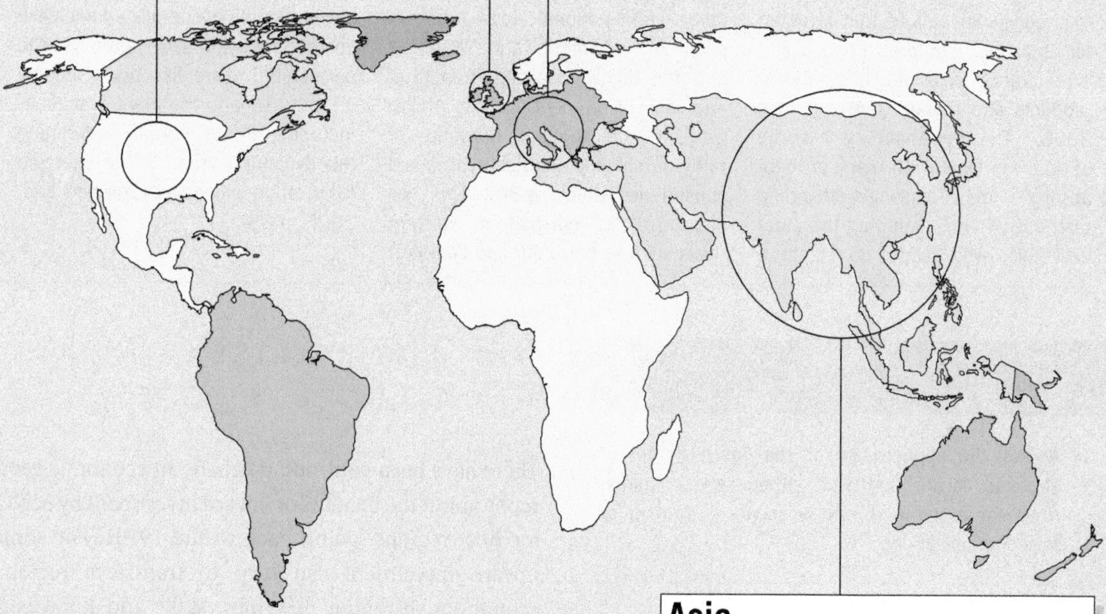

India

In India we have an exclusive franchise agreement with Trent, the retail arm of the Tata Group. We are supporting the development of their Star Bazaar format. We plan to open our first cash and carry store by the end of the year.

Asia

Region	Revenue £m	Stores	Employees
South Korea	4,162	305	22,739
Thailand	2,344	663	34,775
China	844	88	22,668
Malaysia	633	32	9,423
Japan	449	142	4,636
Total	**8,432**	**1,230**	**94,241**

Figure 6.4 Tesco's international operations
Source: Tesco, Annual Report 2010, p.7.

venture into smaller East Asian and eastern European markets. As such, its strategy has been highly regionalized, rather than global, focused on building up market share in a relatively small number of strategic markets.

Tesco's South Korea operation is its largest overseas business, generating the most revenue in 2010 (Figure 6.4). The company strategy there is characterized as one of 'strategic localization' by Coe and Lee (2006), involving substantial adaptation to local market conditions. Tesco's entry

Box 6.3 (continued)

into the South Korean market was facilitated by deregulation and liberalization processes designed to attract increased FDI around the time of the Asian financial crisis in 1997–98. Its South Korean operation was launched in 1999 through a merger with the Samsung Corporation's distribution unit, creating Samsung-Tesco. The stores are branded as Homeplus, further emphasizing the local (Tesco, 2010). This was designed to respond to a consumer culture that is noted for its suspicion of foreign brands and a preference for established national products and brands (Coe and Lee, 2006). The overwhelming majority of staff are local, with many of them absorbed from Samsung's struggling distribution arm, including the chief executive, who sought to create a

hybrid organizational culture, melding Korean notions of staff loyalty and identification with the more 'rational' business practices of Tesco (ibid.). Local sourcing has also been emphasized, involving the establishment of direct procurement channels with local producers and manufacturers.

In these respects, Tesco's strategic localization model can be viewed as a kind of 'third way' between the highly global strategies of certain other trans-national food retailers such as Carrefour and Wal-Mart and the highly localized 'franchise' approach of retailers such as Ahold, who operate through networks of fairly autonomous and self-contained subsidiaries (ibid., p.84). This has allowed it to out-perform western rivals such as Carrefour and Wal-Mart

in claiming an increasing share of the highly competitive South Korean market. In similar fashion, Tesco's operations in other Asian markets have involved a partnership with local firms, such as the CP Group in Thailand and Ting Hsin in China (Coe and Wrigley, 2007, p.352), while its presence in India is based on a franchise agreement with the retail arm of a local corporation. This partnership model has allowed Tesco to territorially embed its operations and achieve operational legitimacy, although the partnership share has been strongly diluted in many countries over time, including South Korea, underlining the dynamic nature of the strategic localization processes (Coe and Lee, 2006, p.85).

Reflect

> Assess the applicability of the Tesco model of strategic localization with other service industries: for example, business services, tourism or telecommunications.

6.6 The impact of MNCs on host regions

There have been continuing debates in economic geography about the benefits of inward investment by MNCs for host regions, going back to the 1970s. For some, inward investment can help to transform regional economies, bringing new jobs, skills and knowledge as well as helping connect local economies up to wider global networks (Table 6.11). For others, however, the effects are more pernicious, leading to a loss of local control over economic development in return for jobs that are often low skilled and always vulnerable to plant closure if MNCs decide to invest elsewhere (e.g. Firn, 1975; Massey, 1984). Here, we consider the effects in different types of region: those in advanced economies; those in developing countries; and those in the transition states of eastern Europe.

Table 6.11 Advantages and disadvantages of MNC inward investment for host regions

Advantages	Disadvantages
Local economy Injection of investment and income opportunities Transfer of global knowledge and management techniques – enhancing competitive advantage in context of globalization Improves skills base through training, employment policies Contributes to region's tax base	*Local economy* Development increasingly dependent on external control Greater vulnerability to closure and job loss Deskilling/downgrading of local economy Production enclaves linked to broader global production systems but with few local benefits Marginalization/displacement of other sectors – economy linked to narrow development trajectory
Local firms New opportunities to supply inward investors 'Piggy-backing' on MNCs into export markets Learning through imitation of best practice in FDIs	*Local firms* Increased competition in local markets Increased competition for labour, land, capital drives up factor prices Become tied in to dominant client firms as suppliers
Local community Upgrading of local infrastructure that may not otherwise have occurred (e.g. transport and communication links) FDI 'social investment' (e.g. in local schools, community services)	*Local community* Disruption/destruction of local culture/society Resource displacement – investment on FDI specific infrastructure rather than community projects/initiatives
Employment New jobs for local people Wages often in excess of average local wage Training and more progressive approach to human resource management	*Employment* Introduction of more coercive management strategies Jobs often low skilled, routine Training limited to company-specific skills and not transferable
Political implications Influx of powerful non-local actors enhances power of region in context of inter-regional competition for state resources	*Political implications* Regional development agenda becomes hijacked by MNC interests Alternative political discourses marginalized – democratic deficit

Source: Adapted from Pavlinek, 2004, p.48.

6.6.1 Branch plant regions in developed economies

Early studies of FDI in developed economies, often informed by a Marxist perspective, suggested that inward investment could lead to relations of dependency, as hitherto locally controlled economies became increasingly subservient to the needs of large foreign corporations. In the 1970s, the term '**branch plant economy**', was coined to reflect how some of the older industrial regions of western Europe and particularly the UK were being repositioned within the wider space economy of capitalism. The concept is defined by functional truncation due to a lack of higher-status activities, such as strategic planning and R&D, limited linkages between MNC plants and the local economy with many components and materials being imported, and a high degree of external ownership and control (Phelps, 1993).

The emergence of branch plant economies in traditional manufacturing regions in the 1970s and 1980s was part of a broader **spatial division of labour** (section 4.3.2). With the decline of traditional industries such as steel production, shipbuilding, textiles and engineering, and the closure of local firms, regional industrial specialization and local ownership were being replaced by a situation in which regions were being recast and repositioned within corporate production systems. In the UK context, the recession of the early 1980s seemed to bear out many of the critics' concerns as several studies indicated that rates of firm closure and job loss were highest in externally owned plants, with foreign owned plants experiencing the greatest levels of decline (e.g. Lloyd and Shutt, 1985; Healey and Clark, 1986).

Controversy over the merits of foreign inward investment as a strategy for regional development in the US and UK was re-ignited during the late 1980s and early 1990s, reflecting a wave of new inward investment into North America and western Europe by Japanese companies and later by other East Asian corporations. Aside from issues of external dependence (see Box 6.4) the amount of money national and regional governments is prepared to spend to attract FDI has been queried by some commentators. When the Nissan car company set up a manufacturing plant in the north-east of England in 1986 (Figure 6.5), it received £112 million from the UK government for its first phase of development (Garrahan and Stewart, 1992). Although it has subsequently expanded to employ over 5,000 employees with an estimated value of £1.2 billion (Conte-Helm, 1999), it is a moot point as to whether the investment would not have been better spent helping to modernize existing firms and protect the jobs of local workers in traditional industries. A later project, set up by the Korean conglomerate LG in 1996 to develop a television and components plant in Newport in South Wales, received over £247 million of public money and made enormous demands on the local infrastructure (Phelps et al., 1998). The promised 6,000 jobs never materialized, however, as LG was severely hit by the Asian financial crisis of 1997–98. Fewer than 1,000 jobs were eventually established at the plant which was closed in May 2003 (http://news.bbc.co.uk/1/hi/wales/3049261.stm).

The negative branch plant stereotype was questioned by some researchers in the 1980s and 1990s (e.g. Munday et al., 1995; Morgan, 1997). Compared with the low-skill assembly line activities of previous US investments, the new wave of inward investment into the UK and US from Japanese and German companies was geared towards the creation of locally integrated industrial complexes that were more firmly tied to their host regions because of the large investments made by MNCs. These investments were seen as more strategic and long term in nature. Instead of being purely concerned with the cost advantages of a particular

Figure 6.5 Nissan factory, Sunderland, north-east England
Source: D. MacKinnon.

Box 6.4

From boom to bust: the changing fortunes of East Asian investment for the UK's old industrial regions

The UK was the dominant location in Europe for inward investment from East Asian MNCs setting up to gain access to the emerging European Single Market, receiving over 40 per cent of the European total during the 1980s and early 1990s (Mason 1994, p.32). The attraction of FDI was central to the approach of the Conservative government of the time, which emphasized the need to create new jobs and introduce modern management practices and attitudes to regions decimated by deindustrialization. At the same time, critics questioned the government's motives, raising questions about the quality of the jobs being created, the degree to which MNCs were territorially embedded in the local economy, and the encouragement of new foreign management with more hostile attitudes towards trade unions (Garrahan, 1986). Nevertheless, there were considerable short-term benefits in terms of employment creation with South

Wales and the north-east of England gaining 13,000 and 11,000 jobs respectively from Japanese investment alone by the mid 1990s (Cumbers, 1999).

Opinions are mixed as to the reasons behind inward investment, but a combination of low labour costs – by EU standards – and government grants appears to have been important (Stone, 1998). What is clear, however, is that few of the investments could be considered to be linked to high technology or high-performance production. Recent research has also questioned the territorial embeddedness of inward investment, finding little evidence that firms were setting down longer-term linkages in the regions with a lack of additional investment after the initial start-up (Dawley, 2007; Phelps *et al.*, 2003).

A reminder of the dangers of over-reliance upon FDI has been brought home in the period since 1998 with the closure of many of the Asian

branch plants, linked to the Asian financial crisis, on the one hand, but also, on the other, to the availability of new sources of cheap labour in eastern Europe, which has resulted in the closure or relocation of earlier branch plants. The collapse of the market for certain electronics components was felt heavily in the north-east of England, where both the German firm Siemens (1,100 jobs) and the Japanese firm Fujitsu (600 jobs) closed semiconductor plants in 1998 (Dawley, 2007). In Scotland, the crisis led to the collapse of plans by Korean firms Chungwa and Hyundai to set up new plants and to the closure of a Mitsubishi plant outside Edinburgh in 2002 (500 jobs). Companies that have relocated from the UK's regions to the Czech Republic include Japanese firm Matsushita (from Wales, costing 1,400 jobs) and US firms Compaq (Scotland, 700 jobs) and Black and Decker (north-east England, 600 jobs) (Pavlinek, 2004, p.53).

location, they were geared to tapping into local skills and 'know-how'.

This more optimistic scenario hinges around a number of organizational changes taking place in MNCs themselves in response to globalization, characterized as a shift away from centralized bureaucratic hierarchies to flatter and more decentralized structures, meaning that more decision-making powers and higher level operations are devolved to local branch plants. Such developments reflect the prerogative of MNCs to be 'globally efficient, multinationally flexible, and capable of capturing the benefits of worldwide learning at the same time' (Dicken *et al.*, 1994, p.30). In these circumstances, it is argued that host regions become home to key forms of knowledge that are

valuable to the firm (Schoenberger, 1994). Learning-by-doing and learning-by-using activities are in effect territorialized through 'the everyday experiences of workers, production engineers and sales representatives' (Lundvall, 1992, p.9), and locationally fixed (at least in the short term) in the sense that they 'remain tacit and cannot be removed from [their] human and social context' (Lundvall and Johnson, 1994, cited in Morgan, 1997, p.493).

As a result of such developments, regions can use the global knowledge networks of MNCs to upgrade and improve their competitiveness. For example, Japanese branch plant investments in the Great Lakes Industrial Belt of the US during the 1980s were seen as instrumental in rejuvenating the regional economy, through the transfer of 'best practice' management and production techniques to local firms (Florida, 1995). In particular, the high-performance manufacturing model employed by Japanese firms, based upon just-in-time production, continuous improvement and team working, was viewed as superior to the assembly line techniques associated with Fordism (see section 3.3.2) (Florida and Kenney, 1993).

The evidence to support such claims is mixed and, as Box 6.4 demonstrates, plants remain vulnerable to closure as corporate strategies and priorities change. In-depth studies of Japanese branch plants, for example, have found the transfer of high-performance manufacturing to their foreign subsidiaries by Japanese MNCs to be the exception rather than the rule, with activities more often than not characterized by the kind of routine assembly work associated with earlier forms of FDI (Danford, 1999).

6.6.2 Export processing zones in developing economies

Concerns about the vulnerability of MNC branch plants in advanced economies are amplified considerably in less-developed economies where host regions often lack the political and social resources (e.g. strong labour unions, strong forms of local democracy and participation, and effective employment and health and safety legislation) for local communities to negotiate the terms of inward investment. The extension of production networks by MNCs into the developing world as part of the NIDL has received much critical scrutiny from researchers concerned at the new forms of exploitation that have emerged. The term '**export platform**' (Smith, 2000) has been used to define production enclaves that are set up by western MNCs (or their suppliers) in parts of the less-developed world to supply markets in advanced economies. Not only do these 'platforms' tend to be totally disconnected from the wider local economy, but they are also often strictly policed and controlled by their national state

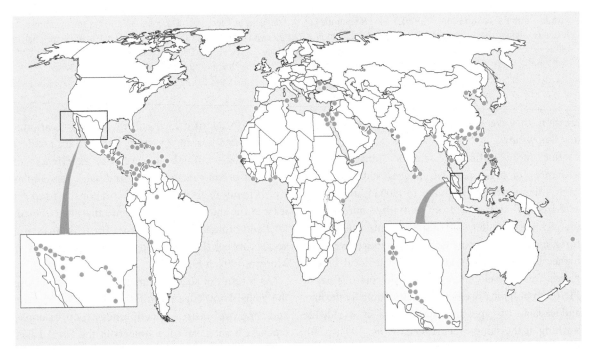

Figure 6.6 Map of export processing zones in developing countries
Source: Dicken, 2003a, p.180.

governments who discourage the formation of labour unions. Such work is commonly associated with the exploitation of women workers in semi-skilled and poorly paid work (Kelly, 2001).

Many developing countries have set up **'export processing zones'** (see Figure 6.6), which effectively offer foreign firms tax and investment incentives to locate there. Typically these will include up to 100 per cent rebates on local taxation, the provision of all infrastructure and the relaxation of the usual rules governing foreign ownership (Dicken, 2003a). Some of the most notorious of these zones have been set up in Mexico along the border with the US. Under the Mexican government's Border Industrialization programme, established in the 1960s, branch plants known as '*maquiladoras*' (literally translated as 'assemblers') have been set up in the region to supply the US consumer market. The plants are known as 'in-bond' plants, which are allowed to import components and materials from the US for the production of finished products, providing that the product is then exported back to the US. Spectacular growth has ensued, with US MNCs attracted by the lower wages and lax environmental and safety conditions over the border. A further stimulus to growth was the setting up of the North American Free Trade Agreement (NAFTA) in 1994 when the removal of all trade barriers with the US, allied to a fall in the value of the Mexican peso, led to further spectacular growth, with plants being established away from the Border region in parts of central Mexico. As such, employment in the major *maquiladora* centres rose from 310,000 in 1988 to an estimated 600,000 around 2000 (Dicken, 2003a, p.80).

Despite the growth in jobs and exports following NAFTA – with Mexican exports increasing from $51.8 billion in 1994 to $166.4 billion by 2000 (Bair and Gereffi, 2001, p.1885) – the *maquiladora* policy has been highly controversial, not least because of the documented evidence of poor and often abusive conditions of employment, especially among women workers who are subject to sexual and physical abuse (Wright, 2001). While there is some evidence of upgrading occurring – for example, Bair and Gereffi found that, as a result of NAFTA, some US clothing retailers had relocated the entire production process to Mexico – the plants

are still vertically organized and controlled by US MNCs, with little evidence of employment conditions improving. Indeed, the vulnerability of Mexico's global connections has been exposed in the period since 2000 with the country's economy being hit hard by competition from China. An article in the *International Herald Tribune* (3 September 2003) estimated that around 500 *maquiladora* plants had closed, with the loss of around 218,000 jobs between 2001 and 2003. A new round of closures occurred from 2008, reflecting reduced consumer spending and industrial production in the US due to the economic downturn (Carrillo and Zarate, 2009).

6.6.3 FDI as regional development panacea in eastern Europe

Nowhere have the merits of FDI as a strategy for the modernization and increased competitiveness of regional economies been made so forcefully as eastern Europe. After the fall of Communism and the entry of the states of central and eastern Europe into the world capitalist economy in 1989, the region experienced an upsurge in FDI from western and predominantly European MNCs. Because of the perceived benefits of FDI, in terms of job creation and industrial modernization, there was fierce competition between the different countries to attract foreign companies. In some cases, this led to substantial subsidies and tax breaks: as much as $5,000 for every job created in the case of some overseas investors in the Czech Republic (Pavlinek, 2004, p.59).

Initially, investment was driven both by the availability of cheap labour and market access, but, in recent years, cost considerations have increasingly driven the strategies of western companies. The effects have been highly uneven between both countries and regions within countries. Generally, the lion's share of FDI has been concentrated in a few regions with four countries dominating investment flows: Poland, the Czech Republic, Hungary and Russia. The general pattern is for FDI to be concentrated in large metropolitan regions, developed industrial regions and those areas that have borders with western European countries (Pavlinek, 2004).

Within areas benefiting from FDI, the effects have been mixed. The positive stimulus of jobs, new investment, western management expertise and contributions to the local tax base has been offset by the displacement of local firms, who find themselves in a position where they are losing their skilled labour to incoming firms that pay higher wage rates. Additionally, with the exception of the few local firms that are acquired by foreign investors, the majority find themselves lacking the capital to modernize so that they face a continuing struggle to survive against better resourced competitors in receipt of government grants. There is also relatively little evidence that incoming firms are becoming firmly embedded within regional economies, with many new start-ups serving as final assembly 'turn-key' plants for components shipped in from western Europe, reducing the scope for local sourcing.

A growing concern among local policy-makers relates to the longer-term sustainability of the FDI wave in the region in the face of cheaper foreign competition, with a number of recent examples of firms relocating production elsewhere (with China being the main beneficiary) as costs have risen (Table 6.12). One strategy of western companies has been to retain ownership of machinery in supplier companies in order to avoid sunk costs (Clark, 1994) and stay 'footloose'. In this case, this would include the training of local labour or the investment in buildings or infrastructure. By owning the machinery but not the plant, and by not directly employing the workforce, firms can avoid such sunk costs. As one director of a local firm supplying a major auto producer noted: 'we could be replaced at any time as we replaced the previous suppliers. It is possible to take away the machinery but impossible to move the building' (Pavlinek, 2004, p.58).

Reflect

> Do you think that the benefits of FDI outweigh the disadvantages for host regions? To what extent would your answer to this question vary between different parts of the world?

Table 6.12 Footloose MNCs: examples of relocation/closure of FDI plants in central Europe

Firm	Location	Nationality	Product	Destination of relocation	Number of job losses	Additional details
Mannesmann	Hungary	German	Steel tubes	China	1,100	
Shinwa	Hungary	Japanese	Electronic equipment	China	500	
Solectron	Hungary	US	Electronics	Romania	unknown	
Flextronics	Hungary and Czech	Singapore	Electronics	China	1,200 in Hungary,	Czech plant opened in 2000, closed in 2002. Labour costs cited as reason
Varta Arku	Czech	German	Telecoms	Unspecified – Asia	2,500 in Czech	Labour costs cited
Massive Production	Czech	Belgian	Light production	China	344	
Takta Petri	Czech	Japanese-German	Steering wheels	Unknown	Unknown	

Source: Adapted from Pavlinek 2004, pp.55–6.

6.7 Summary

In this chapter, we have traced the development of MNCs as important agents of globalization. In particular, we have sketched out the changing patterns of FDI and explained the motivations behind the internationalization of business activity. Contrary to some of the claims about globalization, much MNC activity is driven by market access, although the desire for cheaper labour has also been an important factor, underpinning the emergence of new international divisions of labour in manufacturing and services.

In response to some of the hyperbole in the business studies literature about the placelessness and mobility of MNCs, we have stressed their continued geographical connection to place. While the global economy has become increasingly integrated in recent decades, there are limits to the 'globalness' of MNCs. Even the most powerful of MNCs remain embedded in their own domestic economies in important ways and often lobby their own governments to defend their interests overseas. The geographical imprint of MNCs is also reflected in their operations, which continue to be characterized by cultures and practices that reflect their countries of origin, creating ongoing diversity in the operations of MNCs. Despite the increasingly transnational nature of many firms, national characteristics are still important, meaning that many MNCs have to adapt to the demands of local cultures and institutions in host countries and regions (Box 6.3).

Nonetheless, MNCs have a pronounced impact on host regions, creating relations of dependency in the eyes of some observers, although offering the opportunities for upgrading according to others. Recent processes of reorganization and the growing use of regional and global sourcing networks present MNCs with greater spatial flexibility. For host regions, the vulnerability to competition and MNC relocation seem to be common themes in a range of different circumstances, although it is important not to view regions as passive victims of corporate restructuring. Instead, regional development outcomes are the product of the ongoing interaction between regional actors and MNCs within global production networks (see section 10.4).

Exercise

Using the most recent Annual Reports of two MNCs from different countries listed in Table 6.4 (most reports are now available on the internet), construct detailed corporate geographies. Start by sketching out a map detailing the firms' locations and the type of activities (by sector and function) in different places. Then examine other material (share holdings, number and type of employees, etc.). Having developed as thorough a picture of the geography of each MNC as possible (you will notice that the level and quality of information provided will vary), consider the following questions.

How is each MNC's internal spatial division of labour organized? What does this tell us about the economic and political relationships between employees in different places? To what extent are the two MNCs tied to particular places? Finally, compare the operation of the two MNCs and reflect upon the influence of the home country in shaping the operation of each.

Key reading

Coe, N. and Lee, Y.S. (2006) 'The strategic localization of transnational retailers: the case of Samsung-Tesco in South Korea'. *Economic Geography*, 82: 61–88.
A very instructive case study of Tesco's joint venture with Samsung in South Korea, which discusses the underlying model of strategic localization.

Coffey, W. (1996) 'The newer international division of labour', in Daniels, P. and Lever, W. (eds) *The Global Economy in Transition*, Harlow: Longman, pp.40–61.
A good, though now rather dated, reference for the changing international division of labour and the processes behind it.

Dicken, P. (2007) *Global Shift: Reshaping the Global Economic Map in the 21st Century*, 5th edn, London: Sage, pp.107–37.
An excellent chapter on the evolution of MNCs and the debate on their apparent 'placelessness'. It is also worth reading the following chapter, which covers the organizational forms of MNCs and the networks within which they are located.

Pavlinek, P. (2004) 'Regional development implications of foreign direct investment in central Europe'. *European Urban and Regional Studies*, 11: 47–70.

A valuable analysis of the impact of FDI restructuring on host regions containing useful data on transition economies in central Europe. Provides some good case studies of individual MNC strategies towards foreign investment.

Pike, A., Rodríguez-Pose, A. and Tomaney, J. (2006) *Local and Regional Development*, London: Routledge.
A clear assessment of the impact on MNC investment on local and regional economies which refers to several different examples.

Wright, R. (2002) 'Transnational corporations and global divisions of labour', in Johnston, R.J., Taylor, P. and Watts, M. (eds) *Geographies of Global Change: Remapping the World*, Oxford: Blackwell, pp.68–77.
A good overview of the emergence of MNCs and their role in underpinning successive international divisions of labour.

Websites

http://www.unctad.org
United Nations site containing annual World Investment Reports that provide data on FDI and the trans-nationality index of MNCs.

http://www.ft.com
Financial Times site containing detailed data on MNCs and international business reports.

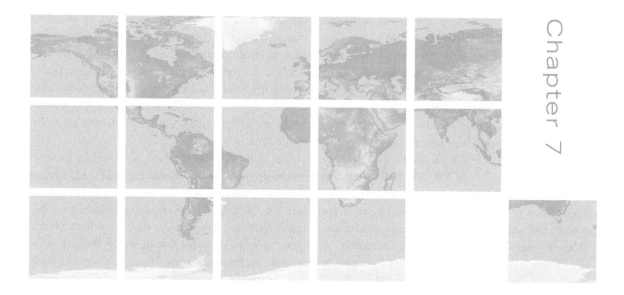

Changing geographies of work and employment

Chapter map

The chapter begins with some basic definitions of work and employment, together with a brief examination of how the relationship between them varies in time and space. It then proceeds to explore some of the key changes to the world of work that have characterized the contemporary era (stemming from the late 1970s), critically assessing different theoretical perspectives on these changes and their geographies. In particular, this part of the chapter considers the shift towards more service-based forms of work in the developed economies, the erosion of more stable forms of work and the development of labour market flexibility. These changes are then framed in a geographical perspective, highlighting their uneven impact upon places. This is followed by a discussion of the implications of these changes for trade unions, while the final section of the chapter looks at the emergence of a 'new labour geography' approach that views labour as an active participant in the construction of the economic landscape.

7.1 Introduction

For many people, regular interaction with the 'economy' takes place through work in its different forms, and our chances of making a decent living are shaped by our relationship to prevailing forms of work and employment. In the developed capitalist economies, the majority of us make a living through forms of paid employment, where we exchange our human labour for a wage. In less developed countries, much of the population is still engaged in non-capitalist forms of work linked to subsistence agriculture. The increasing integration of the world into a single global capitalist economy often threatens such traditional lifestyles, although it also opens up the possibility for some groups to escape feudal or more traditional forms of oppression to work as waged labour. As such, the processes of geographical uneven development and economic restructuring that we have considered in earlier chapters take on particular significance in terms of how they affect our conditions of employment and livelihoods. It is in this sense that economic geography has been defined as the geography of people's attempts to make a living (Lee, 2000, p.195).

In this chapter, our purpose is to examine the nature of employment change in the contemporary economy, exploring in particular the transformation of work that has occurred since the 1970s and the role played by geography in shaping change. We also emphasize that, unlike other factors of production, labour is not passive to processes of economic restructuring but plays a more active role. At both the individual and collective levels (through organizations such as trade unions), labour helps to shape the changing landscape of capitalism, although it is often at a strategic disadvantage because of its relative immobility (Herod, 2001).

7.2 Conceptualizing work and employment

7.2.1 Definitions

Human beings have to perform basic work tasks (e.g. hunting and gathering food, finding shelter, making clothes, looking after and raising children, etc.) to reproduce daily life. In this sense, work is essential to all societies, however primitive or advanced. How work is organized has varied and changed dramatically over time, as societies have developed from early and rudimentary nomadic peoples to the more advanced global

capitalist society of today. In section 3.3 we examined labour as a basic category under capitalism and developed an understanding of how a more complex division of labour emerged with the growth of an industrialized society. The concept of a division of labour allows us to differentiate between different forms of work, some of which are paid and other forms that receive no financial reward.

The determination of which work or jobs are paid and which are not and how much different types of work are paid relates to the social division of labour (section 3.3.2). Under advanced capitalism, a basic distinction can be made between two very different forms of work. Being in **employment,** selling your labour to work for an employer in the formal economy, is usually paid (although see next section), whereas many forms of **household or domestic labour,** typically undertaken by female family members, are not. Arguably, household work is more valuable to the basic reproduction of society than many forms of employment in the formal economy. The fact that it is (usually) not financially rewarded reflects prevailing values and (gendered) power relations within society. As Pahl observes:

> Someone arriving from another planet might be surprised and puzzled by the way that we distinguish between work and employment and the differential rewards that are paid to employees based on the kind of work they do and the kind of person they are. Interesting, creative and varied employment is highly rewarded; dull, repetitive and routine work is poorly rewarded. Men receive more than women, and this is related to social attitudes and conventions more than the actual amount or quantity of work that the individual or the gender category does.
>
> (Pahl, 1988, p.1)

The broader point is that work is highly differentiated and these differences reflect facets of social identity such as gender, ethnicity and age. Up until relatively recently, women were discouraged and actively discriminated against in the labour market (both through government legislation and employer attitudes). Indeed, in most countries, wages for women's employment are usually lower than the equivalent for men. Wage discrimination also continues to apply on the

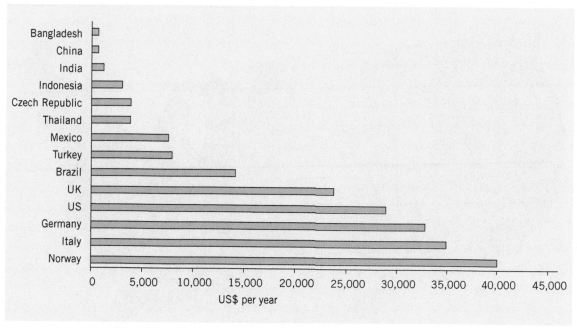

Figure 7.1 Differences in labour cost between selected countries
Source: Adapted from World Bank, 2006, Table 2.6.

basis of ethnicity, religion and even age. While most advanced industrial economies have laws promoting equal opportunities, employers continue to discriminate on the basis of social identity.

Geography is critical in understanding how work and employment are organized. As we have noted in past chapters, the economy under capitalism is characterized by spatially uneven development and the geographical differentiation of work is an important part of this. Through the development of large multi-plant corporations and the construction of a **spatial division of labour** (section 4.3.2), a highly differentiated landscape of employment emerges reflecting variations in the nature and type of work between different places. With the emergence of a global economy and through the activities of MNCs this has been translated into a 'new international division of labour' as noted in Chapter 6 (section 6.3.2). The important point to note here is that these spatial variations in labour (at national and international levels) are, in turn, important for future rounds of capital investment, allowing firms to make location decisions based upon the different types of labour (e.g. rates of pay or levels of skill) available across the economic landscape.

Spatial divisions of labour are reflected in variations in employment conditions between places, most obviously in the wide discrepancies in wages between developed and developing countries (Figure 7. 1), encouraging the relocation of low-wage production by MNCs (section 6.3.2). These differences exist within countries as well as between them. For example, in the US, the average hourly earnings of a manufacturing worker in Dakota were found to be 60 per cent of those in Michigan in the 1990s (Hayter, 1997, p.88). More recent data from the US Bureau of Labor Statistics indicate considerable variation by region in the incidence of unemployment (Figure 7.2). States hit by deindustrialization, such as Michigan in the north, or predominantly rural states heavily dependent on agricultural and other traditional sectors, such as Mississippi or Kentucky, continue to suffer the highest rates of unemployment. However, the recent financial crisis has also hit formerly prosperous states such as Florida and California that have been more exposed to the crash in property prices. Spatial variations in employment do not only amount to wage or unemployment differentials but also include different cultures of labour associated with particular industries (section

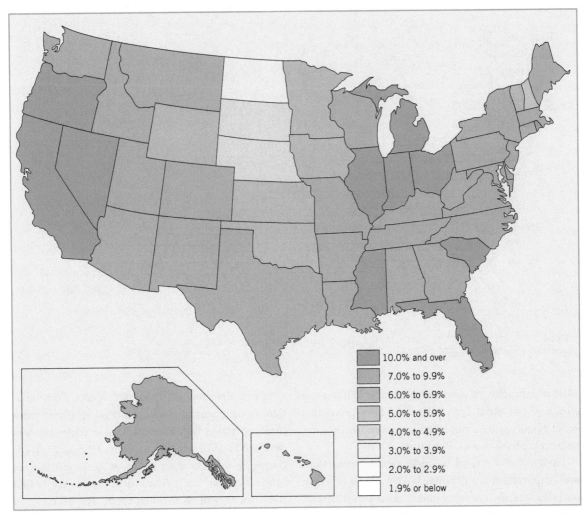

Figure 7.2 Unemployment rates by state, seasonally adjusted, July 2010 (US rate = 9.5 per cent)
Source: Bureau of Labor Statistics, Local Area Unemployment Statistics.

4.3) (see also Peck, 1996). These include differences in working practices, levels of unionization and the way **local labour markets** operate (e.g. through different kinds of recruitment and training strategies).

7.2.2 The transition to industrial work

The forms of employment that we have come to associate with the capitalist economy are relatively recent in historical terms. Prior to the development of the Industrial Revolution in the late eighteenth century, the concept of paid work was relatively marginal to the overall functioning of the economy. The vast majority of the population lived 'off the land' and engaged, to a varying extent, in forms of **subsistence agriculture** (Malcolmson, 1988). Within the pre-industrial economy wage labour was used to supplement other household activities as a way of 'making a living' and would have taken the form of supplying labour to larger farms and estates at particular times of the year, such as during harvesting (Figure 7.3). Some men and women may also have been employed as maids, cooks and other servants within the household of local elites.

Additionally, the idea of 'going to work' in the sense of travelling to a dedicated 'place of work' would have been an alien concept. Work would have been centred upon the household. Early craft work, for example in textile and weaving activities, would have taken place within the home, the origins of the term 'cottage industry'. At a broader geographical scale, work was dispersed across the landscape, in contrast with the massive concentrations of work in large towns and cities that were to develop with the emergence of modern industry and industrial capitalism. In the world's first industrial economy, the UK, there had been only two cities, London and Edinburgh, with a population of over 50,000 in 1750 (Hobsbawm, 1999, p.64). By 1851, there were nine cities, with two of over 100,000 people and over half of the population were now living in urban areas as a result of industrialization.

The transition to an **industrial society** therefore caused massive spatial and social upheaval, and in the process fundamentally transformed the nature of work (see also sections 3.2.4 and 3.3.2), as the renowned Marxist historian Eric Hobsbawm puts it:

> It transformed the lives of men beyond recognition. Or, to be more exact, in its [the Industrial Revolution] initial stages it destroyed the old ways of living and left them free to discover or make for themselves new ones.
>
> (Hobsbawn, 1999, p.58)

For the majority of the working population, this transformation meant the destruction of a predominantly subsistence rural lifestyle, largely controlled by the seasons, to forms of routine, paid work under the strict supervision of employers in densely populated towns and cities.

Figure 7.3 Labouring in a pre-modern landscape: 'The Harvesters' by Hans Brasen
Source: Corbis.

7.2.3 Work and employment in the contemporary global economy

A consequence of the transformation of work was the emergence of a large **industrial working class**, whose sole means of 'earning a living' was employment or waged labour. The appalling conditions faced by labour in the early phases of industrialization (see Chapter 3) gave rise to the emergence of **trade unions** in the late nineteenth century, as collective organizations geared towards defending the interests of the working class (see Box 7.1). Trade unions reflect the inherent conflict between employees, concerned with earning a living through paid employment, and a capitalist class of employers dedicated to producing higher profits and reducing costs.

While employment has become the dominant category of work in the global economy, its geographical incidence is highly uneven. Castree *et al.* have

---

Box 7.1

The emergence of trade unions as key actors in the landscape of capitalism

Harvey's concept of a spatial fix was used in earlier chapters to highlight the role of firms in constructing the geography of the economy. But labour also shapes the economic landscape directly through its own strategies and actions in defending and promoting its interests. The fundamental

interests of labour are focused upon securing improved wages and conditions and the interaction between this and the profit-seeking imperative of capital are of critical importance. This focuses on the organizations set up by workers to promote their interests, principally trade unions. The

formation of unions was the result of struggle by groups of workers to defend their interests and improve their lot, beginning in the industrial heartlands of the UK in the nineteenth century before spreading to continental Europe and the US (Hudson, 2001, p.100). Such struggles were

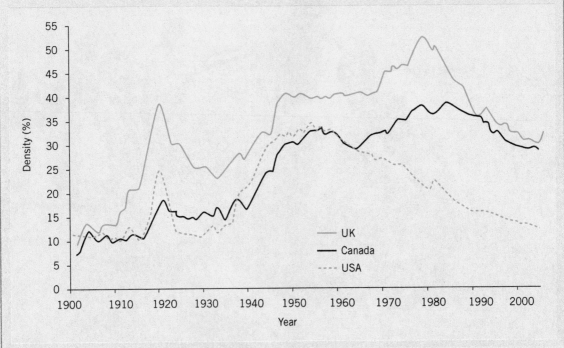

Figure 7.4 Trade union density: Canada, US and UK, 1900–2004
Source: Blanchflower, 2006.

Box 7.1 (continued)

often conducted against employers who were bitterly opposed to trade unions, sometimes supported by the state.

The role of unions is to secure better wages for their members and to promote the interests of labour more generally. As part of a broader labour movement, unions have been closely associated with social democratic and socialist political parties in many countries, with the British Labour Party, for example, closely tied to

trade unions throughout its history. Unions had considerable success in securing better conditions for their members alongside their political allies over the middle decades of the twentieth century. By the late 1960s, growing union strength (see Figure 7.4 for historical trends in Canada, US and UK) in the manufacturing sector in particular was leading to demands for higher wages, as well as a challenge to management's ability to control the nature and speed of

work. Since the late 1970s, however, unions in the advanced industrial countries have faced tougher times, reflected in falling memberships, as many of the heavily unionized traditional industries have collapsed and as governments have placed increased legal restrictions on their activities. Despite such reversals, organized labour remains an important force shaping the geography of capitalism.

---

estimated that three-quarters of world employment (waged labour) is located in only 22 countries with 'almost half of the world's labour force located in four countries: China, India, the United States and Indonesia' (Castree *et al.*, 2004, p.11). Within the global south, the incidence of wage labour varies considerably, reflecting differences in levels of integration within the global economy. In many parts of Africa, Asia and Latin America, forms of subsistence agriculture persist, although attempts to modernize economies have often led to the forced destruction of such traditional ways of 'making a living', without always replacing them with sustainable alternatives. For much of sub-Saharan Africa, in particular, integration into the global economy continues to be defined by the production of primary commodities such as minerals and agricultural products such as coffee and cotton (Chapter 8), with the higher-value-added activities such as manufacturing and processing taking place in Europe, North America or South-east Asia.

Even in those countries where wage labour has grown with industrialization, the conditions of work are often poor, as evidenced by the difference in labour costs between countries (Figure 7.1). Child labour – the employment of children under the age of 16 – is also common place in the Global South, more often than not under extremely exploitative and badly paid working conditions. In China, for example, children are employed to make toys for 12 to 16 hours per day

for a wage of less than 14 cents per hour (Hudson, 2001, p.245). The ILO estimates that there are approximately 246 million child labourers (aged between five and 17) working in the world, 70 per cent of whom are engaged in hazardous work in mining, chemicals industries or with harmful pesticides in agriculture or with dangerous machinery (UNICEF, 2006, p.46). Seventy-three million child labourers are under the age of ten (ibid.).

Although wage labour has spread with the development of globalization, it is important to emphasize that only around half the population at any one time are in paid employment. There are also other forms of work that are unpaid, notably housework, voluntary work and the work of many carers (particularly family members), looking after older people, children and the disabled. Attempts to reduce state welfare provision in many countries since the 1980s have meant that unpaid care work has become increasingly critical both to supporting the economy – through the supervision of children and their 'socialization' for future employment – and in providing for the more disadvantaged sections of society. Unpaid housework is still the dominant form of domestic labour in both advanced and less-developed economies, although the situation is slightly complicated by the existence of paid work for some household tasks (e.g. nannying or cleaning) particularly in many western societies in the US and western Europe, where both members of a household are engaged in full-time paid employment (Gregson and Lowe, 1994).

Alongside paid work, other categories include the self-employed, the definition of which often varies in time and space according to differences in employment laws and regulation. Additionally, a growing number of the global labour force is subject to unemployment or underemployment. The extent of this again varies over time, dependent upon the pendulum of uneven development. A growing number of people in western Europe has been exposed to unemployment in various forms since the 1970s, while after the collapse of Communism unemployment rocketed in eastern Europe during the early 1990s. More recently, the financial crisis of 2008–09 has greatly increased

levels of unemployment in western and eastern Europe and North America with less notable impacts in Latin America, Africa and South and East Asia (ILO, 2010, p.9). Even within countries, the crisis has shown marked variations – a feature that we explore in greater detail in section 9.5.

A final category of work to comment on here is slave or forced labour, which sadly remains a feature of the global economy (see Box 7.2). One of the more pernicious outcomes of globalization has been an increase in the trafficking of people in conditions of forced labour, particularly for the sex trade. Illegal immigrants who lack the status of citizens in their destination countries are particularly vulnerable to highly exploitative employers with an estimated two and a half million

Box 7.2

Forced labour under capitalism

While wage labour has become the dominant form of employment under capitalism, other more exploitative forms of labour continue to exist. In particular, the slave trade, from the fifteenth to the nineteenth centuries, was critical in the emergence of an international system of capitalism and represents one of the most ignominious episodes in European colonial history. As part of colonial expansion, various European states engaged in the enforced transfer of slaves from the African continent to work in the new colonies in the Americas. Although slavery largely predated formal political colonialism, its effect was arguably as devastating for the countries involved. From the late seventeenth century until the nineteenth, it is estimated that somewhere between eight and 10.5 million slaves were transported from West Africa to the Americas to work primarily in tobacco, sugar and cotton plantations (Potter et al., 2004, p.59). Apart from the appalling situation of slavery itself, the conditions of transportation were dreadful and

inhumane; on an average Dutch slaver in the seventeenth century, for example, 14.8 per cent of slaves would die en route, from diseases such as smallpox, dysentery and scurvy (ibid.).

With the abolition of slavery in the nineteenth century, many countries developed systems of 'indentured labour', whereby Asian workers, in particular, were recruited to work in European colonies. Workers would be employed under contract to a single employer for a fixed period (typically between four and seven years) in return for transportation, accommodation and food. In practice, conditions for indentured workers were little better than slavery. Many died during transportation while workers were unable to terminate their contract. Employers, however, had the freedom to sell indentures before the end of the contract period.

Despite the near universal rejection of slavery by states in the modern economy, it sadly persists in various guises. The International Labour Office's 1998 Declaration

on Fundamental Principles and Rights at Work has drawn attention to the continuing use by employers of what it terms 'forced labour' in the global economy. Forced labour is defined as 'all work or service which is exacted from any person under the menace of any penalty and for which the said person has not offered themselves voluntarily' (ILO, 2005a, p.5). Although difficult to accurately measure, for obvious reasons, a conservative estimate is that over 12 million people are in some form of forced labour globally, with Asia and the Pacific Region dominating the trade (ibid., p.12). Two million people are in 'state or military imposed' forced labour, such as prisons, while another growing category is sexual exploitation with over one million 'workers' forced to sell their bodies for sex. Forced labour is a highly gendered affair and often involves children: women and girls account for 56 per cent of total forced labour and 98 per cent of the sex trade.

people engaged in forced labour as a result of trafficking (ILO, 2005b, p.14).

7.3 Changing forms of employment

The nature of employment has changed through time, reflecting broader changes within the capitalist economy. The shift from a craft-based form of production to large-scale factory-based production was critical to the reorganization of the work and the introduction of a more complex division of labour in the nineteenth century (section 3.3.2). More recently, there has been considerable debate about the changes taking place in the contemporary workplace, particularly since the 1970s. In part, these are linked to TNC restructuring and a developing global division of labour as firms seek out cheaper labour resources in the Global South (see Chapter 6) but changes have also been driven by the process of deindustrialization and the emergence of a service-based economy. Restructuring also reflects more deep-seated tensions between employers and employees over both the organization and control of work (**labour process**), and the distribution of the surplus generated by production (i.e. wages versus profits).

Many commentators have viewed the subsequent period from around the mid-1970s to the present day as a new phase of capitalism, characterized first and foremost by a transformation in the nature of employment. We explore two sets of claims here: arguments about a transition from an industrial to a **post-industrial society**; and debates about the shift from **Fordism** (see section 3.2.2) to more **flexible production**. Both have aroused considerable academic controversy and, in evaluating the claims made, it is important to be wary of what we term 'myths at work', and to 'unpick

and deconstruct the myths to show which aspects of them carry credibility and which do not' (Bradley *et al.*, 2000, p.2).

7.3.1 A post-industrial economy

The first set of claims relate to the decline of employment in manufacturing and the shift to a service-based economy. Certainly, in the most advanced industrial economies, there has been a massive shift from employment in manufacturing to services. What this means for the nature of the employment relationship is another matter. The implications of the growth of service work have been the subject of a number of well-known treatises since the 1970s, when manufacturing decline was becoming increasingly apparent. Many have chosen to characterize this as a shift towards a post-industrial society where the whole relationship between work and society has changed (e.g., Bell, 1973; Gorz, 1982; Lash and Urry, 1994). The following claims are typically made about the implications for the nature of work:

> The growth of services has led to an increasing number of people in middle-class, white-collar occupations. Linked to the emergence of high-tech industries, a growing proportion of the workforce is in more educated and professional forms of employment, which involve more autonomy and control in contrast to Fordist mass production.

> In the context of a global economy, advanced economies should prioritize knowledge-based activities for jobs growth. With the advance of global communications and information technology, more routine activities in both services and manufacturing will be increasingly exposed to low-wage competition. Labour market policy should focus upon skills and training.

> Reflecting the increasing interaction with consumers, new social relations are established through service-related forms of work, which involve workers 'performing' scripted roles under the supervision of management. Job prospects and career enhancement are increasingly differentiated by quality of an individual's performance and the individual's ability

to internalize management's wishes (du Gay, 1996; McDowell, 1997).

➤ Traditional class identities are becoming eroded with the decline of the industrial working class that formed the basis of trade union strength. In the post-industrial economy, a new dominant middle class is emerging in the service economy, while the erosion of traditional working-class livelihoods has left a residual underclass of male, unskilled workers

stripped of their traditional identities (McDowell, 2003).

The idea that the post-industrial economy is knowledge intensive and provides plenty of jobs, as long as the workforce has the appropriate skills, is a particularly powerful myth which has been a major element of domestic economic policy in both western Europe and North America since the early 1990s. For example, Robert Reich, President Clinton's Secretary for Labor

Figure 7.5 Call centres: 'an assembly line in the head'?
Source: Sherwin Crasto/Reuters/Corbis.

during his first term of office (between 1992 and 1996), famously claimed:

> The most rapidly growing job categories are knowledge-intensive; I've called them 'symbolic analysts'. Why are they growing so quickly? Why are they paying so well? Because technology is generating all sorts of new possibilities … The problem is that too many people don't have the right skills. (Quoted in Henwood, 1998, p.17)

However, the evidence suggests that this is overstated. In both the US and the UK, for example, the fastest growing job categories in recent years have been in more menial work that does not require high skill or education levels (Henwood, 1998 Thompson, 2004). In contrast, evidence suggests a continuing decline in 'middle level, craft and skilled manual employment' (Thompson, 2004, p.30). In this respect, the discourse of the knowledge economy has been advanced by politicians, academics and policy-makers in the context of the disappearance of relatively well-paid and secure jobs in the manufacturing sector and their replacement by lower-paid, lower-skilled work in service activities. Many of the jobs being created are often very similar to the more routine kinds of work associated with manufacturing assembly lines, where work is strictly controlled and monitored by management (Figure 7.5), often highly intensive, pressurized and monotonous (Beynon, 1997; Henwood, 1998). A good example has been the growth in call centre work, one of the most rapidly growing areas of employment in the UK over the past decade (see below). Because of the routine

nature of much of the work – answering customer calls in a heavily scripted way with little autonomy or creativity – one call centre worker memorably described the job as an 'assembly line in the head' (Taylor and Bain, 1999).

The shift from manufacturing into services has also been accompanied by a growth in female employment relative to male. For some, this reflects the greater personal abilities of women when dealing with customers, which makes them more suitable for service employment. However, it should be noted that many of the jobs created for women are lower paid than those lost to men, while one of the highest growth rates has been recorded in part-time employment (Dex and McCulloch, 1997).

7.3.2 A variegated geography of deindustrialization and services growth

At a more fundamental level, we might challenge the usefulness of the term 'post-industrial' in the context of an integrating global economy where the majority of the world's population is undergoing rapid processes of urbanization and industrialization, while the rich minority experience deindustrialization and a shift towards service work. What is more plausible is that we are seeing the latest phase in an unfolding economic geography of uneven development both between and within the world's major regions. Much of the manufacturing and heavy industrial work is shifting to the

Table 7.1 Deindustrialization in selected developed economies: an uneven global picture

	Share of industrial employment (%)	
	1990	2007
France	29.7	22.6
Germany	38.6[a]	30.0
Japan	34.1	27.9
UK	32.3	22.4
US	26.2	19.8

[a] For West Germany only

Source: OECD in Figures, various editions, available at http://www.oecd.org.

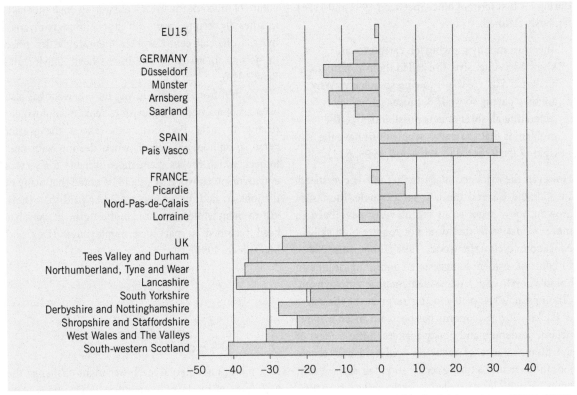

Figure 7.6 Manufacturing employment change in selected European old industrial regions, 1996–2005
Source: Eurostat, Regions – Science and Technology.

global south: already 73 per cent of the world's 3.1 billion workers are in developing countries with an estimated 46 million new workers joining the global labour force each year (Ghose *et al.*, 2008).

In the developed world this equates to a variable geography of manufacturing decline and services growth. While there has been a generalized shift in employment out of industrial activities into services, the rate of change has varied considerably between countries (Table 7.1). Germany and Japan, for example, have been more successful at retaining manufacturing employment than the US and UK. Within countries, deindustrialization has affected areas of traditional and heavy industry (e.g. coal mining, steelmaking and ship-building, textiles and auto production) particularly hard (see section 4.3). However, experiences have once again been very variable between countries. Recent evidence from western Europe's old industrial regions, for example, indicates that French and Spanish indus-trial regions have proved more resilient in protecting

manufacturing employment than their UK and German counterparts (Birch *et al.*, 2010), suggesting that national policy agendas might play an impor-tant role in mitigating processes of global competition (Figure 7.6).

The shift towards service employment has tended to be uneven within countries. In the UK, for example, the geography of employment in the service economy con-tinues to reflect a pronounced spatial division of labour between London and the South-east, and the rest of the economy. In peripheral cities and regions, such as the north-east of England and Scotland, employment growth has been dominated by lower paid and more routine jobs (see Box 7.3), while the lion's share of high-status jobs has been concentrated in the south-east of England. The massive growth in call centre work, in particular, as larger corporations in the business ser-vices sector outsource more routine work to lower-cost locations in the peripheral regions, has almost directly matched earlier phases of spatial reorganization by

Box 7.3

Glasgow's troubled post-industrial transition

During its heyday at the end of the nineteenth century, Glasgow was known as the 'second city of the British Empire', and was synonymous with the growth of the shipbuilding and heavy engineering industries to service the infrastructure and markets of the empire. The twentieth century heralded a long period of decline, culminating in one of the most rapid rates of deindustrialization in western Europe, losing 68 per cent of its manufacturing jobs between 1971 and 2001 (Turok and Bailey, 2004, p.41). Although the city experienced a net increase in employment between 1981 and 2006 of 14,000 jobs, most of these were in low-paid, part-time service work, and full-time employment actually fell by over 34,000 during the same period (Cumbers *et al.*, 2009).

As a consequence, and despite the rhetoric espoused by city elites of a post-industrial renaissance as a city of culture, there is a grim reality of a heavily divided city that has some of the highest levels of deprivation in the UK. The labour market is marked by a growing schism between relatively affluent professional and graduate workers who inhabit the plusher suburbs and fashionable West End districts of the city, and the less skilled and poorly qualified inhabitants of inner city and peripheral housing estates (Turok and Bailey, 2004). Over 40 per cent of households in Glasgow are living below the poverty line, defined as half of the median income (Dorling and Thomas, 2004), while 27.2 per cent of the working age population are defined as economically 'inactive', compared with a UK average of 21.2 per cent (Cumbers *et al.*, 2009, p.11).

Economic deprivation also contributes to severe social and medical problems; in Glasgow this is reflected in appalling differences in health and mortality rates between the more prosperous areas such as the West End and some neighbourhoods in the east of the city where average life expectancy for men is below 60. Research has shown that the city has nine out of the bottom ten areas of the country with the worst premature mortality rates (Shaw *et al.*, 2005), reflecting a distinct 'Glasgow effect' (related to culture and lifestyle) in addition to severe social and economic deprivation.

manufacturing firms (Massey, 1984). Rather than the new environment producing better jobs, for many of the UK's regions it has led to a deskilling and downgrading of work (Hudson, 1989).

7.3.3 The transition from Fordism to flexibility

A second very influential perspective on the changing workplace is associated with the shift towards **post-Fordism** or flexible working (section 4.3.3). Not only have variants of this become dominant within academic and management discourses, but they have also proved very popular with certain politicians, keen to pursue labour market deregulation as a route to competitiveness in the global economy. During the UK's 2005 Presidency of the European Commission, Prime Minister Tony Blair urged his continental counterparts to adopt what he termed the 'Anglo-Saxon model of labour market flexibility' (reported in *The Guardian*,

1 July 2005), while in April 2006 the French Prime Minister was forced to withdraw a new law proposing greater flexibility for employers to 'hire and fire' younger workers in the face of mass public protests.

Fordism as a concept was derived originally from the mass production system introduced into the American automobile industry by Henry Ford. As mass production spread throughout industry in the early part of the twentieth century, the term Fordism became used in a wider sense to encapsulate the system of modern work under capitalism. The Italian Marxist, Antonio Gramsci, was one of the first to use Fordism in this way, associating it with the innovative American model of capitalism that he saw as replacing more antiquated forms. Fordism subsequently has been associated with a particular phase of industrial capitalism and work organization, originating in the early 1900s, but having its heyday between 1945 and 1970 when it spread from North America to western Europe. In employment terms, the key features of Fordism were (Table 7.2):

Table 7.2 Fordist and after-Fordist labour markets

	Fordist–Keynesian	After-Fordist
Production organization	Mass production	Flexible production
Labour process	Deskilled and Taylorized, detailed division of labour	Flexible, functional and numerical
Industrial relations	High union densities; strong worker rights; centralized bargaining	Disorganization of unions; individualized employment relations; decentralized bargaining
Labour segmentation	Institutionalized; rigid hierarchies; large, internal labour markets	Fluid; core-periphery divide; breakdown of internal labour markets
Employment norms	Male, full-time workers; occupational stability and job security	Privileging adaptable workers; normalization of employment insecurity
Income distribution	Rising real incomes and declining pay inequality	Polarization of incomes and pay inequality
Labour market policy	Full employment; secure and high level of male employment	Full employability; ensuring workforce adaptability
Scale characteristics	Privileging of national economy for economic management and labour regulation	De-privileging of national; global economic imperatives; decentralization of labour regulation
Geographical tendencies	Dispersal	Concentration

Source: Peck, 2000, p.139.

> a highly detailed division of labour with considerable deskilling and management control over job tasks (i.e. Taylorism);

> the development of large corporations with complex and extremely hierarchical 'internal labour markets' through which recruitment and promotion were organized;

> relatively high job security, centred around the norm of the white full-time male worker;

> recognition of trade unions as 'social partners' with government and employers in regulating and managing employment conditions, with high levels of union membership and national level collective bargaining.

As part of the response to the economic crisis of the 1970s, European and US firms began to rethink their models of employment organization in order to (i) reduce costs and (ii) create more flexible organizational structures that allowed them to better adapt to increasingly unstable markets arising from global competition (see Table 7.2). For our purposes here we

can distinguish between two types of effect: changes to the way work is organized (i.e. labour process), and changes to the way employers recruit and select workers (i.e. labour market).

Labour process

> The replacement of systems of mass production and Taylorist work practices with new high performance forms of production, most often termed 'lean production' (Womack *et al.*, 1990) or flexible production (Peck, 2000). Heavily influenced by the success of Japanese auto manufacturers during the late 1970s and 1980s, particularly the example of Toyota in out-competing US firms, lean production methods involve slimming down the workforce, a greater attention to quality and the elimination of waste (through quality circles and total quality control and management). As part of the commitment to more efficient methods, just-in-time production methods were introduced, whereby manufacturers keep only the bare minimum of components, eliminating the costly investment in huge stocks associated with

Fordism. This necessitates greater interaction and proximity (both in terms of geographical and managerial distance) with suppliers.

➤ For employees, these new production concepts require a shift away from the deskilled and intensive division of labour associated with Fordist assembly work towards a more multi-skilled and multi-tasked labour force, more able to be flexibly deployed between different parts of the production process or

indeed able to switch to new products as the need arises.

➤ The alienated worker of Fordism, subjected to strong and coercive managerial control, is replaced by a more involved worker, who has greater skills and knowledge of the labour process, has greater autonomy over his/her work.

➤ For some, this brings about the potential for a change in the social relations between management

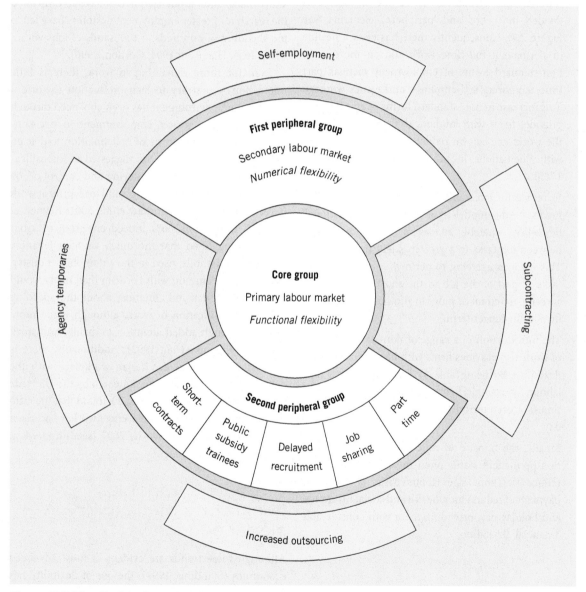

Figure 7.7 The flexible firm model
Source: Allen *et al.*, 1988, p.202.

and worker with a culture of confrontation giving way to trust and cooperation (Oliver and Wilkinson, 1992).

Labour market

➤ Accompanying the shift towards lean production, firms are restructuring their workforces, shifting away from internal labour markets where recruitment and selection are organized within the firm towards dual structures where the workforce is divided into 'core' and 'peripheral' elements (see Figure 7.7). Consequently there has been a decline in permanent full-time work and an increase in non-standard forms of employment such as part time, temporary, self-employed and agency workers. This increase in non-standard forms of employment provides firms with 'numerical' flexibility, whereby the workforce can be reduced or increased in line with fluctuations in product markets (Atkinson, 1985).

➤ In return for job stability, workers in the core segments of the model agree to greater 'functional' flexibility, whereby managers can move them between job tasks to a greater extent than hitherto. This involves agreeing to perform a wider range of tasks as part of the job in the short term, or even agreeing to retrain or move to another job within the firm in the longer term.

➤ The introduction of a range of non-standard forms of work also provides firms with increased temporal flexibility. By being able to draw upon part-time labour, for example, firms can vary shift patterns and times, or even introduce extra shifts into a working day.

➤ Finally, the transfer of much of the workforce to non-permanent status often means that firms can reduce the non-wage labour costs, such as the payment of redundancy, health and social insurance, and holiday pay, providing them with considerable 'financial' flexibility.

7.3.4 The limits to flexibility for workers

While these changes to the nature of employment have been perceived as important by politicians and businesses to enhance competitiveness in the context of a global economy, it is clear that for individual employees the effects can be more pernicious, depending on which side of the core-periphery divide you fall. Various commentators have catalogued how the processes of 'downsizing' and rationalization that have accompanied the search for greater employment flexibility have led to the shedding of hundreds of thousands of jobs within TNCs (e.g., Harrison, 1994; Gordon, 1996).

Even for those remaining in work, there is little impression – contrary to lean production rhetoric – that the work environment has been enhanced through skills upgrading, greater empowerment in decision-making or a greater spirit of collaboration and team working. Rather, research has suggested an intensification of work, increased monitoring and control of job tasks and growing dissatisfaction and insecurity at work (Green, 2001, 2004; Grimshaw et al., 2001; Beynon et al., 2002; Thompson, 2003). Indeed, one group of critics has even suggested that the much vaunted Japanese production methods, used in the automobile industry, have more in common with Fordism than many would admit, being first and foremost about the speeding up and intensification of work, although made more efficient through added attention to quality and stock control (Williams et al., 1992). Additionally, there is little data for a more egalitarian workplace with the data suggesting the opposite. Research by the US trade union confederation, the AFL-CIO, found that the ratio of chief executive pay to the average worker had risen from 1:41 in 1980 to 1:344 by 2007 (see: http://www.aflcio.org/corporatewatch/).

7.3.5 The variable geography of flexibility in a global economy

Although these trends are evident in most advanced economies (Standing, 1999), the use of flexibility has been greater in the Anglo-Saxon world, in countries such as the UK, US and Australia, where successive governments have pursued policies of labour market deregulation and where job protection legislation is minimal. Firms find it easier therefore to 'hire and fire' workers, compared with many countries in continental Europe where employment legislation means that firms have to pay much higher redundancy and other social costs. For example, when the French automobile manufacturer, Peugeot, decided to close its plant near Coventry in the English Midlands rather than one of its domestic plants, trade unionists at the plant contrasted the £50,000 it cost to lay off a British worker compared with as much as £140,000 for a French worker (*The Guardian*, 20 April 2006).

Recent research by the International Labor Office confirmed that, despite increasing global economic integration, there remain wide variations in labour market flexibility. This is reflected in the average length of time that workers spend in a particular job. Once again, the main contrasts exist between a continental European labour market model and the Anglo-American with, for example, the average tenure of American workers being almost half that of larger European countries such as Italy and France (Table 7.3). The differences are even greater when we consider the proportion of the workforce with under a year's tenure: 24 per cent in the US, compared with 11 per cent for Italy. Japan is the other country in the developed world with a tradition of more stable employment, where employers in the large corporate sector have traditionally provided life-time employment guarantees, in return for functional flexibility from employees.

Advocates of labour market flexibility claim that it helps to reduce unemployment by encouraging labour to move from declining sectors and regions to ones that are growing. The US labour market is held up as the model here, because historically it has consistently outperformed that of the EU since the mid-1990s (Sorrentino and Moy, 2002, p.19). The UK, as another model of labour market flexibility, has also outperformed its European neighbours, but it should be pointed out that there are wide variations in experience between European countries with similarly 'rigid' systems of employment protection, while no study has been able to find a decisive link between economic performance and labour market flexibility (ILO, 2005b). Additionally, many of the jobs created in the UK and US are in contingent or temporary work that is highly exploitative (see Box 7.4). Moreover, the most recent evidence suggests that the US and UK have seen much more rapid rises in unemployment than their supposedly over-regulated counterparts, France and Germany. In 2009, unemployment in the US at 9.4 per cent was higher than in both France (9.1 per cent) and Germany (7.4 per cent) (OECD, 2010).

Table 7.3 Average job tenure and tenure distribution for selected OECD countries

	Average tenure (years)	Workers with < 1 year tenure (%)	Workers with > 1 year tenure (%)
Greece	13.6	9.8	52.1
Japan	12.2	8.3	43.1
Italy	12.2	10.8	49.3
France	11.2	15.3	44.2
Germany	10.6	14.3	41.7
Denmark	8.3	20.9	31.5
UK	8.2	19.1	32.1
US	6.6	24.5	26.2

Source: ILO, 2005a, Table 4.1 p.191.

Box 7.4

The growth of the temporary agency

One of the most tangible signs of the new flexible labour market has been the growth of the temporary agency. In the US, it is estimated that one in five jobs created since 1984 has been through a temping agency (Peck and Theodore 2001, p.475). The number of temporary workers in the US grew from 250,000 (0.3 per cent of the workforce) in 1973 to 4.4 million by 1999 (4.3 per cent). Temporary agencies traditionally have been associated with supplying office workers, particularly women, to lower-level administrative positions within firms. However, recent research indicates that they have greatly expanded their remit, both spatially and sectorally. Some temporary agencies are now global multinationals in their own right. The two leading agencies – Adecco and Manpower – both operate in over 60 countries, with combined sales of over $30 billion per annum and the placement of over two million workers (Ward, 2004, p.252).

The greater use of temporary workers by firms and other organizations has meant that the use of agencies has expanded beyond the office to the factory, the warehouse and even the public sector (e.g. in schools and hospitals). In these conditions, Peck and Theodore's (1998) study of temporary agencies in Chicago found emerging divisions within the temporary workforce. For some workers (typically highly experienced administrative workers or secretaries working in the corporate sector) with highly prized skills and attributes, the flexibility provided by temporary work can be empowering. During periods of low unemployment it allows them to choose between employers and make considerable demands in terms of pay and conditions. In the words of one agency manager:

> If you're really good, most companies will hire you on. If [good temps] tell me Friday they want to change placements, Monday I will have a job for them. They're that good. I have to keep them busy or their other agency will take them.
>
> (Manager, small office placement)

For the majority of workers in the temping sector, however, who lack the skills to differentiate them from other workers, the experience is more negative. At the bottom end of the labour market (in areas such as warehouse work, cleaning and construction) many temporary workers are hired, by the day, in impersonal hiring halls, encapsulated by the phrase 'warm bodies delivered on time'. The benefit to the employer of this system is that the disciplining of workers is also outsourced to agencies who are asked to filter out workers with the 'wrong attitude'. In addition, the costs benefits are enormous, revealed in the following quote from an agency manager:

> They'd have to pay $8–9 an hour with benefits, vacation time, sick days, payroll taxes, they'd have to hire someone to pay them ... And their orders fluctuate so much ... Now they can cut 50–100 people in a day. All they have to do is call us.
>
> (Peck and Theodore, 1998, p.661)

From a firm's point of view, using temporary agencies to recruit labour reduces the non-wage costs and increases their numerical flexibility. In this way, labour is subject to a process of 'hyper-commodification' (ibid.), in the sense that employees really are treated as no more than commodities whose labour can be bought and sold on a daily basis. At the same time, the scope for forming unions and organizing for better conditions is greatly reduced by the precarious and ephemeral nature of employment.

Reflect

➤ To what extent is the shift towards a post-industrial economy improving conditions for the average worker?

➤ What are the main elements of labour market flexibility and how do these vary across space?

7.4 A crisis of trade unionism?

7.4.1 The decline of the mass collective worker

Taken together, the shift towards a post-industrial economy and a more flexible workforce is seen as heralding the decline and perhaps ultimate demise of trade unions and the 'mass collective worker' (Hudson,

1997). Five themes, in particular, run through these arguments:

> In the industrial economy, social identities were tied to and constructed through the workplace (e.g. car worker, shipbuilder, steelmaker, etc.). Workers tended to work in the same occupation for most of their working lives and therefore work became an important 'fixing' element in their social consciousness. By contrast, employment in services – for less skilled workers – tends to be more fluid and casualized. People move between jobs more frequently and social identity is less associated with work but more with consumption activities (Lash and Urry, 1994, p.57).

> Subsequently, workers are becoming more individualistic in attitude and less inclined to join trade unions. The fragmented nature of the service economy with large factories giving way to smaller and more dispersed office and retail workplaces make it difficult for unions to organize.

> Trends towards employment flexibility also weaken the position of trade unions. The slimming down of the core, unionized workforce and the creation of more temporary and agency workers undermines traditional union strength in the large single workplaces of Fordism.

> Additionally, increased outsourcing and geographical relocation of work to non-union plants in areas without a culture of union organization further weaken unions, as well as providing employers with increased opportunities to play off workers in one location against those in another.

> Finally, the increased 'feminization' of the workforce also undermines trade unions, as women are less likely to join trade unions than men.

Undoubtedly there is some truth in these claims, but the extent of trade union decline is often exaggerated. Undoubtedly, union membership has declined considerably in most advanced industrial economies since its heyday in the 1970s but the extent of this decline is often exaggerated (Table 7.4). A factor has been the decline of manufacturing, where union strength has traditionally been high, and the growth of service-related employment, where unions have traditionally been weaker. But other factors are also at play. If we consider the international comparative evidence of union decline, what is significant is that decline has been greatest in those countries where employment regulation is weakest and governments have been most hostile to unions (Table 7.4). Union membership in the UK declined faster than most other countries in the 1980s and 1990s but this was in the context of Conservative governments

Table 7.4 Geographical variations in trade union decline for selected OECD countries (measured in terms of union density = % of workforce in union)

	1980	2008
Australia	49.5	18.6
Canada	34.7	27.1
France	18.3	7.7
Germany	34.9	19.1
Italy	49.6	33.4
Netherlands	34.8	18.9
New Zealand	69.1	20.8
Norway	58.3	53.3
Sweden	78.0	68.3
UK	50.7	27.1
USA	19.5	11.9

Source: OECD Trade Union density figures, available at http://stats.oecd.org.

passing anti-union laws restricting the right to strike and encouraging employers to de-recognize trade unions. US, Australian and New Zealand unions similarly have seen their influence declining in the context of neo-liberal economic policies (see Chapter 5), hostile governments and anti-union employment legislation.

In other countries where governments have been less willing to challenge trade unions, they remain a stronger force in the labour market, particularly in some northern European countries such as Norway, Germany and the Netherlands. This does not mean that unions have not been on the back foot in these countries. Indeed, the evidence from Germany suggests that employers have been attempting to withdraw from national systems of national collective bargaining to more decentralized plant- and company-based systems (Berndt, 2000; Zeller, 2000). However, critically, here, as in other western European countries, political support for the idea of social partnership still persists, even if employers are demanding more labour flexibility (Jeffreys, 2001).

It is also worth noting that the situation facing trade unions is different in much of the Global South. In South Africa, trade unions experienced considerable growth in the period between the mid-1980s and 1990s, both in membership and prestige, due to their wider role in campaigns for democratic change alongside their ability to organize in the context of rapid industrialization (Adler and Webster, 1999). Similarly, the trade union movement in Brazil expanded during the 1980s associated with the struggle for democracy, although it has subsequently fallen back in recent years as employers seek to subcontract more work from the formal unionized sectors of the economy to the small firm-dominated informal sectors (Ramalho, 1999). Indeed, in many developing countries, during the 1990s unions have started to come under pressures from processes of corporate restructuring and similar patterns of outsourcing by employers to non-union workplaces as those described for unions in the global north.

7.4.2 The end of trade unions or a crisis of a particular form?

The tendency to write off trade unions as dinosaurs whose days are numbered has been criticized by many commentators who point to the continuing importance of labour action and struggle in the global economy (Herod 2001). Not only are there very different experiences between countries, but even in those places where unions have suffered the worst setbacks, there are signs of revival in recent years. Notably in the UK, the election of a more sympathetic Labour government in 1997 and the passing of new laws providing unions with improved, though still restricted, rights to organize has resulted in a dramatic upsurge in agreements with firms to bargain collectively helping stem the decline in the number of members (Figure 7.8).

More perceptive commentators point to contemporary union decline as representing the crisis of a particular form of trade unionism in time and space, that derived its strength from the economic conditions of the post-1945 period, namely the development of large manufacturing-based workforces, nationally regulated economies and systems of collective bargaining, and the social model of a male breadwinner in permanent employment (Munck, 1999). As these realities have changed, trade unions need to 'change their spots',

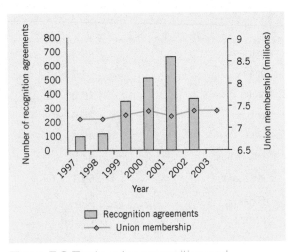

Figure 7.8 Trade union recognition and membership levels in the UK, 1997–2003
Sources: Adapted from Labour Force Survey TUC; Cumbers, 2005.

making themselves more relevant to the dispersed and fragmented economy of services by appealing to non-traditional constituencies such as women and those in non-standard work, who have little social protection. Some have also called for unions to become less narrowly oriented to workplace issues, but instead to work for broader social and political goals, forging alliances with other social movements within local communities on issues such as housing and combating poverty (Wills, 2003) or what has been termed a new social movement unionism (Moody, 1997). Others argue for a new labour internationalism, capable of facing up to the emerging global networks of TNCs by unions themselves becoming less nationally oriented and more capable of operating globally (Waterman, 2000).

Perhaps the most serious issue facing trade unions worldwide is the growth of a massive non-unionized and highly exploited labour force in the global south, stemming from the increasingly complex divisions of labour. The ILO recently noted that the fastest growing area of employment in both Latin America and Africa was in the urban informal economy, accounting for 90 per cent of all jobs in the latter during the 1990s (ILO, 2005b, p.6). Significantly, this sector includes a growing number of 'unregistered' workers engaged in outsourced work linked into global production networks, but with minimal social protection and employment regulation. Not only are such workers subject to considerable exploitation by managers and key customers, but their presence also drives down the wages and conditions of workers in formal employment. Tackling the unorganized workforce is now a common problem for unionized workers in the north and south, which could become a rallying cry for a new labour internationalism.

> ## Reflect
>
> ➤ How well equipped are traditional forms of trade unionism to meet the challenge of globalization?

7.5 New labour geographies

7.5.1 Constructing labour's spatial fix

In the past, geographers have considered the distribution and organization of labour across space and how this shapes the geography of the economy. More recently, however, a new generation of economic geographers has emphasized the active role that labour plays in constructing the global economy. The persistence of measures of employment protection and regulation in all advanced economies (even in the Anglo-American world) and the growth of trade unions in some of the most important economies of the Global South suggests a need to understand how workers, as well as employers, help to organize and regulate the changing landscape of capitalism. As one of the leading lights of the '**new labour geography**', Andy Herod, puts it:

the production of the geography of capitalism is not the sole prerogative of capital. Understanding only how capital is structured and operates is not sufficient to understand the making of the geography of capitalism. For sure, this does not mean that labor is free to construct landscapes as it pleases, for its agency is restricted just as is capital's – by history, geography, by structures that it cannot control, and by the actions of its opponents. But it does mean that a more active conception of workers' geographical agency must be incorporated into explanations of how economic landscapes come to look and function the way they do.

(Herod, 2001, p.34)

While there is a long-established tradition within Marxist geography of studies that explore how workers seek to 'defend place' in the threat of plant closure and industrial restructuring (e.g. Hudson and Sadler, 1986), labour geographers (see Coe and Jordhaus-Lier 2010; Rutherford, 2010) have been pursuing research that highlights the more proactive role of trade unions in creating their own 'spatial fixes'. Critical here, in the context of globalization, is the need for trade unions to rethink their own geographical strategies to respond to the more global production networks being developed by MNCs.

7.5.2 The 're-scaling' of trade union action

The need for trade unions and workers to develop new spatial strategies to deal with the changing economic landscape of globalization has been termed rescaling by labour geographers (e.g. Bergene *et al.*, 2010; Herod *et al.*, 2003). Four themes are evident in the research that is underway on the subject:

➤ Studies exploring attempts by trade unions to develop more effective organizational geographies that allow them to get to grips with the global organization of production (e.g. Cumbers *et al.*, 2008; Wills, 2002). While the labour movement has always seen itself as international – going back to the time of Karl Marx and the International Working Men's Association (often referred to as the First International) in the 1860s – and has long had trans-national structures, such as the International Trade Union Confederation (ITUC) and global union federations (GUFs) that represent particular sectors, effective power has always been held in national affiliate unions (Cumbers *et al.*, 2008). Different spatial strategies are now being advocated to transcend national level organizations to deal with MNCs. On the one hand, many in the trade union hierarchy have prioritized the importance of developing a set of minimal international labour standards (with regard to pay and conditions, gender equality, the right to join unions and the end to child and forced labour) by lobbying governments and key global institutions such as the World Bank and IMF. Others have prioritized the signing of collective agreements with MNCs as a means of improving workers' rights. Since 1994, a number of global union federations has signed Global Frame Agreements with over 40 MNCs, which seek to develop basic standards through each corporation's global production, although there is little evidence that they have been effective in combating MNC power (Cumbers *et al.*, 2008).

➤ In opposition to these 'top-down' internationalist strategies, others have championed a more grassroots approach that seeks to develop independent networks of workers within global production chains (Waterman, 2000; Castree, 2000; Ryland, 2010). This approach tends to be more militant, viewing attempts by union leaders to forge social partnerships at an international level with governments and employers as inevitably compromised. Instead, emphasis is placed upon workers autonomously developing their own networks of communication and struggle that allow them to 'scale up' local disputes to maximum effect (Box 7.5). However, while grassroots initiatives between workers of different countries may result in short-term victories, they can often prove unsustainable without effective support from national and international leaderships (Cumbers, 2004).

➤ A focus upon creative ways that unions are responding to the spatial reorganization of production (Herod, 2001; Holmes, 2004; Tufts, 2007). For example, instead of assuming that post-Fordist geographies of production will inevitably undermine the position of trade unions, researchers have pointed out that much depends upon the strategies developed by unions themselves. Rutherford and Gertler (2002) have observed that one effect of employers introducing more flexible production that requires a skilled labour force will mean a greater dependency upon the workers who have key skills and knowledge about production.

➤ Attempts to organize workers locally in the dispersed and fragmented landscape of the post-industrial service-based economy (Walsh, 2000; Wills, 2005; Tufts, 2007). As labour geographers have been quick to point out, not all capital is mobile. Much service work in particular is tied to place. Restaurants and hotels, cleaning and security work, and employment in public services all have to locate where the customers are and, in particular, within large urban areas. Nevertheless, as we have already documented, such work is often low paid and non-unionized. The geographical dilemmas facing unions in this context are about developing more effective strategies for organizing across a city in multiple workplaces, rather than focusing upon the single large-mass workplaces typical of the industrial economy. Across the US, coalitions of trade unions, church groups and other social activists have achieved considerable success in recent years, in an otherwise hostile industrial relations climate in winning 'living wage campaigns' at the local level (Box 7.6).

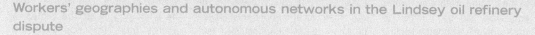

Box 7.5

Workers' geographies and autonomous networks in the Lindsey oil refinery dispute

In June 2009 a strike broke out at the Lindsey oil refinery in north-east Lincolnshire, England (Figure 7.9). The immediate cause of the dispute was the decision of a contracting firm, Shaws, to refuse to consider 51 workers, made redundant following the completion of the first stage of contract work, for vacancies on the second stage of a new construction contract at the refinery.

The dispute had arisen on the back on an earlier strike action in February of 2009, when the 51 workers at the centre of the June walkout had been involved in organizing a protest by local workers at the employment of a foreign workforce, primarily Italian and Portuguese workers. The refusal to consider the 51 workers for the new vacancies resulted in 'unofficial' strike action by all 647 construction

workers at the refinery. The response of the owners of the refinery, French oil multinational Total, with its sub-contractors, was to sack the workers, and to refuse to negotiate until the strike was called off.

In an impressive show of solidarity, hundreds of workers at 17 other oil and gas plants across the UK walked out in sympathy. The dangers that an escalation of industrial action might

Figure 7.9 Lindsey oil refinery
Source: Anthony Ince.

Box 7.5 (continued)

threaten Britain's energy supplies brought the Labour government into the struggle with calls for the two sides to sit down and negotiate an end to the dispute. Within two weeks the employers had been forced to negotiate with the workforce leading to the reinstatement of many of those originally laid off by Shaws.

The dispute was an excellent example of the complex and increasingly contingent geographies of employment in a global economy, but also of the way that grassroots workers can develop their own autonomous geographies of struggle to challenge a powerful MNC. Because

of their positioning within a critical space – the oil refinery network – of the British and European economies, the workers were able to exercise considerable relational power in the way their local actions had much broader spatial implications (see Allen, 2003).

Equally critical, however, was the way that the 'local' workforce was able to scale up its struggle through its own broader networks. The workforce at Lindsey, in common with the engineering construction industry generally, is a mobile and itinerant one. The majority of workers – especially the more skilled workforce

– do not live in the local area and many have experience working at different refineries throughout the UK as well as in the North Sea in the Norwegian and British sectors. From the point of view of the campaign against Total, this was important because the strikers were able to use their existing contacts and networks to scale up the dispute to the national level. Indeed, it was precisely a national escalation of the dispute – in the name of protecting nationally agreed conditions – that was essential to mobilizing workers in other refineries.

Box 7.6

Organizing city-wide: the 'living wage movement' in the US

The 'living wage movement' in the US has been described by celebrated columnist Robert Kuttner as 'the most interesting (and under-reported) grassroots enterprise to emerge since the civil rights movement)' (www.livingwagecampaign.org). Living wage campaigns show that local interventions can still be highly effective in a global economy. They involve the passing of laws by local or city authorities guaranteeing decent wage levels to all workers employed either directly in the public services or by private employers contracted to the public sector. The living wage campaign in the US has cleverly linked pay to public spending, suggesting that the latter should not be used to subsidize 'poverty wages' and its success can be measured by the fact that Republican as well as Democrat city authorities have signed living wage ordinances. In some cities, a 'living wage' ordinance has been

used as a local marketing strategy, in effect using a 'progressive localism' to attract more socially concerned investors.

The first living wage coalition was launched in Baltimore in 1994 when, in the context of deindustrialization and a process of gentrified waterfront regeneration that had done little to deal with the problems of poverty and alienation affecting the traditional workforce, an alliance between trade unionists and religious leaders forced the city government to pass a law requiring a 'local living wage' for public sector workers. Subsequently 122 cities across the US have followed suit and had passed living wage ordinances by April 2006.

The living wage movement has also been gaining ground in other countries. A campaigning group, London Citizens, has been active in promoting the concept of a living wage for the UK capital's low-paid workers. The

campaign has had considerable success since its launch in 2001 and has added an estimated £24 million to the wages of low-paid workers (London Citizens, http://www.london citizens.org.uk). London Citizens lists 85 employers from the private, public and voluntary sectors that currently pay the living wage of £7.60 per hour (£1.87 above the national minimum). They include multinational banks such as Barclays, the accountancy firm Price Waterhouse Coopers, Queen Mary University and the London School of Economics. In addition to paying the living wage, employers must also provide employees with at least 20 days paid holiday plus bank holidays, eligibility for ten sick days per year and access to join a trade union. Perhaps the biggest coup of the campaign to date was persuading the Conservative Mayor, Boris Johnson, to also sign the city government up to paying living wages on its contracts.

7.5.3 Global networks of labour

Alongside the collective agency of labour, it is also important to highlight the way workers at an individual level help to shape the global economy. The most obvious manifestation of this is through the decision to move in search of a better job or way of making a living. In the global economy an estimated 100 million workers are now living away from their countries of origin (Castree *et al.*, 2004, 189). In contrast to previous phases of international migration, what is notable about the current period is the complexity of linkages, although in broad terms movements are from the global south to the north, reflecting the different global geography of income opportunities (see Figure 7.10).

Different kinds of labour migration can also be identified, which reflect the nature of underlying power relations within the capitalist economy and the divisions that exist within the global workforce. On the one hand, we can identify a small group of highly skilled international migrants who form a trans-national capitalist elite, working in the areas of global finance and management. This privileged group of 'workers' operate in a global space of flows (Castells, 2000), located in and moving freely between the headquarters of the world's most powerful TNCs in global cities such as London, New York, Tokyo and Paris (Beaverstock, 2002). They exercise considerable power as the 'movers and shakers' of the global economy whose decisions affect the lives of most of the rest of the world. On the other hand, the vast majority of labour migrants have a less exalted status and are differentiated variously by the extent to which they are temporary or permanent, skilled or unskilled, and voluntary as opposed to 'forced' (see Box 7.1), and legal or illegal.

As Castree *et al.* (2004, p.191) note, despite the rhetoric of globalization and a borderless world, 'national governments actively regulate the international migration of workers', filtering, in the same way that firms do, workers on the basis of their 'desirability'. Not only are migrants increasingly assessed in terms of education and skill levels, but racial and ethnic characteristics typically still inform immigration policy. Racist discourses and nationalistic stances towards immigration often lead to policies that are contradictory from the point

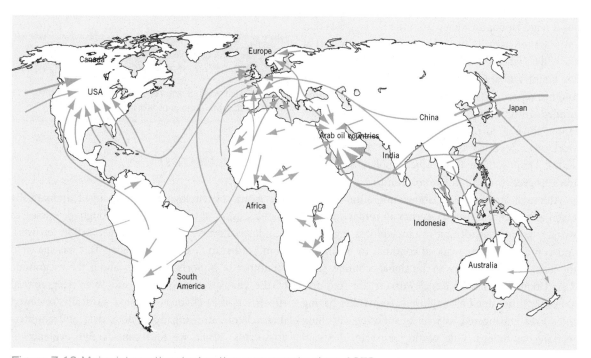

Figure 7.10 Major international migration movements since 1973
Source: Castles, S. and Moller, M., 1998, *The Age of Migration*, 2nd edn, Macmillan, reproduced with permission of Palgrave Macmillan.

Figure 7.11 Illegal immigrants crossing the Mexican–US border
Source: Danny Lehman/Corbis.

of view of capital; for example, tight immigration controls operating during periods of low unemployment and economic boom, when there is a high demand for foreign labour. Evidence from the US suggests that immigration actually fuelled economic growth during the 1990s when around 13.5 million people entered the country (*The Guardian*, 3 December 2002), while another report on the UK estimated that the economy needed 150,000 migrants per year to maintain the labour force at current levels in the face of an ageing workforce (reported in *The Observer*, 8 December, 2002). Despite these facts, official policy still tends towards ever stricter immigration controls.

Although international labour migration can be taken as evidence of worker agency in moving to find work, requiring considerable resourcefulness, ingenuity and courage, migrant workers continue to be among the most exploited groups in the global economy (Box 7.1). Undocumented or illegal workers are routinely abused by unscrupulous employers, sometimes having their lives endangered as well as suffering appalling working conditions. One of the worst incidents in recent times was the deaths of 23 Chinese migrant workers in the UK, by drowning in the Morecambe

Bay area of Lancashire while being employed as cockle pickers (see http://news.bbc.co.uk/1/hi/england/lanca shire/3827623.stm).

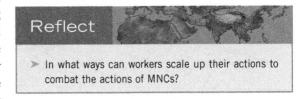

Reflect

➤ In what ways can workers scale up their actions to combat the actions of MNCs?

7.6 Summary

The nature of employment has changed dramatically in the period since the 1970s through processes of deindustrialization and the reorganization of work from Fordist to flexible workplaces. This has spawned a number of important debates about the magnitude of the changes taking place and their geographical effects. In this chapter we have critically reviewed these claims, drawing upon both data and counter-arguments. While we find considerable evidence of a changing workplace, particularly a shift to greater flexibility and insecurity, there is less evidence that these changes have brought positive benefits for the majority of the workforce. There continue to be wide spatial variations and divisions within the workforce, in terms of pay and conditions, and little evidence of a more egalitarian workplace that some commentators predicted with post-Fordism.

We have also highlighted the importance of geography. Continuing variations in the political and institutional framework within which work is embedded between countries and regions continue to account for spatial variation in the adoption of new forms of work. Another key theme is the role of workers and their unions in shaping economic land-scape, as highlighted by the 'new labour geography'. While labour has been on the 'back foot' in processes of global restructuring, due to the increasing mobility of capital, we have highlighted the continuing possibilities for labour action at a number of different scales in the global economy. Some of the more interesting research in economic geography in recent years has been that examining the new spatial strategies being developed by unions in response to globalization.

Exercise

Using Table 7.2 as a starting point and employment tenure as a measure of labour market flexibility, compare the economic performance of two 'flexible economies' with those of less flexible ones.

1. Is there any evidence that labour market flexibility leads to better economic performance and social wellbeing?

2. What other indicators might be used to assess the social impact of flexibility?

3. Does the use of such statistics provide us with adequate knowledge of geographical variations in the performance of labour markets?

4. What other methods might be used to understand the different experience of flexibility for individual workers?

5. How have the different national models of labour regulation performed in the recent economic downturn?

Key reading

Bergene, A.-C., Endresen, S.B. and Knutsen, H.M. (eds) (2010) *Missing Links in Labour Geography*, Aldershot: Ashgate.
An excellent set of articles relating to current debates on labour geography with many of the leading protagonists in the field. The collection includes both recent advances in theories concerned with labour agency as well as a wider variety of empirical studies on union efforts to organize in the Global North and South.

Bradley, H., Erickson, M., Stephenson, C. and Williams, S. (2000) *Myths at Work*, Cambridge: Polity.
A thorough and critical analysis of various theories about changes in the contemporary workplace. As the title suggests, the book usefully unpacks the various myths about employment transition with particularly valuable chapters on flexibility and the 'end of trade unionism'.

Castree, N., Coe, N., Ward, K. and Samers, M. (2004) *Spaces of Work: Global Capitalism and Geographies of Labour*, London: Sage.
An excellent general text on labour geography with the first two chapters providing useful introductions to the spatial variations in the experience of work in the global economy and key concepts in understanding employment relations. In later chapters there is a wealth of empirical data and maps to help illustrate essays.

Pahl, R. (ed.) (1988) *On Work: Historical, Comparative and Theoretical Approaches*, Oxford: Blackwell.
A little dated but contains a number of very useful chapters about the history of work from the pre-industrial period through to the industrial. It is particularly good at under-standing the changing social and technical divisions of labour through time, evaluating changes in gender and class relations.

Peck, J. and Theodore, N. (2001) 'Contingent Chicago: restructuring the spaces of temporary labor'. *International Journal of Urban and Regional Research*, 25(3): 471–6.
A detailed empirical case study of the workings of the flexible labour market for those at the sharp end of the US economy. The paper offers real insights into how the labour market is being restructured and how the meaning of flexibility is very different for firms and employees.

Useful websites:

http://www.//ilo.org
International Labor Organization's website: the UN body that carries out research into global labour issues. Produces data and research papers on a range of labour issues from trade unions membership to trends in labour flexibility.

http://www.icftu.org
International Confederation of Free Trade Unions. The main website for the umbrella body that represents the international trade union movement.

http://www.aflcio.org
The website for the main US trade union federation.

http://www.//tuc.org.uk
British trade union confederation site.

http://www.bls.gov/
US Bureau of Labor Statistics.

http://www.//nomisweb.co.uk
UK Official Labour Market Statistics.

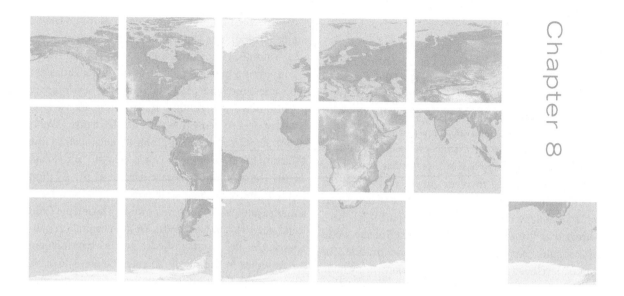

Geographies of development

Chapter map

After a brief introduction to the issue of development, we consider the meaning of development as a term and its history as a concept in section 8.2. This is followed by a review of the main theories of development that have influenced policy and practice. We focus on the modernization school of the 1950s and 1960s, the dependency theories of the 1960s, the neoliberal model of recent decades and grassroots development. Section 8.4 focuses on changing patterns of development, emphasizing the increased economic divergence between different parts of the developing world that has become apparent in recent years. We then review the growth of social movements in developing countries, highlighting examples of local groups resisting particular projects. Current development issues and challenges are discussed in section 8.6, focusing on trade, aid and climate change. This is followed by a brief summary which highlights the main points covered in the chapter.

8.1 Introduction

Development is defined by the *Concise Oxford English* dictionary as a 'gradual unfolding, fuller working out; growth; evolution … ; a well-grown state, stage of advancement; product; more elaborate form …' (quoted in Potter *et al.*, 2008, p.4). In the context of economic policy, the term conveys a sense of positive change over time, applied to a particular country or region. Such change involves growth or progress, as countries become more prosperous and advanced. As an economic and social policy, development has been directed at those 'underdeveloped areas' of the world that require economic growth and modernization. The 'underdeveloped world', which we are concerned with in this chapter, consists of Africa, Latin America and Asia, representing the main focus of international development policy since the 1950s. The division between this periphery and the core of developed countries in Europe, North America, Japan and Australasia can be expressed as a geographical divide between the global north and south (Figure 8.1)

The problems of developing countries in the Global South have attracted much attention in recent years, with a number of campaign groups and social movements raising concerns about global inequalities. At the same time, and partly in response, development organizations and political leaders have focused on issues of poverty reduction and climate change (World Bank, 2009), setting high-profile targets such as the Millennium Development Goals (Hulme and Scott, 2010). The promotion of development as the solution to the problem of large-scale poverty in Africa, Latin America and Asia is not new, however. During the **Cold War** (late 1940s to late 1980s), western capitalist countries, led by the US, established large-scale development programmes in the newly liberated colonies to 'save' them from the 'evils' of communism, stressing the need for economic progress and modernization. While post-war development was driven by Cold War geopolitics, the recent interest in global poverty and development is shaped by opposing perspectives on **globalization**. For pro-globalists, globalization is a long–term process that will eventually lift all people out of poverty, so long

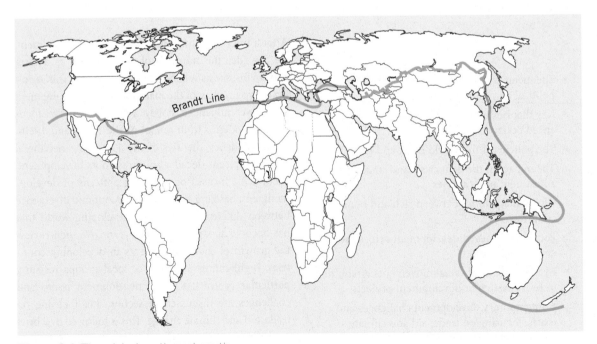

Figure 8.1 The global north and south
Source: Adapted from Knox *et al.*, 2003, p.24.

as their governments pursue market-oriented policies. For 'counter-globalization' groups, however, neoliberal globalization is the problem, not the solution, and the global economic system requires major reform if not outright revolution.

8.2 The programme of development

The countries of Africa, Latin America and Asia were designated as the 'underdeveloped world' in the 1950s in a context of Cold War geopolitics and decolonization. As part of its efforts to break the old imperial trading blocs and create a world open for international trade and investment, the US sought to encourage economic growth and modernization in the newly liberated colonies. In his inaugural address in 1949, President Truman emphasized the need to tackle poverty through the promotion of development:

We must embark on a bold new program for making the benefits of our scientific advances and industrial progress available for the improvement and growth of underdeveloped areas … For the first time in history, humanity possesses the knowledge and the skill to relieve the suffering of these people … I believe that we should make available to peace-loving peoples the benefits of our store of technical knowledge in order to help them realize their aspirations for a better life … What we envisage is a program of development based on the concepts of democratic fair-dealing … Greater production is the key to prosperity and peace. And the key to greater production is a wider and more vigorous application of modern scientific and technical knowledge. Only by helping the least fortunate of its members to help themselves can the human family achieve the decent, satisfying life that is the right of all people.

(http://www.bartleby.com/124/pres53.html)

This 'program of development' was not only an inherent good, overcoming the 'ancient enemies' of hunger, misery and despair, it would also provide a necessary bulwark against communism, viewed as a disease to which those living in poverty were particularly prone.

As such, the former colonies in Africa, Latin America and Asia became key sites of the Cold War struggle, as the US and Soviet blocs competed for influence and power.

The notion of the '**third world**' was invented as a label for the 'underdeveloped areas', in contrast with the 'first world' of western democracies and the 'second world' of communist states in the USSR and eastern Europe. The term was applied to vast areas of the globe, reflecting a kind of negative labelling of them as backward and in need of assistance from outside. It is a view that has been perpetuated over the decades, supported by powerful images of poverty and underdevelopment. As the development geographer Morag Bell remarks, this dramatic image masks a less clear-cut reality:

What is the geography of the Third World? Certain common features come to mind: poverty, famine, environmental disaster and degradation, political instabilities, regional inequalities and so on. A powerful and negative image is created that has coherence, resolution and definition. But behind this tragic stereotype there is an alternative geography, one which demonstrates that the introduction of development into the countries of the Third World has been a protracted, painstaking and fiercely contested process.

(Bell, 1994, p.175)

The presentation of the 'third world' as a monolithic bloc is highly simplistic and misleading, ignoring a complex and diverse geography. Development is never straightforward, involving a range of policies and programmes devised by main development organizations, and received or 'consumed' by the nations, communities and households that they are aiming to assist. It has resulted in different outcomes in different places, shaped by a range of locally specific factors such as social attitudes, environmental conditions, labour skills and farming practices. It is important to recognize that economic development policy is not confined to the so-called 'third world' as underdeveloped areas in developed countries have also been the subject of state-sponsored programmes and initiatives (sections 5.2 and 5.5.4).

Following Truman's crucial speech and the hardening of Cold War divisions, a development industry emerged in the 1950s and 1960s, funded by the US and other western capitalist countries. This was defined by a shared belief in economic planning, modern technology and outside investment as the key drivers of change. Academic disciplines, particularly economics, played a significant role in providing expert knowledge about the process of development and the conditions shaping it. International organizations such as the World Bank, International Monetary Fund (IMF) and United Nations (UN), established as part of the new post-war order, were charged with the responsibility of promoting development in poor countries. Driven by an ideology of modernization and development, such organizations sought to introduce modern knowledge and investment in order to overcome the 'ancient enemies' that held back progress in the underdeveloped lands. The UN, for instance, designated the 1960s as the 'decade of development'. Other key players in the development field include the governments of developing countries and **non-government organizations**

(NGOs) working in the development field. NGOs are organizations, often of a voluntary or charitable nature, which make up the so-called 'third sector', belonging to neither the private nor public sectors. Oxfam is a good example of an NGO focused on development (Box 8.1). Others include Christian Aid, Save the Children and the Red Cross.

Economic aspects of development were heavily stressed in the 1960s and 1970s as policy focused on the need to foster investment and growth at the national level, assuming that this would generate increased incomes and employment opportunities for individuals. Development indicators concentrated on the material wellbeing of a country or region, with gross domestic product (GDP) and gross national product (GNP) by far the most commonly used measures of development. This corresponds to the quantitative change section of Figure 8.2. Since the late 1970s, however, the notion of development has broadened to include more qualitative aspects of change,

Box 8.1

The development of Oxfam

Oxfam has grown to become one the most important NGOs operating in the field of development. It is an independent British organization, registered as a charity, working with partners, volunteers, supporters and staff of many nationalities. It had a total income of £308.3 million in 2008–09 and employs 3,423 staff working overseas and 2,151 staff in Great Britain as well as over 20,000 volunteers working in more than 750 shops (Oxfam, 2009, p.39, 58–9).

Oxfam's origins lie in the Second Wold War when the Oxford Committee for Famine Relief was set up in response to famine in Greece in 1942. The committee remained in existence after the war, focusing on 'the relief of suffering in consequence of the war' in Europe in the late 1940s. After 1949, the scope of its

operations expanded to encompass the world. In the 1960s, Oxfam's income trebled, reflecting growing concern for the world's poor. Support for self-help schemes whereby communities improved their own farming practices, water supplies and health provision became the major focus of activity. Community involvement and control remain key principles of Oxfam's work.

As well as famine relief and self-help schemes, Oxfam campaigns to raise awareness of global poverty and its structural causes in terms of the debt burden, unfair terms of trade and inappropriate agricultural policies. Its current activities focus on three main areas: emergency response to disasters; development work that addresses poverty; and campaigning for changes in policy.

In 1995, Oxfam opened an office in Washington, DC in order to lobby the World Bank, IMF and UN more effectively, while it launched a UK poverty programme in 1996. Oxfam has identified five main priorities for the 2007–10 period: the right to sustainable livelihoods, allowing poor people to achieve food and income security and secure employment; the right to basic social services such as education, particularly for girls; the right to life and security, focusing on providing an improved response to conflict and natural disasters; the right to be heard, helping poor and marginalized people to influence decisions; and the right to equity, covering gender and diversity (Oxfam, 2007).

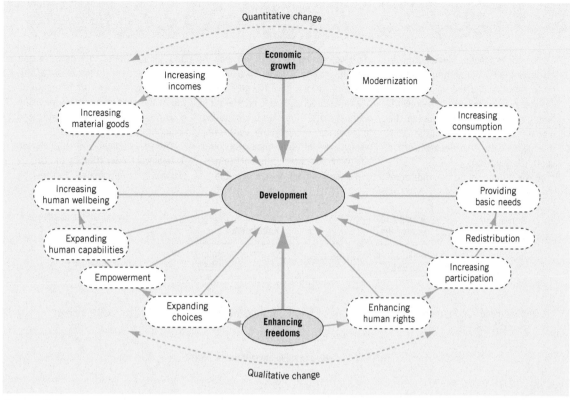

Figure 8.2 Changing conceptions of development
Source: Geographies of Development, 2nd edn, Potter, R.B., Binns, T., Elliott, J.A. and Smith, D., 2004, p.16. Pearson Education Limited.

encompassing broader social and political goals such as quality of life, choice, empowerment and human rights (Potter *et al.*, 2008, pp.16–17). The work of the Nobel Prize-winning, Indian-born economist, Amartya Sen, has been particularly important in advancing the idea of 'development as freedom'. Through factors such as better education, increased political participation and free speech, working alongside the process of economic growth, people are liberated from 'unfreedoms' such as starvation, undernourishment, oppression, disease and illiteracy. This broader conception of development is reflected in the evolution of development indicators with more emphasis now focused on assessing social and political aspects of development alongside the economic dimension (Box 8.2).

Box 8.2

Measuring development

As indicated above, the main economic measures of development are GDP and GNP, usually expressed on a per capita basis. GDP is a measure of the total value of goods and a service produced within a country, while GNP also includes income generated from investments abroad, but excludes profits repatriated by foreign MNCs to their home countries. GDP and GNP were the key measures employed by international development agencies in the post-war era, but began to attract criticism from development activists and analysts in the late 1960s and

Box 8.2 (continued)

1970s for neglecting social aspects of development. They remain important, however, providing a useful summary measure of development, emphasizing the divide between the global north and south, and growing divergence between the regions of the south (Section 8.4).

A wide range of social indicators of development was published in the 1970s and 1980s, focusing on issues such as poverty, education, health and gender. The proliferation of these social indicators threatened to generate considerable confusion, however, as different measures could be used to show different things, and it was almost always possible to find some statistics to 'prove' a particular argument (Potter *et al.*, 2008, p.9). What seemed to be required was some kind of summary measure constructed out of key economic and social indicators.

The development of the **Human Development Index (HDI)** by the United Nations Development Programme

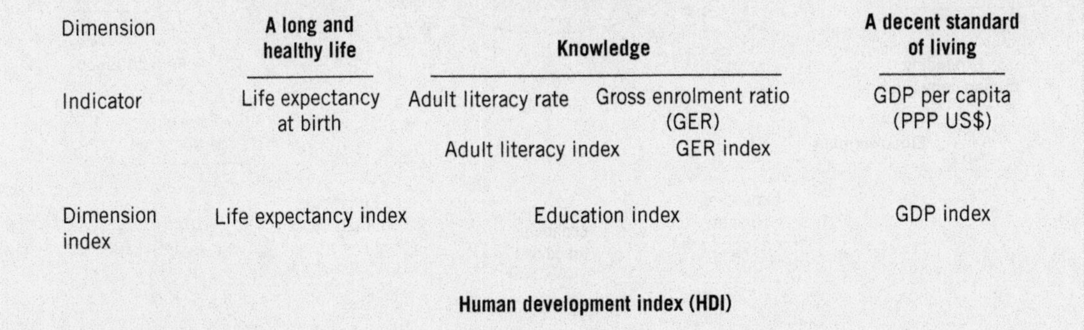

Dimension	**A long and healthy life**	**Knowledge**		**A decent standard of living**
Indicator	Life expectancy at birth	Adult literacy rate Adult literacy index	Gross enrolment ratio (GER) GER index	GDP per capita (PPP US$)
Dimension index	Life expectancy index	Education index		GDP index

Human development index (HDI)

Figure 8.3 Calculating the Human Development Index (HDI)

Source: Calculating the Human Development Indices, p.340 from 'International cooperation at a crossroads: aids, trade and security in an unequal world', by Human Development Report. By permission of Oxford University Press, Inc.

Figure 8.4 The UNDP Human Development Index

Source: UNDP, 2005, pp.219–21.

Box 8.2 (continued)

(UNDP) – published annually since 1990 in the *Human Development Report* – has met this need, becoming widely used and adopted. The HDI 'measures the overall achievement of a country in three basic dimensions of human development – longevity, knowledge and a decent standard of living' (UNDP, 2001, p.14). The specific measures used are life expectancy, educational attainment (adult literacy and combined primary, secondary and tertiary enrolment) and GDP per capita in US dollars (Figure 8.3). A separate ratio is calculated for each of the three dimensions, where the actual value is divided by the maximum possible one (for life expectancy, 85; for GDP per capita US$40,000), giving a value between 0 and 1. The HDI is then calculated as the simple average of the three dimension indices (UNDP, 2004). In recent years, is has been supplemented by the development of other related indexes, including the Human Poverty Index, the Gender Related Development Index and the Gender Empowerment Measure (GEM). Countries with HDI scores above 0.8 are classified as high human development; ones with scores between 0.5 and 0.8 as medium human development; and ones with scores below 0.5 as low human development (Figure 8.4).

Reflect

➤ To what extent should development be led by external northern agencies such as the World Bank, IMF and NGOs?

8.3 Theories of development

8.3.1 The modernization school

The **modernization school** approach was dominant in 1950s and 1960s, shaping and informing the efforts

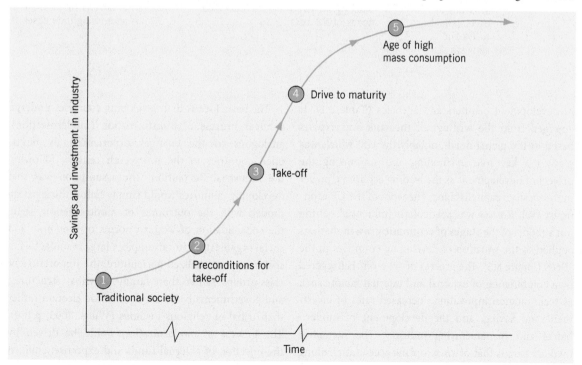

Figure 8.5 Rostow's stages of economic development
Source: Geographies of Development, 2nd edn, Potter, R.B., Binns, T., Elliott, J.A. and Smith, D., 2004, p.91. Pearson Education Limited.

Table 8.1 Theories of development

Theory	Key theorists	Time frame	Main argument	Recommendations	Problems
Modernization theory	Rostow	1950s and 1960s	Development occurs in distinct stages; developing countries undergo linear process of modernization, akin to developed countries in the nineteenth century	Need for external funds and expertise, along with modern planning and investment methods, to generate growth	Eurocentric; ignores structure of world economy; growth fails to alleviate poverty
Structuralism and dependency theory	Prebisch, Frank	1960s and 1970s	Metropolitan core of world economy exploits 'satellites'. Development and underdevelopment are opposite sides of the same coin	Import substitution or withdrawal from world economy	Crude and static view of global relationships. Undermined by experience of E. Asian NICs
Neoliberalism	Lal and Balassa in development circles	1980s and 1990s	Developing countries should reduce state intervention in the economy and embrace the free market	Economic reform through SAPs, free trade, promotion of exports	Focus on growth failed to alleviate poverty; SAPs and privatization led to cuts in public services, and the introduction of user charges
Grassroots approach	Emerged from activities of development NGOs and activists	1970s–	Development agencies should focus on meeting the everyday needs of the poor	Local participation and self-help should be encouraged	Limited funds. Tends to alleviate the symptoms of poverty rather than addressing the causes

of development planners and agencies (Table 8.1). It emerged from the writings of theorists and experts based in the global north, notably the US, which was playing a key role in funding and supporting the project of development as the leading economic power in the western capitalist bloc. The work of the US economist Walt Rostow was particularly influential, setting out a model of the **stages of economic growth** that was applied to the situation of developing countries in the 1960s (Figure 8.5). The process of 'take-off' is triggered by a combination of external and internal factors such as technological innovation, increased rates of investment and saving, and the development of modern banks and manufacturing industry. The metaphor invoked here is that of an aeroplane accelerating along a runway before gathering power to become airborne (Power, 2008, p.188).

The basic idea is that developing countries undergo a linear process of transformation (modernization), analogous to the changes experienced by developed countries in the nineteenth century following the Industrial Revolution. The assumption was that developing countries would simply follow this existing model, with the outcomes of modernization being the consolidation of western norms of economic and social organization (for example, a large manufacturing sector, commercialized agriculture, the importance of class-groups rather than family or tribal structures, and governments based on democratic election rather than tribal or religious loyalties (Willis, 2005, p.189). The process of modernization would be driven by the injection of external funds and expertise, coupled with national government intervention and planning to mobilize resources and stimulate investment. Economic growth was paramount, generating increased income and employment opportunities which, it was assumed, would 'trickle down' to the poorest groups in society.

From this perspective, the problems of under-development are internal to the countries concerned, reflecting the attachment to traditional values and the absence of modern technology and scientific knowl-edge. As such, the solution was to import these factors from outside, giving western experts and agencies a key role in assisting development. When linked to ini-tiatives and reforms carried out by developing country governments, this would generate the momentum required for 'take-off'. The **Eurocentric** assumption that western values and methods are always superior to those found in developing countries has been criticized as arrogant and condescending.

The focus on internal factors means that the struc-ture of the world economy, particularly in terms of the relationships between richer and poorer countries, was neglected by modernization theorists. It was simply assumed that it is just as easy for developing countries to develop as it was for developed countries in the nine-teenth century. This ignores the fact that the latter were not required to compete with a group of already indus-trialized countries that dominated world production of manufactured goods and services. As such, the poli-cies followed by Europe and North America were not necessarily appropriate to the problems confronting developing countries in the 1960s and 1970s.

8.3.2 Structuralism and dependency theory

The idea that the existing structure of the world economy was impeding the development of 'third world' coun-tries was a key point of departure for radical critics of modernization theory in the 1950s and 1960s. These critics adopted a Marxist approach, focusing on the

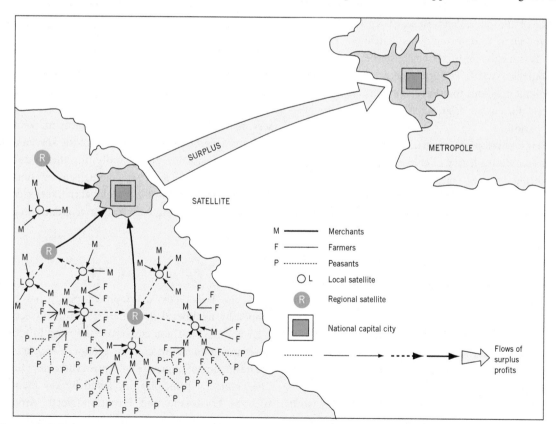

Figure 8.6 Dependency theory

Sources: Geographies of Development, 2nd edn, Potter, R.B., Binns, T., Elliott, J.A. and Smith, D., 2004, p.111. Pearson Education Limited. Adapted from R.B. Potter, *Urbanisation in the Third World*, COOP, 1992. By permission of Oxford University Press.

political economy of economic development. In particular, they sought to apply Marx's insights about the historical and geographical unevenness of capitalist development to the experience of developing countries in the post-war period (Power, 2008, pp.188–9). In contrast with the modernization school, **structuralist or dependency theories** emanated from the global south, being particularly associated with a group of theorists and activists based in Latin America.

This approach can be termed structuralist for their characteristic focus on the structure of the world economy, particularly the relationship between developed countries and developing countries (Table 8.1). It focused on the mode of incorporation of individual countries into the world economy, viewing this as a key source of exploitation. From this perspective, the causes of poverty and underdevelopment are external to developing countries, stemming from the relationship between them and the wider world economy. According to Andre Gunder Frank, the most influential of the dependency theorists in the 1960s, the metropolitan core exploits its 'satellites', extracting profits (surplus) for investment elsewhere (Figure 8.6). Colonialism was a key force here, creating unequal economic relations that were then perpetuated by the more informal imperialism characteristic of the post-war period.

Many developing countries have found it difficult to overcome the legacy of colonialism, involving them exporting primary commodities (agricultural goods, minerals, fuels, etc.), while the European powers export manufactured goods. Over time, the price of the former has tended to fall relative to the latter, reducing the export earnings of developing countries and making them less able to pay for imports. At the same time, MNCs based in Europe and North America have been able to repatriate the profits from their plantations and factories in developing countries to their home countries, paying low wages and having very few links with the host economy in which they are operating.

The implications of dependency theory are that developing countries should protect themselves against external forces or even withdraw from the world economy altogether. Development and underdevelopment are opposite sides of the same coin; closer links between core and periphery merely widen the economic

gap between them. The milder form of structuralism associated with Raul Prebisch and the UN Commission for Latin America in the 1950s advocated that countries should follow protectionist policies rather than simply embracing external assistance (Willis, 2005, p.191). In particular, they should focus on **import substitution industrialization** (ISI), seeking to develop domestic industries to produce goods that are currently imported, using import tariffs to protect them against competition from the established industries of Europe and North America (section 5.4.1). For the radical version of dependency developed by Frank, however, the solution was withdrawal from the global economy and the creation of alternative forms of society based on socialism (Willis, 2005, p.193).

Dependency theories have been heavily criticized, relying, again, on a simple divide between 'core' and 'satellite' economies. Their view of global relationships is static and crude, assuming that the patterns established under colonialism will inevitably persist. This assumption was undermined by the experience of rapid economic growth in the **NICs** of East Asia such as South Korea, Taiwan and Singapore in the late 1970s and 1980s (Box 5.3). The success of these countries was based on policies of **export oriented industrialization** (EOI), where industrial development is based on serving international rather than domestic markets. As such it seemed to undermine some of the key foundations of dependency theory, indicating that countries could overcome the legacy of colonialism. By contrast, the notion of withdrawal from global markets seemed impractical, promising only further economic stagnation and decay.

8.3.3 Neoliberalism

Since the late 1980s, development policy has been shaped by **neoliberal** (free market) theories (sections 1.2.1 and 5.5.1). These emphasize the need to reduce government intervention in the economy, encouraging the development of private enterprise and competition. The neoliberal principles developed by key theorists such as Friedmann and Hayek in North America and western Europe were translated into development thinking in the early 1980s, following the counter-revolution in development economics when neoliberal ideas overturned the Keynesian orthodoxies of the 1960s and 1970s. Key figures included Depak Lal and Bela Balassa, both based at leading US universities, who argued in favour of free trade and the application of standard economic (neo-classical) principles to the developing world (Peet and Hardwick, 1999, pp.49–50).

Neoliberal principles have underpinned the so-called **Washington Consensus**, which has structured economic development policy since the early 1990s (section 5.5.2). Key strands include a focus on low inflation, the reduction of barriers to trade, openness towards foreign direct investment (FDI), the liberalization of the financial sector and the privatization

Box 8.3

The debt crisis

The **debt crisis** has been a key factor shaping north–south relations over the past 25 years, with many developing countries struggling to service loans originally taken out in the 1970s. It first came to attention when Mexico defaulted on its loans in August 1982, with other Latin American countries such as Brazil and Argentina also experiencing major problems. Much of sub-Saharan Africa and parts of Asia have also been severely affected. The debt crisis has not only seriously undermined development efforts, requiring developing countries to spend their limited export earnings on servicing debts, it has also threatened the viability of the world financial system, forcing northern governments and bodies such as the IMF and World Bank to intervene. The impact of debt on development was emphasized by development activists and groups such as Jubilee 2000 in the late 1990s and early 2000s, feeding into the 'Make Poverty History' and Live 8 campaigns, which brought the issue into the public consciousness around the 2005 G8 summit in Gleneagles, Scotland.

The origins of the debt crisis lie in the interactions between three sets of factors:

➤ The borrowing of large sums by developing countries from northern banks and institutions

in the 1970s. Following the sharp rise in oil prices that occurred in 1973, the world economy was awash with funds invested by the oil-exporting companies. These so-called 'petrodollars' were lent by leading banks and government agencies to developing countries to pay for oil imports and to finance large-scale industrial development programmes.

➤ Most of the loans were made for a period of five–seven years, denominated in US dollars and subject to floating interest rates (Corbridge, 2008, p.508). In the late 1970s and early 1980s, interest rates rose markedly, following the introduction of monetarist policies in the US and UK particularly. For example, the London Inter-Bank Offered Rate, the main index of the price of an international loan, rose from an average of 9.2 per cent in 1978 to 16.63 per cent in 1981 (ibid.). As a result, developing countries were faced with hugely increased debt repayments and the world economy was plunged into a recession

➤ The collapse of commodity prices in the early 1980s, 'such that in 1993 prices were 32 per cent lower than in 1980; and in relation to the price of manufactured goods, they were 55 per cent lower than in 1960' (Potter *et*

al., 2008, p.369). This amounted to a serous deterioration in the terms of trade for developing countries, reducing their export revenues relative to the price of imported goods. As a result, they were faced with a 'scissors' crisis of declining export revenues and mounting debt repayments in a strengthening dollar (Corbridge, 2008, p.508). This is what led to a number of countries defaulting, threatening the viability of the international financial system.

The scale of the debt crisis seemed greater in Latin America in the 1980s, but the difficulties facing developing countries in sub-Saharan Africa, particularly, have become apparent since the early 1990s. While Latin American and Asian countries often have the largest absolute debts, it is African countries that are worst affected in terms of the relationship between debt and GDP. Africa's indebtedness is estimated to have trebled to around $US216 billion between 1980 and 1999, increasing from approximately 28 per cent of its GDP to 72 per cent, compared with 40 per cent for Latin America (Potter *et al.*, 2008, p.369). In some cases, annual repayments to creditors – around $15 billion every year – outweigh expenditure on education and health.

Box 8.4

The development of SAPs

The central aim of SAPs is debt reduction through reform packages designed to enable countries to pay their debts while maintaining economic growth and stability. Agreement to introduce a SAP became an essential pre-condition for developing countries to obtain finance from the World Bank, IMF and other private donors, a pre-condition that few countries were in a position to refuse. As a result, they rapidly spread across the Global South. The first SAP was introduced in Turkey in 1980 and 187 had been negotiated for some 64 developing countries by the end of the decade (Potter *et al.*, 2008, p.299).

SAPs were based on four main objectives (Simon, 2008, p.87):

➤ The mobilization of local resources to foster development.
➤ Policy reform to increase economic efficiency.
➤ The generation of foreign revenue through exports, involving diversification into new products as well as expansion of established ones.
➤ Reducing the economic role of the state and ensuring low inflation.

The measures required to achieve these objectives are generally divided into two types (ibid.):

1. Stabilization measures, which were immediate steps designed to address the economic difficulties facing developing countries in the short term, providing a foundation for longer-term measures.
➤ A public sector wage freeze – to reduce wage inflation and the government's salary bill.
➤ Reduced subsidies on basic foods and other commodities, and on health and education – to lower government expenditure
➤ Devaluation of the currency – to

make exports cheaper and more competitive, and to deter imports.
2. Adjustment measures, which were to be implemented as a second phase, having a longer-term impact. Their objective was to ensure the structural adjustment of the economy, creating a platform for future growth.
➤ Export promotion through incentives for enterprise (including increased revenues and access to foreign currency) and diversification.
➤ Downsizing the civil service through the retrenchments following a programme of rationalization to reduce 'overstaffing', duplication, inefficiency and cronyism.
➤ Economic liberalization – relaxing and removing regulations and restrictions on economic activities. Examples include import tariffs and quotas, import licences, state monopolies, price fixing, subsidies and restrictions on the repatriation of profits by overseas firms.
➤ Privatization through the selling off of state enterprises and corporations.
➤ Tax reductions to create stronger incentives for individuals and firms to save and invest.

SAPs proved highly controversial. Their impacts have often been harsh, with ordinary people rather than elites bearing the brunt of the adjustment costs (ibid., p.88). In general, large traders, merchants and rural agricultural producers have benefited from increased export opportunities, often at the expense of the urban poor, who have suffered from the abolition of food subsidies and reductions in public expenditure. The effects of SAPs have been felt through cuts in public services, privatization and

the introduction of user charges for services. In Ghana, for instance – regarded as a 'star pupil' by the World Bank and IMF – the introduction of user charges for health services in 1985, in a climate of failing wages and increased poverty, led to a drop in hospital and clinic visits of 25–50 per cent in urban Accra and of 45–80 per cent in certain rural areas (Konadu-Agyemang, 2000, p.474). The privatization of water in Ghana has led to higher water rates in a situation where 35 per cent of the population already lack access to safe water. There is also evidence that economic reform has benefited foreign interests over domestic producers. Large-scale privatization and liberalization of Ghana's lucrative gold mining sector has benefited outside MNCs, attracted by tax breaks and incentives with 70–85 per cent of the industry now foreign-owned (Ismi, 2004, p.17). Minimal environmental regulations have led to widespread contamination and degradation, while 30,000 people were displaced during 1990–98 (ibid.). According to many critics, SAPs have reinforced inequality and poverty by reducing access to crucial social services and opening up developing economies to outside interests such as MNCs, although some local traders and farmers have benefited (Simon, 2008).

SAPs were refined in the late 1980s and 1990s, taking better account of local needs and circumstances with the longer-term adjustment element rebranded as Economic Recovery Plans (ERPs) (ibid., p.90). Since 1999, SAPs have been replaced by **poverty reduction strategies** (PRSs), which require national governments to produce a comprehensive plan for reducing

Box 8.4 (continued)

poverty, requiring consultation with the World Bank, IMF, NGOs and local communities. The World Bank has stated its commitment to work more closely with NGOs to foster democracy and local empowerment within developing countries (Potter *et al.*, 2008, p.303). Kenya was one of the first countries to experience the shift from SAPs to PRSs after IMF support was suspended in 1998 because of the government's failure to adhere to the terms of its SAP. The resumption of support was made dependent on the production of a PRS by the government during 2000, with the emphasis on economic reform prompting widespread criticism from development NGOs and activists in Kenya (ibid.). As this indicates, central elements of the neoliberal Washington consensus model remains in place, having essentially been augmented by the new focus on poverty reduction and 'good governance' (section 5.5.2) (Sheppard and Leitner, 2010).

of state enterprises. The World Bank and IMF have played a crucial role in requiring developing countries to sign up to this policy agenda. In the context of the debt crisis, developing countries were often in desperate need of further financial assistance from those organizations (Box 8.3). This allowed the World Bank and IMF to set conditions requiring developing countries to reform their economies. These reforms have become known as **structural adjustment programmes** (SAPs), encompassing a range of measures requiring countries to open up to trade and investment and reduce public expenditure (Box 8.4). Developing countries should seek to compete in the global market through the development of competitive export sectors, with the experience of the East Asian NICs often cited in support of this argument. In a similar fashion to modernization theory, neoliberalism is based on the imposition of a set of externally derived solutions from the global north through mechanisms such as SAPs (Willis, 2005, p.194).

8.3.4 Grassroots development

This approach is rather different in nature and orientation from the other theories discussed above. Instead of advancing a set of overarching prescriptions about development policy, based on abstract economic analyses, it is directly concerned with the practical problems and needs of poor people in developing countries, the ultimate targets of development aid (Table 8.1). While the previous models are designed and implemented in a top-down fashion, focusing on the national scale, grassroots development, as the term implies, is focused on the local level. In addressing local needs in a 'bottom-up' manner, this approach is based on the observation that increased economic growth does not necessarily reduce poverty; in some cases; growth may wider inequalities within society, benefiting some groups at the expense of others. NGOs are closely associated with grassroots strategies, often working in partnership with local agencies and groups (Townsend *et al.*, 2004).

This approach to development has evolved from the '**basic needs**' strategies of the 1960s and 1970s. These focused attention on the everyday lives of the poor, reflecting concerns that these were being neglecting by orthodox modernization policies. Recent work shares the same concern for identifying and meeting such needs in relation to food, shelter, employment, education, health, etc. The ethos is one of 'helping people to help themselves', involving small-scale projects that directly benefit individuals, families and households, supporting local services and livelihoods. In urban areas, these are often based on the large informal sector of the economy, supporting the efforts of small-scale producers and traders to survive. An example of an urban grassroots initiative is a partnership of an NGO and two community-based organizations in Mumbai, India, known as *The Alliance*, which aims to organize and mobilize poor people in slum areas to pursue better housing and sanitation facilities (McFarlane, 2004). In rural areas, farming is usually the focus of attention, with the establishment of schemes offering targeted assistance and support. An example is Oxfam's cow loan scheme, whereby families are given cows from

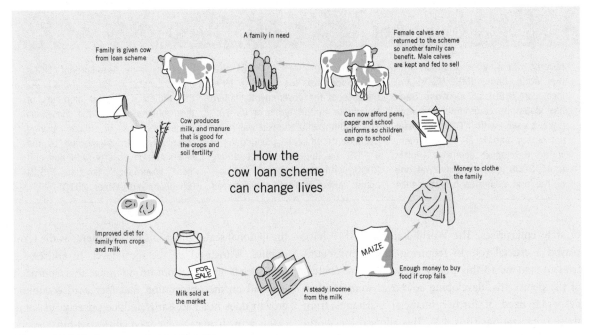

Figure 8.7 Oxfam cow loan scheme
Source: Willis, 2005, p.196.

the loan scheme, fertilizing crops and providing milk, which not only provides sustenance for the family but can also be sold to generate an income, allowing them to purchase food and clothes (Figure 8.7) (Willis, 2005, p.196). Calves are returned to the scheme to be loaned out to other families or sold.

The main constraint on grassroots development is the limited resources available to NGOs and other bodies. While the number of NGOs has increased significantly since the 1970s, their resources are still outweighed by the scale of problems confronting them. Furthermore, as much of their money is derived from donations by individuals and governments, projects often reflect the concerns of donors rather than the priorities of local people. At the same time, the proliferation of small-scale local projects can sometimes be associated with a lack of collaboration and integration between agencies and initiatives. Grassroots development could be said to focus on alleviating the symptoms of poverty rather than addressing the underlying causes. This helps to explain some of the World Bank's and IMF's reluctance to embrace such approaches, concentrating instead on economic reform programmes. Such a view is rather harsh, however, with NGOs and others doing much

valuable work in meeting local needs and highlighting the failure of mainstream initiatives to reach those most in need. The adoption of explicit poverty reduction strategies by the World Bank and others in recent years suggests that such concerns have filtered into the mainstream agenda.

Reflect

➤ Which of the theories offers the most appropriate model for developing countries? Do elements of different theories need to be combined? If so, which ones? Justify your answers.

8.4 Patterns of development

8.4.1 An unequal world

Since the 1970s, despite many setbacks and difficulties, some modest progress has been made in improving conditions in poorer countries. Whilst these improvements are overshadowed by the sheer magnitude of the gap between rich and poor countries, their significance should not be minimised. Key outcomes include:

> Average incomes in developing countries almost doubled, from $US1300 to $US2500 between 1975 and 1998 (UNDP, 2001, p.10).

> The number of people in developing regions living on less than $1.25 a day has fallen from 1.8 billion in 1990 to 1.4 billion in 2005 (United Nations, 2010, p.6).

> Since 1970, for the world as a whole, life expectancy has increased by nine years, adult literacy by 20 points and income per capita by US$3,800 (Molina and Purser, 2010, p.1).

> Globally, the total number of deaths of children under five has fallen from 12.5 million in 1990 to 8.8 million in 2008 (United Nations, 2010, p.26).

> Access to mobile telephones continues to expand in the developing world, with cellular subscriptions per 100 people reaching 50 per cent by the end of 2009 (ibid.).

At the same time, huge global inequalities remain. Gross national income (GNI) per capita, for instance, ranged from an average of 39,345 ($US, PPP) for high-income economies to 2,789 for low-income and middle-income economies in 2008, using the World Bank's categories (World Bank, 2009, p.379). Much of the reduction in extreme poverty is a result of rapid growth in populous countries such as China and India, with the use of global aggregates masking slower falls or even increases in other regions (Hulme and Scott, 2010, p.7). At the same time, measures of international inequality take no account of the distribution of income within countries, which seems to have become more unequal in recent decades (UNDP, 2005).

Some of the progress made in recent years is being undermined by the effects of the global economic crisis and substantial rises in food and fuel costs from 2006 (Hulme and Scott, 2010). The economic crisis has slowed growth in developing countries due to reductions in demand for exports, trade and investment, leading to sharp drops in employment-to-population ratios and reduced output per worker (United Nations, 2010). The crisis is likely to have increased unemployment and poverty with the largest negative effects occurring in sub-Saharan Africa, southern Asia,

Southeast Asia and Oceania (ibid., p.11). Higher food prices caused the proportion of the population suffering from undernourishment to start to rise again in sub-Saharan Africa, having been falling slowly from 1990 to 2006 (Hulme and Scott, 2010, p.7)

8.4.2 Divergence between developing regions

It is important to recognize that not all developing countries are equal in terms of economic development. Since 1960 there has been increased divergence between the major regions of the developing world, symbolized by the contrasting performance of East Asia and sub-Saharan Africa (Figure 8.8). The different regions were close together in 1960, with their income standing at about one-ninth/one-tenth of the high-income economies belonging to the Organization for Economic Cooperation and Development (OECD). The main exception was Latin America and the Caribbean, which enjoyed significantly higher income levels, at about one-third/one-half of the OECD level. East Asia and the Pacific experienced strong growth from the early 1970s, progressing to almost one-fifth of OECD income levels. The position of southern Asia worsened in the 1960s and 1970s, before improving significantly from the early 1980s.

At the same time, a dramatic decline occurred in sub-Saharan Africa, which plunged from about one-ninth of OECD countries' income in 1960 to one-eighteenth in 1998. In absolute terms, the GDP per capita of sub-Saharan Africa grew by 36 per cent between 1960 and 1980 but fell by 15 per cent in the 1980–90 period (Ismi, 2004, p.11). The divergence between the performance of East Asia and Africa is highlighted by the contrasting experiences of Ghana and Malaysia (Box 8.5). Although this is not shown in Figure 8.8, the position of many transition economies in the former Soviet bloc also worsened in the 1990s.

The decline of sub-Saharan Africa in the two 'lost development decades' of the 1980s and 1990s requires some explanation (Hulme and Scott, 2010, p.7). While the efforts of the World Bank and IMF have focused on internal inefficiencies and failures, external relationships are also crucial. The region's decline certainly does not reflect its isolation from the world economy;

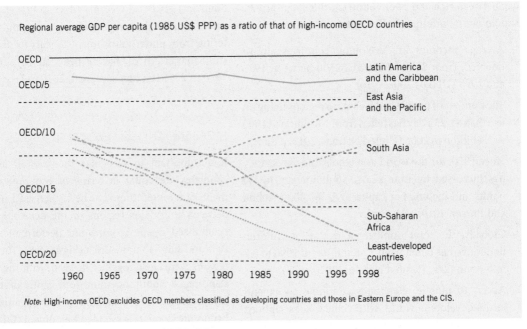

Regional average GDP per capita (1985 US$ PPP) as a ratio of that of high-income OECD countries

Note: High-income OECD excludes OECD members classified as developing countries and those in Eastern Europe and the CIS.

Figure 8.8 Income trends by region, 1960–98
Source: UNDP, 2001, p.16.

Box 8.5

Contrasting experiences of development: Ghana and Malaysia

Both Ghana and Malaysia are former British colonies, which gained independence in 1957 and 1961 respectively. Both had fairly typical colonial economies, based on the export of raw materials, particularly cocoa and gold in Ghana and rubber, tin and timber in Malaysia. While poor, both were probably better placed than many other former colonies, with Malaysia enjoying higher incomes than many of its neighbours and Ghana regarded as one of the best prospects in sub-Saharan Africa in terms of the high value of exports, a well-developed infrastructure and a relatively skilled and educated workforce (Khan, 2002; Konadu-Agyemang, 2000).

Since 1960, however, the economic fortunes of the two countries have diverged dramatically, with Malaysia experiencing sustained economic growth while Ghana has suffered prolonged stagnation and decline (Doyle, 2005), before undergoing something of an upturn over the last few years. Malaysia was ranked 66 in the UN's Human Development Index in 2007, with a GNI per capita (PPP US$) of 6,970 compared with Ghana, ranked 153 with a GNI per capita of 670, although it is a few places above the bottom category of 'low human development' (UNDP, 2009, pp.168–69; World Bank, 2009, p.378). Ghana remains dependent on the export of raw materials, particularly cocoa and gold, while Malaysia has diversified successfully, producing a range of industrial goods, including cars, and advanced services.

How can this contrasting economic performance be explained? The role of the state is clearly of central importance. Malaysia has enjoyed four decades of political stability, with one politician, Dr Mahathir Mohamad, occupying the position of Prime Minister for an unbroken 20 years. This has offered a stable climate for investment and growth, with the government pursuing a consistent policy of rapid export-led growth and poverty reduction (Khan, 2002). At the same time, it retained the flexibility to adapt the mechanisms through which this policy was implemented as global economic conditions changed. Ghana, by contrast, soon lapsed into political instability with the first of a series of military coups occurring in 1966 (Doyle, 2005). Corruption and economic mismanagement became rife, coupled with an overvalued currency and unfavourable terms of trade

Box 8.5 (continued)

(Konadu-Agyemang, 2000). Malaysia retained control of its economic destiny, while high levels of debt and inflation, coupled with falling export revenues, meant that Ghana has been subject to a succession of World Bank and IMF-sponsored adjustment programmes. These have had the effect of reducing the access of the poor to social services and increasing foreign ownership and control of the economy, although Ghana's economic performance has improved since the late 1990s (OECD, 2006).

instead, trade expanded considerably in the 1990s, as did FDI, reflecting the impact of SAPs. But African exports remain based on raw materials for which prices have been falling. Copper, for example, accounts for 70 per cent of Zambia's exports and coffee for 73 per cent of Burundi's (BBC, 2004). Non-oil commodity prices dropped by an average of 35 per cent on average since 1997 (Ismi, 2004. p.11), leading to declining terms of trade as export revenues fell relative to the price of imported goods. At the same time, under SAPs, Africa's external debt has actually increased by more than 500 per cent since 1980 (ibid., p.12). Alongside cuts in public expenditure, this results in Africa spending four times more in debt servicing than health. At the same time, HIV/AIDS has had a devastating impact, with sub-Saharan Africa accounting for an estimated 60 per cent 28 million of the 39.5 adults and children living with HIV worldwide (Potter *et al.*, 2008, p.214).

Reflect

➤ Is it still meaningful to talk of a 'third world' in view of the increased economic divergence between developing regions and the end of the Cold War?

8.5 Resisting development: the growth of local social movements in developing countries

Orthodox forms of development based on the promotion of economic growth and open markets have often failed to reach those groups most in need of assistance.

As we saw, neoliberal models of development based on the principles of deregulation, liberalization and privatization have often reduced the access of the poor to vital public services and increased externally owned MNCs' control over local resources and services. As part of this project of development, western values and expertise have been privileged over local knowledge and culture (Routledge, 2005, pp.211–12). The agency of local people in developing countries, viewed simply as passive subjects of development, is often ignored by mainstream approaches. This provides the context for assessing the activities of local **social movements** in resisting particular aspects of neoliberal globalization and the development projects undertaken by individual states. The activities of these groups have helped to fuel the growth of the counter-globalization movement (section 1.2.1), attracting the interest of researchers and activists in northern countries.

According to the political geographer Paul Routledge (2005, p.214), local social movements in developing countries are highly diverse, incorporating 'squatter movements, neighbourhood groups, human rights organizations, women's associations, indigenous rights groups, self-help movements amongst the poor and unemployed, youth groups, educational and health associations and artists' movements'. Such groups articulate an 'environmentalism of the poor', concerned with defending local livelihoods and maintaining communal access to local resources in face of their commodification or appropriation by the state and corporations and with freeing people from extreme poverty and domination by other groups (for example, large landowners or private logging companies).

Social movements in developing countries can be seen as an expression of conflicts between different groups in society over the control and use of space.

While local groups wish to retain control of local economic resources such as land, forests and water, utilizing them to meet their material needs on a day-to-day basis, states and private interests often want to exploit them for economic gain, threatening the basis of local livelihoods (Bebbington and Bebbington, 2010). Deforestation schemes, mineral extraction projects and the construction of dams for hydro-electric power and irrigation are often the focus for conflicts which can be seen as examples of '**resource wars**' between the different parties. Many of the local resistance movements are place-based, asserting local values and identity in the face of the knowledge and power of states and private corporations. One of the effects of policies of privatization and liberalization has been to make it easier for external corporations and interests to gain control over local resources, resulting in increased conflict with local groups. While most movements are locally-based, their struggles have often assumed a wider dimension, involving imaginative use of the media and internet to gain international support.

While social movements have been formed in a range of developing countries and regions in developing countries, including southern and South-east Asia and sub-Saharan Africa, some of the most interesting examples are found in Latin America, highlighting the unequal social relations between poor and indigenous peoples, on the one hand, and the state and private capital, often in the shape of externally owned MNCs, on the other. One of the best known is the Zapatista guerrilla movement in the Chiapas region of Mexico, which represents the interests of the indigenous Mayan people. On New Year's Day 1994, a group of masked guerrillas emerged from the jungle, seized the local state capital and declared war on the Mexican state. The Zapatistas were protesting against poverty and the exploitation of local resources, problems that were enhanced by reforms associated with the North American Free Trade Agreement (NAFTA) – between the US, Canada and Mexico, completed in 1994. This fuelled the growth of intensive agriculture for international markets, leading to the emergence of a small group of wealthy farmers and a large class of landless Indian labourers (Routledge, 2005, p.215). The Zapatista movement employed a range of tactics designed to gain international media attention

alongside their armed insurgency, engaging in a prolonged 'war of words' with the Mexican government through the pronouncements and communiqués of their mysterious spokesman, Subcomandante Marcos.

Another example is the *Movimento Sem Terra* (MST) in Brazil, a mass national social movement of an estimated 1.5 million members (http://www.mst.org.br). Founded in 1984, MST is active in 23 out of Brazil's 27 states. It is made up of mainly landless labourers and peasants from rural areas, many of whom have been dispossessed and displaced by agricultural reforms, mechanization and land clearance. The context for the growth of the movement is the highly uneven distribution of land ownership in Brazil, with most of the land owned by fewer than 50,000 people, whereas 4 million peasants share less than 3 per cent (Routledge, 2005, pp.216–17). While land reform remains the main goal of the MST, it has come to recognize that the struggle

Figure 8.9 An MST protest march
Source: Luciney Martins.

is not only directed against the Brazilian *latifundio* (system of large estates with a single owner), but also the neoliberal economic model, campaigning vociferously against the proposed Free Trade Area of the Americas (FTAA) (http://www.mst.org.br). The strategy of the movement involves targeting large unused private estates and illegally squatting and occupying the land. After areas are 'secured' in this way, the MST resettles large numbers of people on these sites, builds schools and houses, and begins farming. Over 600,000 people have been resettled since 1991, leading to considerable violence as large landowners and their private armies

have attacked and killed squatters (ibid.). The MST has organized several marches and congresses in the capital city of Brasilia to publicize its agenda for agricultural reform (Figure 8.9). Its campaigning continued despite the election of 'Lula' da Silva of the left-wing Workers' Party to the Brazilian presidency in 2002 and again in 2006, since progress on agrarian reform remains limited, reflecting the government's continued adherence to elements of the neoliberal model (Jacobskind, 2010).

A third example concerns indigenous and environmental movements protesting against resource

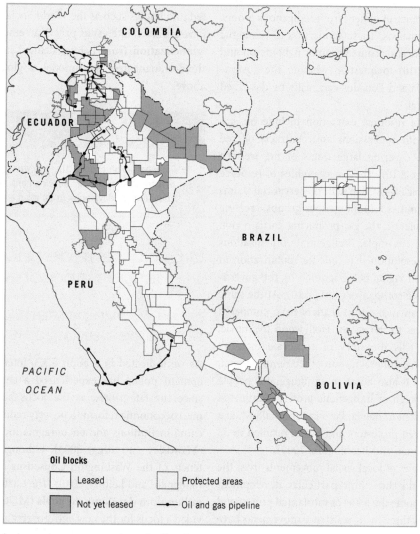

Figure 8.10 Hydrocarbon concessions in the Amazon
Source: Bebbington, 2009, p.14.

extraction in the Andean region. In recent years, governments of different political complexions in Bolivia, Ecuador and Peru have sought to encourage the expansion of extractive activity as a key pillar of macroeconomic strategy, viewing this as means of income generation that can be used for poverty reduction and social investment (Bebbington and Bebbington, 2010, p.11). While the Peruvian state espouses a conventional neoliberalism, emphasizing private interests, the leftist governments of both Bolivia and Ecuador have privileged the role of the state in governing the extractive process and redistributing the surplus for social ends. At the same time, all three governments share an underlying authoritarianism that is manifest in a growing intolerance of dissent from indigenous groups and environmentalists. These similarities in economic policy and political discourse prompt Bebbington and Bebbington (2010) to question whether the governments of Bolivia and Ecuador can really be described as 'postneoliberal'.

Key forms of resource extraction include oil, gas and mining, with concessions and contracts offered for the former covering large tracts of the western Amazon (Figure 8.10). These geographies of resource exploitation typically overlap with the territorial claims of local communities and indigenous groups, creating a basis for conflict. In the gas-producing Tarija region of Bolivia, for example, indigenous organizations have issued a resolution that calls for mobilization in response to their rights being ignored by the government and private companies (ibid., p.10). At the same time, Ecuadorian indigenous leaders have compared their experiences to the recent Hollywood hit movie, *Avatar*, which tells the story of corporate interest's exploitation of the mineral resources of a remote planet, Pandora, which is also home to an indigenous people (ibid.). A campaign of indigenous protest culminated in a tragic incident in Bagua, Peru in June 2008 when a stand-off between protestors and police resulted in 33 deaths (Bebbington and Bebbington, 2009).

These examples of local social movements from the developing world show how particular development projects can become the focus of substantial protest and resistance. In all three cases, resistance movements have emerged when local livelihoods have been undermined and local cultures threatened by economic reforms and large-scale development projects. They illustrate the continuing scope for 'bottom-up' action and initiative in an increasingly globalized economy, indicating that local people should be seen as active players in the development process rather than passive subjects of the state and international agencies. While most of these movements are place-based, mobilizing local culture and identity to animate their campaigns, they have become increasingly active in appealing to international audiences and in forging links with similar movements in other countries, facilitated by new communications technologies such as the internet. This has fed into bodies such as the World Social Forum, which have attracted interest as part of an emerging model of **'globalization from below'**, defined in opposition to the dominant neoliberal model of 'globalization from above'.

> **Reflect**
>
> ➤ To what extent does the growth of local social movements resisting development indicate that conventional development policies are failing?

8.6 Current development issues and challenges

8.6.1 Development strategies and goals

As we indicated in section 8.3.3, international development policy has experienced a shift of emphasis since the late 1990s, as the focus has moved from macroeconomic reform to poverty reduction, strengthening institutions and encouraging local participation (Mawdsley and Rigg, 2003), although many of the tenets of the Washington Consensus remain in place (Sheppard and Leitner, 2010). The establishment of the **millennium development goals** (MDGs) in 2000 provided a focus for the new anti-poverty agenda. Adopted by the UN in 2000, the MDGs identified a number of objectives, setting specific targets to be achieved by 2015, against which progress can be measured and monitored on an annual basis:

1. Eradicate extreme poverty and hunger.

2. Achieve universal primary education.

3. Promote gender equality and empower women.

4. Reduce child mortality by three-quarters.

5. Reduce maternal mortality by two-thirds.

6. Combat HIV/AIDS, malaria and other diseases

7. Ensure environmental sustainability.

8. Develop a global partnership for development.

Progress towards these goals is commonly described as mixed, with substantial improvement in some areas matched by slower improvement, or even regression, in others (Hulme and Scott, 2010). As we saw, the target of halving extreme poverty between 1990 and 2015 is on course to be met, largely due to sustained growth in China and India, particularly in the 1990s, before the MDGs were established. In relation to education (goal 2), the universal enrolment target is close to being met, but not in sub-Saharan Africa and south Asia. The gender equality goal (3) is likely be met in part, with the developing regions as a whole approaching gender parity in educational enrolment, although large regional gaps remain, particularly at primary level (United Nations, 2010, pp.20–1). Substantial progress has been made in reducing child mortality (goal 4), but three-quarters of countries remain 'off-target' (Hulme and Scott, 2010, p.6). The reduction of maternal mortality shows the least progress of all, with deaths remaining very high in sub-Saharan Africa and southern Asia. While the spread of HIV- and AIDS-related deaths has peaked (goal 6), the disease remains a huge problem in sub-Saharan Africa. While the setting of global goals has provided a clear overall target for development organizations, it tends to mask significant geographical divergence, as the above summary indicates, involving a lack of progress in certain regions (Hulme and Scott, 2010, p.12).

8.6.2 Trade and the politics of globalization

Trade has become one of the key battlegrounds of globalization in recent years, ever since protestors disrupted the WTO's meeting in Seattle in December 1999. Throughout the post-war period, GATT/WTO (Box 1.2) operated through successive 'rounds' of trade negotiations, leading to agreements between member-states to reduce tariffs (charges imposed on imported goods). Until the 1980s, GATT agreements concentrated on trade in manufactured goods in particular, with restrictions on agricultural goods and primary commodities remaining in place. This left many developing countries dissatisfied with a system that they felt was dominated by powerful northern states, requiring them to open up their markets to manufactured goods, while they were confronted with barriers to the export of the primary goods in which they enjoyed a comparative advantage.

The **Uruguay Round** which ran from 1986–94 was the most ambitious and wide-ranging of the GATT agreements. For the first time, it incorporated not only agriculture, textiles and clothing but also new agreements on services:

➤ The General Agreement on Trade in Services (GATS)

➤ Trade-Related Aspects of Intellectual Property Rights (TRIPs)

➤ Trade-Related Investment Measures (TRIMs).

The Uruguay Round led to an average tariff reduction of one-third by developed countries (Dicken, 2003a, p.586). The agreement on services again seemed to reflect the interests of the north, representing an area in which it enjoyed a strong competitive advantage over the countries of the Global South. In general, the Uruguay Round failed to resolve a number of tensions between developed countries and developing countries. One of the effects of the TRIPs agreement, for instance – pushed hard by northern multinationals in sectors such as pharmaceuticals to give stronger protection to their patent rights – was to make essential life-saving drugs more expensive for poor people in developing countries (Stiglitz, 2003).

The **Doha Round** of trade negotiations, launched in Qatar in late 2001, was presented as the 'development round' by the WTO, EU and US, who argue that a new trade agreement represents the best way to reduce global poverty. The implication is that the inequities stemming from the Uruguay Round would be addressed, with developed countries opening up their markets to agricultural goods and textiles, while industrial tariffs would be slashed further. At the same time, the EU and the US aim to advance the liberalization of trade in services, introducing new rules for investment, competition, government procurement and trade facilitation (the so-called Singapore issues). If agreed, these new rules would open up the public sector, in particular, to trade, with countries prevented from favouring domestic suppliers of services over foreign companies.

Meanwhile, developing countries remained angry about continuing restrictions that developed countries imposed on trade in agriculture and textiles through **subsidies**, favouring domestic producers (Box 8.6). The subsidies which the EU and US offer to farmers result in the dumping of surplus products on world markets, driving down prices for farmers in developing countries (Oxfam, 2005, p.10). The $3 billion plus subsidies that the US offers to its 25,000 cotton farmers, for example, has undermined the livelihoods of more than 10 million farmers in West African countries such as Benin, Chad and Mali (*The Economist*, 2003). Instead of launching a new round of trade talks, most developing countries wished to concentrate on implementation issues left over from the Uruguay Round, particularly

Box 8.6

Trade reform and development: the case of the EU sugar industry

One of the main barriers to economic development in the world's poorest countries is their lack of access to the agricultural markets of the Global North. Within the current Doha Round, the agricultural protectionism of the EU and US has become a major sticking point. The EU's Common Agricultural Policy (CAP) favours European producers through a range of subsidies and restrictions on imports. Sugar is one of the most protected sectors in European agriculture, with EU prices set over 300 per cent higher than the world market price in March 2002 (Gibb, 2006, p.6). A small proportion of sugar from third-world countries outside the EU is imported under special quotas offering preferential access.

The EU's trade relations with the developing world were governed by the Lome Convention, which ran from 1975 to 2000. This offered the African, Caribbean and Pacific (ACP) group of states, made up of former European colonies, duty-free access to the European market for most

of their exports. It effectively discriminated against other developing countries not part of the ACP group. As became increasingly apparent during the 1990s, such discrimination was not compatible with the WTO agenda of trade liberalization, prompting reform. In response, the EU has divided the existing ACP grouping into 'least developed countries' (LDCs) and 'developing countries', extending Lome trade preferences to all the former, while seeking to negotiate reciprocal free trade agreements with the latter. There are only nine developing countries that were not previously members of the ACP group (see Gibb, 2006, p.4 for listings). Under its 'everything but arms' (EBA) system, introduced in March 2001, the EU introduced duty-free access for all exports from LDCs (49 countries) with the exception of arms, rice, bananas and sugar.

The prospect of LDCs gaining free access to the European sugar market under a three-year transitional arrangement sparked vigorous

lobbying by European producers to block the reform. As a result, the final version watered down the access offered to LDC sugar exporters, using a quota system to restrict access over an extended transitional period from 2001/2 to 2008/9. The burden of adjustment arising from this limited new access is placed squarely on the developing countries who lose the preferences that allowed them to supply quota sugar to the EU. 'In short, poor countries are paying for the access privileges offered to the very poor' – 'an extraordinary outcome for an initiative designed to help the world's poorest countries' (ibid., pp.9, 14). In the case of sugar, then, the EU's reform of its trade regime offers only minimal access to one group of poor countries at the expense of another. In late 2005, EU agriculture ministers, under pressure from the WTO and other bodies, agreed a 36 per cent cut in sugar prices, sparking protest from sugar-producing LDCs affected by the price cut (BBC, 2005).

the failure of developed countries to reduce agricultural subsidies.

Against this backdrop, trade negotiations were always going to be difficult. Talks broke down at the Mexican resort of Cancun in September 2003. Basically, developing countries were not prepared to discuss EU-led proposals on new rules for investment and competition without large-scale reduction of agricultural subsidies and textile quotas, which the EU and US were unwilling to accept. The emergence of organized coalitions of developing countries prepared to take on the EU and US has been particularly significant. The so-called G21 group, led by India, China and Brazil, played a key role in advancing the interests of developing countries in Cancun, refusing to bow to pressure. The Doha Round remained stalled at the time of writing (December 2010), although negotiations continue.

The substantial difficulties experienced by the Doha Round raise important questions about global economic regulation. A new assertiveness by developing countries in the face of continuing trade inequities has thrown the GATT/WTO model into doubt. There is little sign of any resolution on the underlying issues dividing developed and developing countries. The EU and US remain unwilling to eradicate subsidies in agriculture and textiles, while continuing to push for further liberalization of trade in services and industrial goods. Without a commitment to the former, developing countries will not agree to the latter. Without major concessions from the EU and US, it seems unlikely that an agreement will be reached. Even if a deal is eventually struck, developed countries are likely to gain more from access to developing country markets than developing countries will from agricultural liberalization

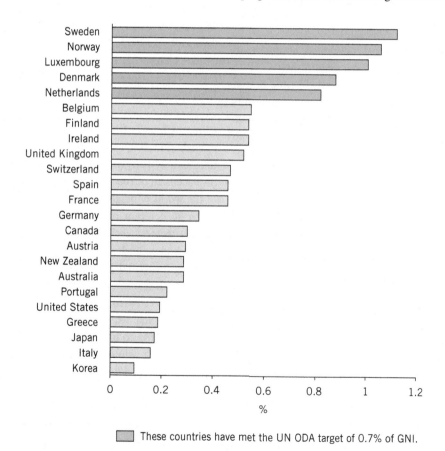

Figure 8.11 Overseas development assistance in 2009 as a percentage of GNI
Source: http://webnet.oecd.org/oda2009.

(Wilkinson, 2009), reproducing the existing trade inequities between the Global North and South.

8.6.3 Aid to developing countries

Aid is defined as official development assistance offered by developed countries to promote economic development and poverty reduction (Burnell, 2008). It has been an important aspect of development policy throughout the post-war period, although levels of assistance have often fallen short of what development activists and campaigners have called for. At the same time, aid has been used to further donors' strategic and political interests, by, for example, tying it to the purchase of goods and services from that country. During the 1980s and 1990s, a particularly high proportion of aid from the UK was tied, reaching 74 per cent of bilateral aid in 1991 (Potter *et al.*, 2008, p.373), although all UK aid has been fully untied since April 2001.

In 1970, the UN urged developed countries to commit the equivalent of 0.7 per cent of their GNP to foreign aid by 1975, a demand reiterated by the Brandt Commission in 1980 and subsequent global summits in the 1990s (Potter *et al.*, 2008, p.370). At the 2005 G8 summit in Gleneagles, world leaders agreed an additional $50 billion a year in aid to developing countries by 2010, representing 0.36 per cent of estimated GNI with $25 billion going to Africa. Moreover, the EU countries pledged to reach a collective aid target of 0.56 per cent of GDP by 2010 and 0.7 per cent by 2015. Actual expenditure on aid has risen considerably to a total of US$119.6 billion in 2009, representing 0.31 of GNI, fairly close to the overall target of 0.36 per cent. Figure 8.11 shows that only five members of the Development Assistance Committee (DAC) of the OECD committed more than the 0.7 target in 2009 with several other EU countries unlikely to meet the 0.56 of GNI target by 2010, although some have made considerable progress since 2004.

8.6.4 Climate change and development

The need to respond to climate change has become increasingly accepted by government and policy-makers in recent years, although climate sceptics reamin influential. The emission of so-called greenhouse gases, particularly carbon dioxide, is identified as a key cause of global warming with increased emissions over the past couple of centuries largely reflecting the burning of fossil fuels (Potter *et al.*, 2008, p.257). While climate change remains an abstract question for many people in the developed world, its impacts are already being felt in many developing countries. This reflects the fact that developing countries are more exposed to the impact of change and less resilient to climate hazards, because of their dependence on ecosystem services and natural resources for production, particularly in the agricultural sector. It is estimated that they will bear 75–80 per cent of the costs of climate change (World Bank, 2009, p.5). As such, the need to address climate change has become a key concern of development organizations, providing the focus for the *World Development Report 2010*, for instance (ibid., 2009).

According to the Inter-Governmental Panel on Climate Change's 4th Report and the Stern Report, both published in 2007, the impacts of global warming, in the absence of any mitigating action, would include: the displacement of 100 million people by floods; one in six of the world population experiencing water shortages; up to 40 per cent of species becoming extinct; more intense tropical cyclones; and increases in precipitation in high latitudes and decreases in most subtropical land regions (cited in Potter *et al.*, 2008, p.259). The effects on development will be regionally differentiated, involving: increased risk of drought in sub-Saharan Africa, coupled with the spread of tropical diseases such as malaria to higher altitudes; rising sea levels in southern Asia, where they are predicted to submerge much of the Maldives and to inundate 18 per cent of Bangladesh's land; increased water scarcity in the Middle East and North Africa, the world's driest region, which will undermine agriculture; and the melting of the tropical glaciers of the Andes, leading to reduced water availability and the threatening of hydropower which supplies more than half the energy of many

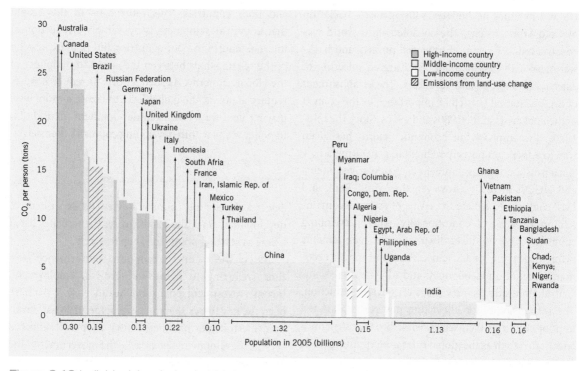

Figure 8.12 Individuals' emission in high-income countries and developing countries

Sources: Emissions of greenhouse gases in 2005 from WRI, 2008, augmented with land-use change emissions from Houghton, 2009; population from World Bank, 2009.

South American countries (World Bank, 2009, p.6). These predictions are based on rises of 5°C or more compared with pre-industrial times, focusing attention on the need to stabilize temperatures at around 2°C above pre-industrial levels, still a highly ambitious target in view of current emissions (ibid.).

Questions of adaption to climate change raise huge issues of environmental and social justice, since the developing world is likely to experience much of the impact, while developed countries are responsible for the bulk of emissions by person (Figure 8.12) and have enjoyed most of the benefits of industrial growth since the nineteenth century. This contradiction is heightened by population growth, with the world's population predicted to reach 9 billion by 2050, including an additional 2.5 billion in developing countries, placing huge pressures on ecosystems and natural resources (World Bank, 2009, p.40). Expecting poor countries to sacrifice development to conserve resources without reducing emissions from developed countries is, of course, hugely unjust. As such, proposed responses to climate change are based on proportional reductions in emissions and support for adaptation measures in developing countries. Recent progress has been limited, however, with the Copenhagen Summit in December 2009 failing to reach a legally binding agreement, although further talks are scheduled for Mexico and South Africa in 2010 and 2011 respectively.

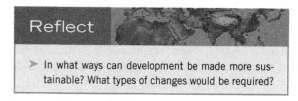

Reflect

➤ In what ways can development be made more sustainable? What types of changes would be required?

8.7 Summary

The development of the former colonies in Africa, Asia and Latin America can be seen as a key political and economic project of the post-war era, launched by the

US and its allies in the late 1940s against a backdrop of Cold War tensions. The so-called third world was constructed as a realm of universal poverty and backwardness with development requiring an injection of external knowledge and resources. The establishment of SAPs assumed particular importance in the context of the debt crisis in the 1980s and 1990s. Since the mid-1990s, the emphasis on economic reform has given way to a focus on poverty reduction, generating a new consensus among development agencies, governments and NGOs. In reality, however, the World Bank and IMF continue to dictate the terms of assistance, maintaining a regime of 'conditionality', where developing countries have to meet their specific requirements before qualifying for assistance. A commitment to economic liberalization, openness and adjustment remains central alongside the new focus on poverty reduction. The responsibility for alleviating poverty is allocated to poor countries themselves, while broader structural constraints such as the global rules underpinning trade and investment are largely ignored.

Despite some improvements over the last 30 years, global inequalities remain massive. Within the Global South itself, processes of uneven development have resulted in a growing divergence between regions. Sustained growth in East Asia and improvements in southern Asia contrast with economic decline in sub-Saharan Africa and stagnation in Latin America. According to the UNDP (2005, p.34), 'the war against poverty has witnessed advances on the eastern front, massive reversals in sub-Saharan Africa and stagnation across a broad front between these poles'. In recognition of the problem of global poverty, world leaders agreed the MDGs in 2000, committing themselves to reaching a series of targets by 2015. While considerable progress has been made towards reaching these targets, the reliance on global aggregates masks substantial regional differentiation and the effects of the global economic crisis and increased food and fuels costs have impeded progress since 2006. A new trade deal seemed little closer in 2010 than at the start of the Doha 'Development' Round in 2001, while the Copenhagen Conference failed to produce a binding international agreement to reduce carbon emissions, reflecting underlying conflicts of interests between rich and poor countries. This returns us to the insight that uneven development is not simply a reflection of internal failures in poor countries, but is an expression of the relationships between the core and periphery of the global economy. As such, efforts to reduce poverty require action at the global level to create a more level playing field for sustainable development, in addition to policy shifts within developing countries themselves.

Exercise

Select a developing country. Review its experience of development since the 1980s, using the websites listed below as starting points for your research.

Assess the country's economic performance over time (referring to figures on growth, employment, income, investment, education, health, etc.). What have been the key forces shaping development? What development strategies has the country adopted? Who has set the development agenda – the government, the World Bank/IMF, foreign MNCs, NGOs or domestic interests (e.g. landowners, industrialists, traders, workers, peasants)? Have development strategies been informed by any of the theories discussed in section 8.3? Are there any examples of local social movements resisting particular development projects? How would you describe the country's development prospects at the present time?

Key reading

Desai, V. and Potter, R. (eds) (2008) *The Companion to Development Studies*, 2nd edn, London: Arnold.
Contains a number of short overviews of various development themes and issues written by leading authorities. Extremely comprehensive, dealing with a wide array of topics including the meaning of development, the main theories, rural development, urbanization, industrialization, the environment, gender and population, health and education, violence and instability and key agents of development.

Mohan, G. (1996) 'SAPs and development in West Africa'. *Geography*, 81: 364–8.
A brief and accessible review of the main impacts of structural adjustment programmes in West Africa. Provides a concise definition of SAPs and highlights their effects on trade and

economic policy, poverty, health and welfare, employment and gender roles, resource extraction and the environment and the state and democracy.

Potter, R., Binns, T., Elliott, J.A. and Smith, D. (2008) *Geographies of Development*, 3rd edn, Harlow: Pearson.
Probably the best contemporary textbook on development issues in geography. Provides a comprehensive and integrated treatment of development which covers the main theories and the historical legacy of colonialism; assesses the role of population, resources and key institutions; and examines spaces of development within developing countries.

Power, M. (2008) 'Worlds apart: global difference and inequality'. In Daniels, P.W., Bradshaw, M., Shaw, D. and Sidaway, J. (eds) *Human Geography: Issues for the 21st Century*, 3rd edn, Harlow: Pearson, pp.180–202.
An engaging discussion of development issues, emphasizing the scale of global differences and inequalities. Covers the key theories, institutions and history of development, highlighting its unequal impact on households and individuals within developing countries.

Routledge, P. (2005) 'Survival and resistance', in Cloke, P., Crang, P. and Goodwin, M. (eds) *Introducing Human Geographies*, 2nd edn, London: Arnold, pp.211–24.
A useful account of the growth of local social movements against particular projects. Covers two of the three examples outlined in section 8.5, stressing both the place-based origins of the movements and the increasing global links between them.

Willis, K. (2005) 'Theories of development', in Cloke, P., Crang, P. and Goodwin, M. (eds) *Introducing Human Geographies*, 2nd edn, London: Arnold, pp.187–99.
A very clear summary of the main theories of development, covering the modernization school, dependency theory, neoliberalism and grassroots development. Contains useful bullet point summaries of each approach.

Useful websites

http://www.oxfam.org.uk/index.htm
The website of one of the leading British development NGOs, providing details of its history and current strategies, and highlighting some of the key projects and campaigns it is undertaking. Contains information on how you can support Oxfam's activities through volunteering and fund-raising.

http://hdr.undp.org/
The UNDP site provides access to information about human development in the form of current statistics, the annual *Human Development Report* and details of how these measures such as the Human Development Index are compiled. Contains interactive maps and tools illustrating human trends and outcomes.

http://www.worldbank.org/
The official World Bank site contains a wide range of information, reports and projects. Key resources include the annual *World Development Report*, World Development Indicators and the speeches of the President and other senior figures, giving an indication of current thinking. Data profiles of countries and country groupings are also available.

http://www.mstbrazil.org
The website of Brazil's Landless Workers' Movement, the *Movimento Sem Terra* (MST). Contains a range of information about the movement's history, objectives and campaigns.

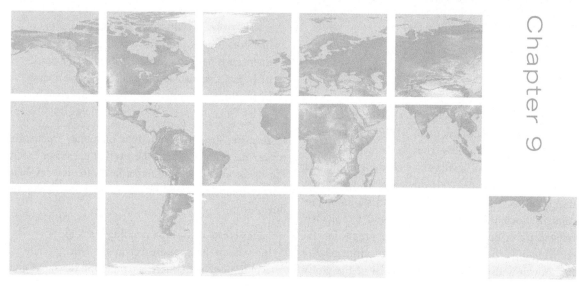

Chapter 9

The uneven geographies of finance

Topics covered in this chapter

- The role and functions of money.
- Money and finance under capitalism.
- Geographies of money.
- Financial globalization.
- Financial crises and cycles.

Chapter map

We begin the chapter with a discussion of the role and functions of money; before moving on to explore the development of money under capitalism. We then consider the changing geographies of money and the emergence of a global financial system, emphasizing how financial globalization has accentuated patterns of geographical unevenness. The latter parts of the chapter explore the increasing tendencies towards 'crisis' under financial globalization and examine how these resulted in the recent financial crisis and economic downturn.

9.1 Introduction

The financial sector is one of the fundamental building blocks for the global capitalist economy. It plays a critical role as the provider of money and credit for new investment and ultimately economic growth but, at the same time, as the recent financial crisis has demonstrated, it is also prone to considerable volatility. This volatility appears to have been accentuated in recent years by the revolution in information technology and the development of an increasingly globally integrated economy. These factors have together resulted in a situation of financial hypermobility, whereby billions of dollars are traded instantaneously within and between the leading financial centres on the world's currency markets. For all this, geography still matters in understanding the workings of money and finance. The global financial system is characterized by **uneven development** involving the increased concentration of key activities in leading financial centres, on the one hand, and the marginalization and exclusion of poorer countries and regions, on the other.

9.2. The role of money

Money, finance and credit are critical to the functioning of the global capitalist economy. Although money as a medium of value and exchange dates back to the earliest large human civilizations in the Middle East and Egypt more than 3,000 years ago, and was clearly important as a signifier of wealth and status in pre-capitalist societies, its pre-eminence under capitalism reflects its critical role in providing credit for the securing of surplus value or profit, and ultimately for the growth and geographical expansion of the global economy. If new technology and innovation provide the catalyst for change, and profit is the motor for the expansion of the capitalist economy, money and its credit form are the lubricants.

9.2.1 The functions of money, its regulation and the role of central banks

As we indicated in section 3.2.3, money is a measure and store of value. It can be anything that is accepted as payment for goods. In ancient human societies, particular commodities or items that were more durable were often used as forms of money: grain, weapons and even cattle were accepted as forms of exchange before precious metals such as silver and gold were used in the minting of coins.

In addition, money has taken on an increasingly important second function as a medium of exchange and circulation. Money is crucial to the functioning of the economy, allowing the trading of goods but also providing credit, whereby those individuals and groups with a surplus can provide loans for new investments and products. Credit in this way becomes critical to economic growth and the evolution of economies. As one economist puts it:

The function of the financial system as a whole is financial intermediation, the pooling of financial resources among those with surplus funds to be lent out to those who choose to be in deficit, that is to borrow ... With financial intermediation, investors in new productive activities do not themselves have to generate a surplus to finance their projects; instead the projects can be financed by surpluses generated elsewhere in the economy.

(Dow, 1999, p.33)

In practice, these two functions of money frequently come into conflict, where the over-extension of credit during economic boom times leads to the devaluation of financial assets during an economic crisis or downturn:

banks create credit by extending loans to individuals, firms and government. During economic growth, there is temptation for banks and other lending institutions to create large volumes of credit. Any sudden loss of confidence in the quality of credit, as happened in the Wall St crash of 1929 ... can trigger financial panic and devaluation as depositors, investors and financial institutions flee from credit monies and seek out safer havens.

(Pollard, 2008, p.359)

A third related function of money is as a universal equivalent. In other words, money enables us to assess the relative value of commodities. This, as Marx recognized, has geographical significance, for it allows the potential annihilation of space over time. As we show later in the chapter, the transformation of money from material to electronic form has dramatically accelerated this time–space compression (Harvey, 1989a) as part of the globalization processes.

The tensions and contradictions in the functions of money have resulted in governments setting up regulatory authorities and/or **central banks** to manage the conflicts inherent in the restriction and expansion of money. In the UK, this role has been carried out traditionally by the Bank of England, a formerly independent institution that was nationalized in 1946, and given operational independence in 1997. In the US, central bank duties are undertaken by the Federal Reserve (or 'The Fed'), while the establishment of the single currency (the euro) in the EU required the setting up of the European Central Bank. Conventionally, central banks use interest rate policy to try to influence the supply of money and credit in an economy: raising interest rates makes borrowing money more expensive and therefore is a means of restricting credit and vice versa. Ultimately, central banks are known as 'lenders of the last resort' in the sense that they have the key strategic role of ensuring that the supply of credit is maintained during periods of crisis.

9.2.2 The operation of money in a capitalist economy

The classic mainstream economics explanation of the operation of money is that money is created largely by central banks and then loaned out through commercial banks, who are required to keep a particular ratio of reserves to loans. Money supply is therefore seen as 'neutral' and not a factor in the creation of booms or economic crises. However, the mainstream view is disputed by 'heterodox' economists who stress both the role of credit emanating from commercial banks in the creation of money and the dangers of imbalances occurring between credit and monetary deposits. Furthermore, many Marxists, post-Keynesians and other heterodox scholars argue that, rather than money being somehow neutral, it is both active in the shaping of the economy and socially constructed (Ingham, 1996, 2004; Gilbert, 2005), through the activities of banks and financial institutions and their decision-making power to mediate the flow of funds going to businesses and individuals (Box 9.1).

Marx long ago recognized that commercial banks and investment houses control the circulation of funds and have the power to 'create capital' through provision of credit. As the following quote demonstrates, Marx also recognized that, instead of money markets operating through perfect competition, over time the competitive dynamics of the market place would lead to a concentration of finance into fewer and larger organizations with immense market power:

Box 9.1

Alternative theories of money supply: the monetarists versus the post-Keynesians

Following the work of John Maynard Keynes, a school of post-Keynesian economics developed in the UK and US from the 1940s onwards. Not to be confused with the neo-Keynesian and new Keynesian schools, post-Keynesian economists, who include distinguished figures such as Joan Robinson, Michal Kalecki and Nicholas Kaldor, dismiss the idea that the economy will have a tendency to return to full employment and market equilibrium without government intervention. They place a strong emphasis upon the role of aggregate demand in the functioning of a market economy and in the role government should play in intervening in and stabilizing the economy during economic downturns.

Post-Keynesians have been particularly influential in relation to understanding the workings of money in the modern capitalist economy, with their emphasis upon the way banks expand the supply of money through credit. Post-Keynesian monetary circuit theory suggests that money is generated internally within the economy and therefore cannot be controlled by central banks themselves restricting the money supply, a policy associated with the monetarist theories of US economist Milton Friedman. Friedman's basic point was to reiterate the mainstream perspective, known as the 'quantity theory of money', whereby the value of money is like any other commodity, and varies in accordance with the quantity in circulation. To control inflation, and the devaluation of money, governments needed to restrict its supply. The failure of monetarist strategies to control inflation in the early 1980s led to most governments and central banks implicitly accepting the post-Keynesian critique by using the setting of interest rates as a means of attempting to control inflation and the flow of money in an economy. The financial crisis and subsequent responses to it seem to bear out the post-Keynesian view of the futility of monetarist policy and the dominant role played by commercial banks in the development of credit money (see below).

Talk about centralisation! The credit system, which has its focus in the so-called national banks and the big money-lenders and usurers surrounding them, constitutes enormous centralisation, and gives to this class of parasites the fabulous power, not only to periodically despoil industrial capitalists, but also to interfere in actual production in a most dangerous manner – and this gang knows nothing about production and has nothing to do with it.

The Acts of 1844 and 1845 are proof of the growing power of these bandits, who are augmented by financiers and stock-jobbers.

(Marx, 1894, pp.544–5, cited in Keen, 2009)

Furthermore, banking and investment activities can have damaging effects on the rest of the economy, for example in promoting speculative activity over investment in more productive capital, or, in times of financial crises (as is the case in the current period), restricting finance to perfectly well-functioning firms that prevents normal business activities and can cause job losses and closures.

Reflect

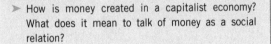

➤ How is money created in a capitalist economy? What does it mean to talk of money as a social relation?

9.3 The changing geographies of money

From the outset, money and finance have been tied to the international expansion of capitalism as an economic system. However, we can recognize various stages in the development of the geography of money since the advent of the modern capitalist economy (Table 9.1). In the early phases of capitalist development in the late eighteenth century, industrial innovations were largely financed by local and regional banks. As capitalism developed further, growing concentration of the banking sector led to the emergence of national financial systems. In practice, this varied across different countries. For example, in countries such as the UK, the financial sector became heavily oriented around London and a national system whereas in Germany the banking system continues to be more decentralized with regional banks (often owned by regional governments) continuing to play an important role in the financing of companies.

The growth of more nationally oriented financial systems resulted in a growing internationalization of financial activities as banks took advantage of investment opportunities in overseas markets. Deindustrialization domestically often went alongside successful geographical diversification in foreign markets. Since the 1970s, the financial system has undergone an even more

Table 9.1 Geographical development of the financial system

Geography	Local/regional	National	Global
Periodization	Eighteenth and early nineteenth centuries	Late nineteenth century to 1970s	1970s onwards
Phase of development	Industrialization	Mature national economy, growth of services	Post-industrial, economy of flows, hyper-capitalism
Objectives of finance	Local manufacturing firms	National firms but also growing international investments	Increasing separation between financial and 'real' economy; growth of derivatives, futures, SIVs, etc.
Financial characteristics	Emergence of regional and national banks	Concentration of banking in national institutions, growth of capital markets	Growing internationalization of banks, emergence of hedge funds
Type of finance, capital	Loans, risk capital, profit	Increasing importance of share capital	Capital and credit markets

Source: Adapted from Martin 1994, Table 11.1, p.256.

profound set of changes, as national banking structures have become increasingly integrated with global financial flows. As part of this, there has been an increasing separation between the financial system and what we might term the 'real' economy of productive activities. An increasing amount of financial activity is tied to speculative investment and new financial products such as derivatives (see below).

9.3.1 The globalization of the financial system

One obvious indicator of the increasing globalization of the financial sector has been the massive increase in global financial flows; daily turnover on the major foreign exchange markets grew from over $10–20 billion in 1973, about twice the value of trade, to an average of $1.5 trillion in 2004, over 90 times the value of trade (Pollard, 2008, p.366). In this way, it is possible to talk of a virtual economy or what Thrift (2002) refers to as a 'phantom state' operating alongside the more 'real' economy in which products are manufactured in specific locations. This **hyper-mobility of capital** – made possible, in part, by advances in electronic technology and the ability to shift billions of dollars from one part of the globe to another at the touch of a button – is seen as having its own disciplining effect on national economies and states, encouraging policies that are pro-business and fiscally conservative (e.g. low taxes or low inflation) in order to attract mobile funds and prevent capital flight.

Three factors have been critical to the globalization of the financial system.

First, the deregulation of financial activities through neoliberal policies. Conventionally, the financial system was tightly controlled by governments through a range of detailed rules and restrictions that separated different kinds of activities – for example, banking and insurance – and limited entry of firms into the financial sector, particularly foreign ones. But an unprecedented 'crumbling of walls' has taken place since the 1970s, through major deregulation programmes (Dicken, 2007). These involved the abolition of exchange and capital controls – which essentially limited trade in foreign exchange markets and the volume of capital that could be exported – and the removal of barriers

between different financial activities and limits on entry of foreign firms:

> A series of changes in the United States since the 1970s has both eased the entry of foreign banks into the domestic market and facilitated the expansion of US banks overseas … In the United Kingdom, 'the so-called 'Big Bang' of October 1986 removed the barriers which previously existed between banks and securities houses and allowed the entry of foreign firms into the Stock Exchange. In France the 'Little Bang' of 1987 gradually opened up the French Stock Exchange to outsiders and to foreign and domestic banks.
>
> (Dicken, 2007, pp.388–9)

In the US, a final important piece of legislation was abolition of the Glass-Steagall Act (1933) in 1999, which removed most of the remaining barriers between retail and investment banking in the US, allowing more speculative activities into markets such as housing and real estate.

Second, the development of advanced communications technologies. Computers have transformed payment systems, through allowing electronic money to be moved around the world at great speed. The introduction of chips (microprocessors) allows customers to pay for their purchases using plastic in the form of credit and debit cards. Technological developments have facilitated 24-hour trading on a global basis, exploiting the overlap between the trading hours of the world's major financial centres. In essence, we live in an age of increasingly dematerialized electronic money or what some refer to as a 'cashless' society (Martin, 1999). Above all, ICT has greatly increased the pace of financial activity, allowing almost instantaneous trading between distant centres. Finance is a crucial agent of time–space compression, obliterating the friction of distance in terms of the rapid movement of capital and information across space (Harvey, 1989a).

Third, the number of financial and monetary products has grown massively. Of particular significance is the rise of new financial instruments known as **derivatives**, which are traded and have made it easier to move money across the globe. The term encompasses a highly complex range of instruments, but these are essentially 'contracts that specify rights/obligations

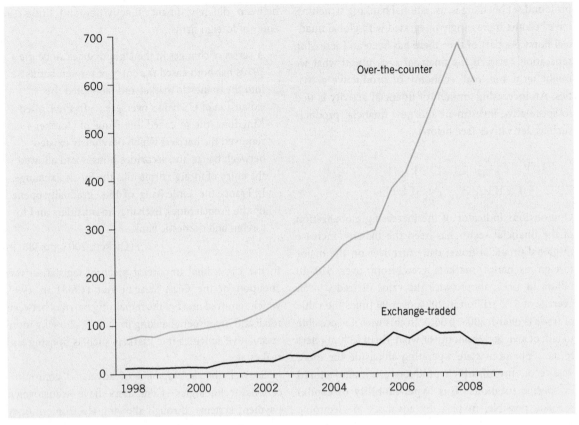

Figure 9.1 International derivatives markets
Source: Bank for International Settlements.

based on (and hence derived from) the performance of some other currency, commodity or service used to manage risk and volatility in global markets' (Pollard, 2005, p.347). Such risk and volatility are associated with changes in prices of commodities, currencies and instruments such as interest rates. As Figure 9.1 shows, after a period of continuous and rapid expansion from the mid-1990s, the value of the derivatives trade fell sharply in 2008 as the financial crisis struck.

The basic forms of derivatives include: 'simple futures (where an agreement to buy a commodity at a given date and given price is made, allowing purchasers to purchase a degree of certainty)' (Tickell, 2000, p.88); swaps (where the parties to an agreement exchange particular liabilities, for example interest repayments, usually through an intermediary such as a financial institution); and options 'which, in exchange for a premium, allow buyers to trade assets

at a predetermined price at a point in future') (ibid.). Derivatives are traded in two ways: on organized and regulated exchanges such as the London International Financial Futures Exchange or the Chicago Mercantile Exchange and 'over-the-counter' through direct agreements between the parties concerned. Since the early 1980s, derivatives have become symbolic of a highly monetized and financially driven world economy, in which ICT has rendered financial instruments increasingly mobile and separate from flows of material goods (Tickell, 2003).

9.3.2 Finance and uneven development

Although globalization has brought the dramatic compression of space and time to the financial sector, it has not resulted in the end of geography (cf. O'Brien, 1992).

Instead, it has resulted in processes of both agglomeration and geographical exclusion, reflecting a broader tendency towards uneven development. In this context, place still matters – or in some senses is becoming more important – to the operation of the financial system. As French *et al.* have recently suggested: 'an understanding of the historical geography of money and finance is essential to an apprehension of the volatile financial world in which we currently find ourselves' (2009: p.289).

While world financial markets may be global in their scope and almost frighteningly hyper-mobile in

Table 9.2 Average daily turnover of equities trading 2005 (millions of $US)

Exchange	Average daily turnover	Region % of world total	Exchange	Average daily turnover	Region % of world total
North America		50.0	Swiss Exchange	3,817.7	
Amex	2,394.1		Vienna (Austria)	187.4	
Bermuda	0.3		Warsaw (Poland)	121.2	
Mexican Exchange	222.3		**Middle East**		0.2
NASDAQ	40,026.7		Tehran (Iran)	32.3	
NYSE	56,052.6		Tel Aviv (Israel)	199.9	
Toronto (TSX Group)	3,587.6		**Africa**		0.4
South America		0.4	JSE South Africa	803.9	
Buenos Aires (Argentina)	27.2		**Asia/Pacific**		17.0
Lima (Peru)	10.6		Australian	2,668.2	
Santiago (Chile)	75.2		Colombo (Sri Lanka)	4.6	
São Paulo (Brazil)	663.8		Hong Kong	1,879.6	
Europe		31.0	Jakarta (Indonesia)	171.3	
Athens (Greece)	260.5		Korea	4,862.1	
Budapest (Hungary)	95.5		Malaysia	208.9	
Deutsche Bourse (Germany)	7,452.5		Mumbai (India)	633.4	
Euronext (Amsterdam, Brussels, Lisbon, Paris)	11,308.2		National SE India	1,253.7	
Irish	266.5		New Zealand	83.2	
Ljubljana (Slovenia)	5.4		Osaka (Japan)	883.4	
Istanbul (Turkey)	790.8		Philippines	28.4	
Italian exchange	5,053.4		Shanghai (China)	985.6	
London (UK)	22,530.6		Shenzhen SE (China)	637.4	
Luxembourg	1.3		Singapore	465.8	
Malta	0.6		Taiwan	2,370	
OMX[1]	3,760.6		Thailand	390.4	
Oslo (Norway)	926.4		Tokyo (Japan)	18,292.7	
Spanish Exchange (BME)	6,141.6				

[1] OMX includes exchanges in Copenhagen, Stockholm, Helsinki, Riga, Tallinn and Vilnius.
Source: Adapted from World Federation of Exchanges, at http://www.iasplus.com/stats

Table 9.3 Ranking of global financial centres, September 2010

Rank	City
1	London
2	New York
3	Hong Kong
4	Singapore
5	Tokyo
6	Shanghai
7	Chicago
8	Zurich
9	Geneva
10	Sydney
11	Frankfurt
12	Toronto
13	Boston
14	Shenzhen
15	San Francisco
16	Beijing
17	Washington, DC
18	Paris
19	Taipei
20	Luxembourg

Source: Z/Yen Group, The Global Financial Centres Index 8, p.9.

their effects, they are centred upon a few key nodes in dominant world cities. The massive asymmetries of global finance are illustrated in Table 9.2, which shows the concentration of equities trading in North America, western Europe and Asia and the virtual absence of the rest of the world. Geographical concentration is even more profound when one considers the dominance of a small number of cities in their respective world regions: particularly London and New York, but also Hong Kong, Singapore, Tokyo and Shanghai, as the financial importance of Asia grows (Table 9.3).

The paradox of globalization in this sense is that financial markets are dominated by the decisions of a small number of market traders and analysts, operating out of a select group of cities. As Thrift (2002, p.39) puts it: 'what we find is a system where people are often in interaction with only four or five other people at a time – on a trading floor or a bank branch'. These geographies of financial power are deeply entrenched. As French *et al.* (2009) note, London and New York have been the dominant financial centres in the global economy for over a century (Figure 9.2), despite various crises and downturns.

While groups such as financial traders, brokers, investors and executives have benefited hugely from financial globalization, not least through the payment of massive bonuses linked to financial transactions (see below), most of the world's population has been

Figure 9.2 Wall Street
Source: D. MacKinnon.

Box 9.2

Geographies of financial exclusion

Financial exclusion can be defined as 'these processes that serve to prevent certain social groups and individuals from gaining access to the financial system' (Leyshon and Thrift, 1995, p.314). It is a key part of the broader problem of poverty or social exclusion defined by low incomes, unemployment, a lack of skills, bad health, low educational attainment, high-crime environment and family breakdown (Kempson *et al.*, 2000, p.7). Financial exclusion is based on income having a cumulative effect, compounding and accentuating the difficulties facing low-income individuals and groups, widening the group between them and the more affluent sections of society. It is estimated that 1.5 million people in the UK, for instance, do not have a bank account, while a further 4.4 million (20 per cent) exist on the margins of the financial system, operating little more than a basic bank account (ibid.). There is an important geographical dimension to financial exclusion in the sense that people are denied access to financial products and services because of the social and economic characteristics of the area in which they live, attracting the interest of economic geographers

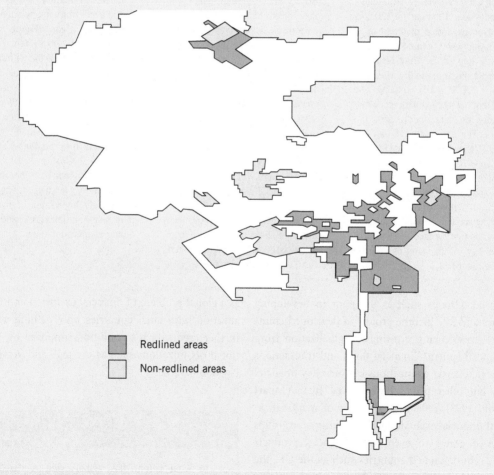

Redlined areas

Non-redlined areas

Figure 9.3 Redlining in Los Angeles
Source: Pollard, 1996, pp.1209–32.

Box 9.2 (continued)

(Leyshon and Thrift, 1995). Inner-city neighbourhoods in the US and UK are typically characterized by financial exclusion.

A key point to appreciate is that lending to the poor is a higher-risk enterprise for financial institutions than lending to more affluent groups, since a low income means that the former are less able to repay loans. In countries such as the UK and US, the 1990s saw a redrawing of the boundaries of financial markets in 1990s (Leyshon and Thrift, 1995). A spate of business insolvencies, mortgage defaults and personal bankruptcies in the recession of the early 1990s, following an expansion of credit in the 1980s, plunged the financial services industry into crisis. They responded through a 'flight to quality' (Kempson et al., 2000, p.16), moving away from lending to high-risk customers to focus on safe middle-class customers. The same process has been evident following the financial crisis of 2007–08 as lenders have made it far more difficult for people to access loans and mortgages. Due to its sheer magnitude, this 'credit crunch' is affecting some middle-income groups as well as the low paid.

Two key mechanisms of exclusion can be identified.

1. The closure of bank branches. Established banks have sought to reduce costs in face of increasing competition, for example from supermarkets and large retailers offering services such as 'cash back,' and through the expansion of telephone and internet banking particularly. One of the principal ways in which banks have reduced costs is by closing a large number of branches, also fuelled by mergers between banks. A 36 per cent reduction in the number of banks and building societies in Britain occurred between 1989–2003 (French et al., 2008). The pattern of closure was geographically uneven, focusing particularly on low-income areas in urban environments.

2. **'Redlining'** which is defined as cases 'where goods and services are made unavailable, or made available only on less than favourable terms, to people because of *where they live*' (Squires quoted in Pollard, 1996, p.1209). 'Redlining' is a product of the risk assessment procedures employed by financial institutions where computer-based analysis has replaced the personal assessment of bank managers and insurance underwriters. Contemporary methods classify geographical area-based postcodes according to a variety of social and economic factors. Typically, low-income inner city areas are regarded as high risk and are 'redlined' or marked out on a map. As a result, people living within these areas are denied access to credit and insurance services or, more commonly, are charged very high premiums. This process is most advanced in the US, where mainstream financial infrastructure has been withdrawn from many deprived inner city areas (Figure 9.3), although it is also evident in the UK. In such areas, the withdrawal of mainstream financial services has resulted in the growth of unregulated '"second tier" financial sector[s]: pawnbrokers, moneylenders, cheque-cashing firms and hire purchase shops' (Tickell, 2005, p.250). Such enterprises tend to be highly exploitative, lending money at cripplingly high rates of interest and using intimidation and violence against defaulters.

left behind. Groups such as the poor in developing countries and low-income groups in developed countries are experiencing growing marginalization from the financial system. Financial flows and transactions have created closer connections between key financial centres but other places have moved further apart (Leyshon, 1995), reflecting processes of marginalization and **financial exclusion** where low-income groups are denied access to even basic financial products (Box 9.2). In developed countries such as the UK, the problem of financial exclusion has attracted growing attention from governments in recent years The **debt crisis** afflicting many developing countries can be seen as a global problem of financial exclusion and marginalization with such countries only gaining access to further credit on the conditions imposed by international organizations such as the IMF and World Bank (section 8.3.3).

Reflect

➤ To what extent has globalization rendered the financial system 'placeless'? In what ways does geography continue to shape financial markets?

9.4 Financial crises and cycles

As we observed in earlier chapters, capitalism is prone to uneven development in time and space. Periods of economic growth are punctuated by periods of stagnation, decline and even crisis as over-expansion of the system eventually leads to the point where firms are producing more than the market can absorb. The financial system plays an important part in this uneven dynamic, with the historical evidence suggesting that it is prone to the same kinds of crises and cycles that characterize the rest of the economy (Minsky, 1975, 1986; Wolfson, 1986). Indeed, **financial crises** are often constitutive of broader economic crises, because of the critical role of the financial sector in providing credit for development. An oversupply of credit can lead to excessive speculation, increasing levels of debt and over-inflated asset prices (e.g. housing). Such circumstances create financial bubbles that eventually

burst in a variety of different circumstances as the laws of financial gravity take effect.

Globalization appears to have greatly exacerbated the crisis tendencies in the financial system with an acceleration of crises in time and space. Since 1980, financial crises have been 'both endemic and contagious' (Harvey, 2005, p.94). Although much attention has recently focused on the 2007–08 financial crisis, which began in the US and quickly spread to the rest of the world (see below), on average 25 per cent of the world's countries suffered some form of banking crisis every year between 1986 and 2001 (Palma, 2009, p.849). The increasing degree of financial mobility in the world economy and the deregulation of economies by national governments following neoliberal policies seem to be generating greater instability in the world economy, with international investors now able to shift funds from one country to another in search of higher returns. This has had devastating effects in particular places. As David Harvey succinctly puts it: 'the "herd mentality" of the financiers (no one wants to be the

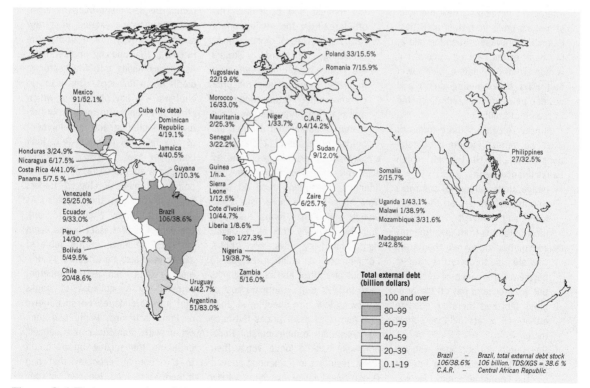

Figure 9.4 The geography of the debt crisis in the 1980s
Source: Corbridge, *Debt and Development* in Harvey, 2005, p.95.

last one holding on to a currency before devaluation) could produce self-fulfilling expectations' (2005, p.94). The opening up of an economy to greater foreign trade can lead to a gradual influx of speculative investment, which in the short term can bring prosperity. However, when international investors lose confidence in an economy there is a rapid outflow of such 'hot money', leading to a financial crisis, the collapse of the banking system and knock-on effects for the rest of the economy.

A case in point was Mexico, which, as a major oil exporter, suffered an economic and financial crisis in 1982 due to the collapse in the oil price. The government also defaulted on the debts (largely with US financial institutions) that it had run up in the economic downturn of the 1970s. Although Mexico was at the epicentre of the crisis, it spread to many other developing countries (Figure 9.4) which had been encouraged to borrow heavily from western financial institutions in the 1970s to finance development projects, only to find themselves unable to service their debts in the early 1980s (Harvey, 2005).

In the second half of the 1980s, and under pressure from the IMF to address its massive debts, the Mexican government cut its tariff barriers to overseas trade and privatized 90 per cent of its state-owned enterprises

Box 9.3

Minsky's financial instability thesis

Minsky follows the post-Keynesian view in recognizing that banks are profit-driven institutions that have an interest in developing new methods of making profit from their activities. Financial innovations therefore are at the heart of the banking system if it is not strictly regulated: banks will find ways of expanding credit and develop new forms of credit financing in search of profits.

Minsky recognizes three forms of what he terms 'income–debt relations' (1992, p.6) through which banks finance credit:

➤ hedge units refers to borrowers who are able to pay off the initial loan and interest from their income returns from investment;

➤ speculative units are borrowers who are able to cover the interest payments from income received, but are unable to pay off the principal debt and therefore need to either sell assets held or borrow more;

➤ Ponzi units are borrowers who are unable to pay either the interest or the principal from income returns on investment. In this

situation, the borrower must rely on the appreciation of the value of the underlying asset to repay the interest and principal. This often requires the selling on or further borrowing of money which, according to Minsky, 'lowers the equity of a unit, even as it increases liabilities and the prior commitment of future income. A unit that Ponzi finances lowers the margin of safety that it offers the holders of its debts' (1992, p.7).

Minsky recognized that, over time, as economic boom conditions develop, financial institutions tend to shift from hedge to the more risky speculative and Ponzi forms, meaning that the economy acts as a 'deviation amplifying system' (ibid.). Furthermore, if inflation begins to get out of control and central banks raise interest rates, the extra cost of borrowing will turn speculative units into Ponzis. If asset prices fall once an investment bubble bursts, Ponzi borrowers can no longer repay their loans, resulting in a financial crisis if Ponzi-style finance has been widely used across the system.

Minsky's financial cycle model recognizes five stages: displacement, boom, euphoria, profit-taking and panic. Displacement happens when investors become excited in a new sector of activity; for example, the internet, the housing market, etc. As money floods into the sector, a credit boom develops. This leads to euphoria – a key phase – in which credit is extended to more risky borrowers, including the development of new financial products (e.g. junk bonds in the 1980s, 'sub-prime' mortgages in the 2000s). At some point, the more intelligent traders begin to cash in their profits. As more institutions attempt to sell, the danger is that asset prices begin to fall. Confidence in the value of assets may also be affected by dramatic events such as the apparent vulnerability or collapse of banks that have overloaded on speculative or Ponzi schemes. Panic can then set in with dramatic consequences, including the seizing up of credit to the 'normal' economy that had hitherto remained innocent of any dealings with the speculators.

(Dicken, 2003a, p.192). The result was a dramatic influx of foreign investment, which had the effect of driving up the value of the peso to levels that made much of Mexican industry uncompetitive. Eventually the government was forced to devalue the currency, leading to an outflow of funds and a new financial crisis in 1995. In both cases, the crises spread to other countries as financial markets became nervous about investments in developing countries. Major global financial crises have now affected every world region since 1980, the most devastating prior to the current downturn being those affecting Asian and former Soviet economies in the late 1990s.

9.4.1 Understanding the crisis tendencies in the global financial system

As recent events have shown, economic orthodoxy has a poor record in both understanding and predicting the tendency for financial markets to veer between boom and bust. Famously, in response to the 2007–08 financial crisis, the British Queen, Elizabeth II, who was estimated to have lost around £25 million of her personal fortune, asked the pertinent question to academics at the London School of Economics, 'Why did no one notice it?' (reported in *The Daily Telegraph*, 5 November 2008).

One response would have been that non-mainstream economists such as John Maynard Keynes and Hyman Minsky have long warned of the inherent instability of financial markets. After having been dismissed as a maverick during the 1980s and 1990s, Minsky's theorization of the tendency for unregulated financial markets to create havoc in the wider economy are now being taken seriously again. His 'financial-instability hypothesis' was first developed in the 1960s. Its central message is that the accumulation of debt by financial institutions, as a result of increasingly over-exuberant speculative activities, eventually produces a crisis which can have knock-on effects for the rest of the economy, with the potential to produce a full-blown recession (see Box 9.3). In the words of one recent commentator: 'bankers, traders and financiers periodically [play] the role of arsonists, setting the entire economy ablaze' (Cassidy, 2008).

Reflect

➤ Consider the relevance of Minsky's theory to the 2007–08 financial crisis.

9.5 The 2007–08 financial crisis and the global recession

Minsky's insights regarding the crisis tendencies of the financial system appear particularly prophetic in anticipating the 2007–08 crisis. At the time of writing (September 2010), the global economy has yet to recover from the worst economic downturn since the 1930s' Great Depression and a number of major financial institutions in North America and western Europe have either gone bankrupt or been taken over by the state. Only substantial government intervention across the world to prop up the banking system and stimulate demand has staved off a meltdown of the global economy. One key indicator has been a collapse in world trade, which declined by over 10 per cent in the period 2008–09 (Figure 9.5). While it is still too early to assess the full implications of the downturn, there is no doubting the scale of the crisis facing the global economy and its origins in the activities of the global financial sector.

What is significant about the current downturn – unlike other financial crises in the past 30 years – is that the effects are global, affecting all countries, although to varying degrees. The trigger for the crisis, beginning in late 2006, was the growing number of households in the US's so-called **'sub-prime' housing market** defaulting on their mortgages. Many of the mortgages provided by banks to this market were of the Ponzi type, since the borrowers could repay neither the interest nor the principal, relying on rising house prices, which lenders assumed would continue indefinitely. As 'sub-prime' borrowers defaulted, house prices started to fall, threatening a wider financial crisis. Thirty years ago, given the essentially regional nature of US banking, this might have been contained, even within the US, to a handful of states and regions. In 2007–08, however, because of

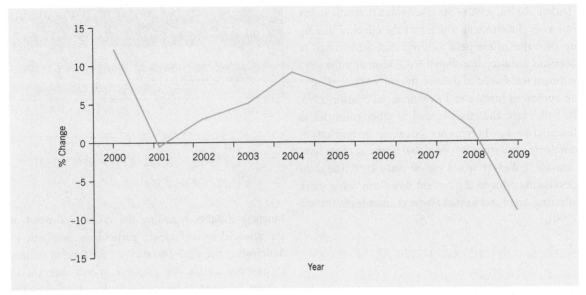

Figure 9.5 Annual change in world trade, 2000–09
Sources: WTO, *International Trade Statistics* (various).

the complex ways in which the US domestic housing market has become entangled in more geographically extensive financial networks, the crisis spread like a virus to the rest of the global economy.

The increasing integration of global financial markets has meant that financial institutions from many other countries were caught up in the US sub-prime housing crisis. A crisis that started out in the US in one sector of the housing market has rapidly spread to other sectors and other parts of the world. The German company, Deutsche Bank, for example, was one of the first firms forced to take action to safeguard two of its own property based US funds (Blackburn, 2008). Other European banks were similarly caught out by the US housing downturn. Leading French bank *Société Générale* lost $7 billion through the activities of one 'rogue trader' while Northern Rock became the first British bank to suffer a run on its assets, following news of its over-exposure to the sub-prime crisis. Only government intervention to nationalize the bank saved it from collapse (section 1.1).

The main reason for the globally contagious nature of the sub-prime crisis relates to the increasing connections between global financial centres and national retail banking markets. Whereas the domestic banking sectors in most countries had operated relatively autonomously from the world of high-finance up

until the 1980s, deregulation blurred the boundaries between international investment activities and the more mundane banking practices involved with serving local and national markets. Key financial centres such as London, New York, Singapore and Hong Kong as well as secondary banking centres such as Zurich, Frankfurt, Boston and Dublin were the key hubs through which national banking operations became embedded in wider global networks (Klagge and Martin, 2005).

9.5.1 The makings of a financial crisis

While the US sub-prime crisis was the trigger for a broader economic downturn, the financial crisis had deeper origins of which three interlinked factors are critical:

➤ the increasingly speculative and high-risk activities of the global financial system following the deregulation of the sector in the 1980s and 1990s, as predicted by Minsky;

➤ growing imbalances in world trade, most notably between the US and UK, running increasingly large trade deficits, and export-driven economies such as China, Germany and Japan;

➤ mainstream policy orthodoxy, which encouraged market deregulation on the basis that it would contain risk but actually achieved the opposite.

In the first place, what Crotty (2009) has termed the 'new financial architecture' of the post-1980 era of neoliberalism and market deregulation has resulted in a set of 'flawed institutions and practices of the current financial regime' (ibid., p.564) that have led to:

➤ greater risk-taking;

➤ increased financial innovations, which have become ever more complex and opaque rather than conveying 'perfect information' on risk and return;

➤ a lack of proper scrutiny and regulation, which meant that many of the more risky assets were held off balance sheet in an increasingly large 'shadow economy' (Blackburn, 2008);

➤ the wider geographical dispersal of highly risky and unstable assets to other institutions and countries because of the increasingly integrated global financial system. For example, the repackaging and selling on of mortgage liabilities from the sub-prime sector, although instruments such as credit default swaps (a form of derivative) became so complex, involving so many different parties and intermediaries, that it became impossible in many cases to make any rational assessment of asset worth and risk.

High-risk speculation by the financial markets was accompanied by a growing imbalance in the global economic system between debtor nations, such as the US and UK, running large current account and trade deficits, and export-based or creditor nations such as Germany, China and Japan. In the US and UK, economic growth from the mid-1990s through to 2007 was largely financed on the basis of a massive build-up of private debt, encouraged in particular by the US Federal Reserve's policy of keeping interest rates and the cost of borrowing low. Credit was largely provided by overseas governments buying US treasury and corporate bonds, especially China (Blackburn, 2008; Wade, 2010). Ultimately, though, this was not enough to prevent the dollar from falling in value, as market traders recognized the increasingly weak position of the US economy, leading to the Federal Reserve raising interest rates, which caused a hike in mortgage rates

and subsequent defaulting problems in the sub-prime market.

A further important factor behind the financial crisis has been the mainstream economic ideology of free markets and the view that markets are essentially stable phenomena that are best left to their own devices and self-regulation. Crucially, these ideas have been absorbed uncritically by most mainstream politicians and the business media to become hegemonic – to use the term popularized by the Marxist theorist Antonio Gramsci – as a kind of accepted 'common sense' of the age. The financial crisis has exposed the underlying fragility of these ideas, but mainstream orthodoxy may be more difficult to shake off because it tends to reinforce the powers of political and economic elites and allows continuing income inequalities (see Palma, 2009). Indeed, it is notable how quickly the economic discourse shifted, once immediate financial crisis had been averted, from blaming the banks for the financial crisis, to the need to reduce public sector spending which had grown dramatically in many countries between 2007–10, partly as a result of the rescue of the banks and the stimulus packages.

One the ways in which the financial crisis spilled over into a global economic recession in 2008–09 was through the so-called 'credit crunch', whereby banks dramatically reduced their lending to businesses and individuals in response to their losses and in order to start repairing their balance sheets. This reaction affects firms that afford loans as well as those that can't, as the reduced availability of credit – in stark contrast with its widespread availability during the boom period – led to a contraction of the economy. In addition, households had become heavily indebted through excessive borrowing to fund consumption, with UK household debt rising from 109 per cent of disposable income in 1999 to 176 per cent in 2007 (Dolphin, 2009, p.21). As a result, households are having to reduce expenditure to bring it into line with income, which may itself have been squeezed by the downturn. In contrast with previous downturns caused by the use of monetary policies (increased interest rates) to control inflation, the current recession is of the balance-sheet variety, caused by excessive debt and the subsequent credit crunch. Such recessions tend to be longer and deeper than those caused by monetary policy, as balance sheets take years to repair, meaning that the recovery is likely to be slow and faltering (ibid.).

9.5.2 The uneven effects of the recession

The economic downturn in the US as the world's leading consumer of other countries' exports has had dramatic knock-on effects on a global scale. World trade plummeted dramatically – faster than economic growth – as the leading consumer nations entered recession (Figure 9.5). While China's growth figures remain relatively strong, buffered by increased state spending on infrastructure projects, other poorer countries that are over-reliant upon raw materials and basic commodities for export have seen a dramatic downturn as prices collapse. The Commonwealth of Independent States (CIS) witnessed the biggest fall in GDP (−6.7 per cent in 2009) of any regional bloc as a result of its dependence on world markets for its vast oil and gas resources (IMF, 2009). Other countries that have suffered considerably include smaller states such as Iceland, Ireland and the Baltic states, which have undergone processes of market liberalization and opening up of their economies to global competition and financial flows since the late 1980s (Table 9.4). These are countries that had previously enjoyed rapid rates of growth during the early 2000s, all hailed to varying degrees as success stories of free market globalization.

In some cases, the economic downturn has been devastating in its suddenness and intensity. Countries such as Lithuania, Armenia and Romania, that had still been enjoying relatively rapid growth rates in 2007–08, suffered dramatic reversals as the crisis effects rippled out from the epicentres of the world's financial markets. In the developed world economies, Ireland and Iceland stand out as the countries most affected. In Ireland's case, overdependence in the domestic economy on a boom in construction and housing seems to have been the main factor in the country's devastating turnaround from 'Celtic Tiger' – as late as 2007 enjoying some of the highest economic growth rates in western Europe of 6 per cent – to one of Europe's weakest economies. Unemployment rose from 6 per cent in 2006 to 11.7 per cent in May 2009.

At a national level, the US sub-prime crisis itself has taken a particular geography. The states worst affected by housing repossessions or foreclosures as a result of people defaulting on mortgages are those that were home to the most extreme housing boom, such as Florida, and western states such as California and Nevada (Figure 9.6). The broader impact of economic downturn has also taken particular geographical forms. The most recent data available suggest that, outside of the states directly impacted by the housing market collapse, some of the worst effects have been felt in more traditional 'problem regions' such as the industrial Midwest, which suffered deindustrialization during the earlier recessions in the 1970s and 1980s (Figure 9.6). Similar trends have also been apparent in the geography of the recession in the UK (Box 9.4).

Table 9.4 Ranking of countries most severely affected by global economic downturn (2008–09 figures estimated by IMF)

Ranking	Country	GDP decline (2008–09)	GDP performance (2007–08)
1	Lithuania	−18.5	+3.0
2	Latvia	−18.0	− 4.6
3	Armenia	−15.6	+6.8
4=	Estonia	−14.0	−3.6
4=	Ukraine	−14.0	+2.1
6	Moldova	−9.0	+7.2
7	Seychelles	−8.7	+1.9
8	Iceland	−8.5	+1.3
9	Romania	−8.5	+7.1
10	Ireland	−7.5	−3.0

Source: IMF 2009, Tables A1–5, pp.169–76.

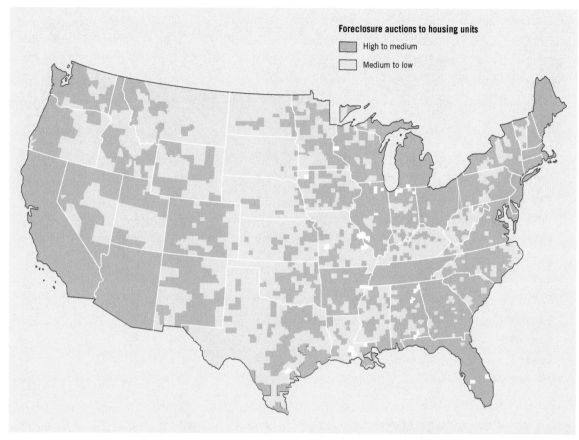

Foreclosure auctions to housing units

High to medium

Medium to low

Figure 9.6 The geography of foreclosures in the US as of May 2009

Box 9.4

The geography of the economic downturn in the UK

While many observers initially predicted that the financial crisis would produce a 'white-collar recession', affecting London and the south-east of England the most, it has been traditional industrial areas in the north of England, Northern Ireland and the West Midlands that have suffered the worst effects of the associated recession (see Table 9.5). Indeed, the knock-on effects of the crisis have impacted heavily upon manufacturing firms and regions that rely upon export markets and the availability of credit, which has been constrained as banks try to rebalance their books. Geographically, the recession has exacerbated longer-term patterns of spatial inequality. At the more local level too, analysis has revealed that within northern English city regions, the most deprived neighbourhoods have been hardest hit by recession (Dolphin, 2009).

Although it is too early to understand fully these geographical variations, there is an apparent paradox that those places at the heart of the financial crisis, such as London, appear to be more resilient than regions and cities that are still more reliant on manufacturing activity. The recession has largely reinforced existing geographies of uneven development, so that those places that have suffered through long processes of deindustrialization without creating sufficient new forms of employment seem to have been the most vulnerable to the recent downturn. There is certainly some more important work to be done by economic geographers in getting to grips with these issues.

Box 9.4 (continued)

Table 9.5 Regional unemployment rates in the UK (based on JobSeekers' Allowance claimants)

Region	Unemployment rate (%)		Change
	March 2008	March 2009	
Northern Ireland	2.7	6.2	3.5
North-east	3.9	7.3	3.4
West Midlands	3.4	6.7	3.3
Yorkshire & Humber	2.9	6.0	3.1
Wales	2.8	5.8	3.0
North-west	3.0	5.8	2.8
East Midlands	2.4	5.1	2.7
East	1.9	4.3	2.4
Scotland	2.5	4.7	2.2
South-east	1.4	3.5	2.1
London	2.6	4.7	2.1

Source: Dolphin, 2009, p.7.

Reflect

➤ Why has the 2008–09 recession had such a severe effect on traditional industrial regions in countries such as the UK and US. How does this compare with the geography of previous recessions?

9.6 Summary

Money and finance have always been critical to the expansion of capitalism, through the provision of credit for investment and development. Over time, the evolution of capitalism has seen dramatic changes in the nature and form of money and its relation to broader processes of economic development. Recent decades have seen a massive expansion of financial activities relative to the rest of the economy. This is associated with the growth of an increasingly globally integrated financial system, as financial services have become the most advanced form of time–space compression (see Harvey, 1989a). The increasing hyper-mobility of finance, facilitated by neoliberal deregulation and ICTs, seems to have unleashed a period of increasing instability in financial markets. Indeed, the financial volatility of the 1990s and 2000s can be seen as the latest chapter in the unstable history of capitalism, which has been characterized by cycles of growth and crisis (see section 3.2.4).

Rather than the annihilation of geography predicted by some (O'Brien, 1992), the globalization of the financial system has created distinct forms of geographical connection and differentiation. In particular, financial globalization has accentuated existing patterns of uneven geographical development, fuelling the growth of leading global cities, which act as key hubs in linking local and national financial systems to broader and increasingly dynamic global flows of money. By contrast, other parts of the global economy in both the Global North and South are becoming distanced and excluded from financial flows. Within the global economy as a whole, relations between the centres of financial power and poorer countries and regions have become increasingly uneven. This is also evident within individual countries such as the US and UK where the financial crisis and recession have had the most severe impact on traditional industrial regions

Exercise

Drawing upon two different countries, evaluate the effects of the 2007–08 financial crisis and associated recession in terms of economic growth, employment and geographical relations between different cities and regions.

1. How well do different national financial systems respond to the broader global crisis?

2. In what ways do the geographical effects of the recession vary between the two countries?

3. What kind of government policies are being used to deal with the economic downturn?

4. What are the geographical effects of these policies?

Key readings

Crotty, J. (2009) 'Structural causes of the global financial crisis: a critical assessment of the "new financial architecture"'. *Cambridge Journal of Economics*, 33: 563–80.
A good account of the forces behind the financial crisis. See also the other papers in the special issue of the *Cambridge Journal of Economics* by Wade and Palma in particular.

Dolphin, T. (2009) *The Impact of the Recession on Northern City-Regions.* Newcastle upon Tyne: IPPR North. Available at http://www.ippr.org.uk/ipprnorth/publicationsandreports/.
An excellent analysis of the impact of the recession on city-regions in the north of England, which contains a splendidly concise and accessible account of the financial crisis and recession.

French, S., Leyshon, A. and Thrift, N. (2009) 'A very geographical crisis: the making and breaking of the 2007-8 financial crisis'. *Cambridge Journal of Regions Economy and Society*, 2: 287–302.
A good early analysis of the geography of the financial crisis which links the cultural and discursive aspects of the crisis to wider economic and geopolitical processes.

Harvey, D. (2005) *A Brief History of Neoliberalism*, Oxford: Oxford University Press, Chapter 4.
An excellent account of the uneven development of neoliberalism and financial globalization and its geographical effects.

Pollard, J. (2008) 'The global financial system: world of monies', in Daniels, P., Bradshaw, M., Shaw, D. and Sidaway, J. (eds) *An Introduction to Human Geography*, 3rd edn, London: Prentice Hall/Pearson, pp.358–75.
A good overview of the workings of the financial system and its geographies.

Useful websites

http://www.imf.org
The website of the International Monetary Fund, the global organization responsible for securing financial stability. Useful in indicating how this body has responded to the financial crisis of 2007–08.

http://www.thecityuk.com/
The website of The CityUK, the body responsible for promoting the British financial services sector. Contains key facts and figures and up-to-date reports on the industry.

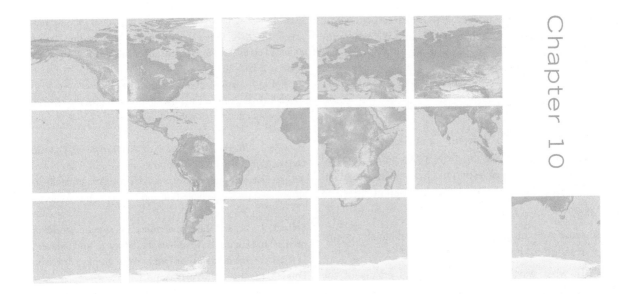

Chapter 10

Commodity chains and global production networks

Chapter map

The chapter begins by emphasizing how places are connected by flows of goods and services, among other things. We then go on to assess the global commodity chain (GCC) approach, highlighting the key distinction between buyer-driven and producer-driven chains. This is followed by a discussion of the global production networks (GPN) approach, which has been developed by economic geographers, identifying the main similarities and differences between the two approaches. In section 10.4, we consider the implications of the GPN's approach for understandings of regional development, involving the adoption of a relational perspective that stresses the interaction or 'strategic coupling' between regions and GPNs.

10.1 Introduction

In recent years, economic geographers and others have emphasized the importance of the wider social relations and networks in which places and regions are entangled. Flows of different kinds of material – goods, services, money, information and people – are of particular importance here, creating a range of connections between places. With globalization, the volume and intensity of these connections has increased considerably, focusing attention on the interactions between places and this global 'space of flows' (Castells, 1989). While we focused on flows on money in the last chapter, here we are concerned with flows of goods and services between places.

The production and consumption of particular goods and services or commodities typically involves a complex chain of actors, spread across different countries and regions. These include producers such as farmers or labourers in plantations, mine or plantation owners, sub-contractors, transport operators, distributors, retailers and consumers. The **global commodity chain** approach aims to capture these relations (Box 1.5). The production and distribution of commodities reproduce patterns of uneven development between places with low-value activities, often confined to poorer regions in the global south, while higher-value ones are typically located in wealthier places in the global north, reflecting, to a considerable extent, the legacy of colonialism. The process of **commodity fetishism** means that consumers in the Global North are typically concerned with the price and appearance of the goods that they buy, obscuring the relations of production and distribution associated with these goods. This commodity fetishism is often reinforced by advertising.

By contrast, recent research by economic geographers, sociologist and anthropologists has been concerned to overcome commodity fetishism by uncovering the complex geographies of commodity production and distribution, revealing webs of interdependencies that connect different places within the global economy. As the following quotation from David Harvey illustrates, even routine, everyday activities like having breakfast rely on complex sets of spatial relations:

Consider, for example, where my breakfast comes from. The coffee was from Costa Rica, the flour that made up the bread probably from Canada, the oranges in the marmalade came from Spain, those in the Orange juice came form Morocco and the sugar came from Barbados. Then I think of all the things that went into making the production of these things possible – the machinery that came from West Germany, the fertiliser from the United States, the oil from Saudi Arabia ... it takes

very little investigation for the map of where my breakfast came from to become incredibly complicated. I also find that literally millions of people all over the world in all kinds of different places were involved just in the production of my breakfast. The odd thing is that I don't have to know that in order to eat my breakfast. Nor do I have to know it when I go shopping in the supermarket. I just lay down the money and take whatever it will buy.

(Harvey, 1989c, p.93)

It is these kinds of complex connections that the geographer Michael Watts (2005, p.530) is referring to when he describes the commodity as a bundle of social relations. By uncovering these relations as suggested by the Harvey quote above, advocates of the commodity chain approach have argued that it is possible to trace the 'life' or 'biography' of a commodity.

10.2 The commodity chain approach

The GCCs approach originally emerged out of world systems research (Table 10.1), defining the commodity chain as 'a network of labour and production processes whose end result is a finished commodity' (Hopkins and Wallerstein, 1986, p.159). Commodity chains involve the transformation of material and non-material inputs into finished goods or services. Four basic stages of inputs, transformation, distribution and consumption have been identified (Figure 10.1), with each stage adding value to the previous stage. The approach is applicable to both services and manufacturing, with even the production of manufactured goods relying on a series of 'non-material' inputs or services such as procurement, insurance, legal services, advertising and marketing (Bryson, 2008).

Much of the literature on GCCs focused on the issue of **governance**, referring to how chains were organized, controlled and coordinated by 'lead' firms. Here, the sociologist Gary Gereffi (1994) drew an important distinction between what he called producer-driven and buyer-driven commodity chains (Table 10.2). **Producer-driven commodity chains** are typical of industries which are highly capital and technology

Table 10.1 The global commodity chain and global production network approaches compared

	GCC	GPN
Main disciplines	Economic sociology, development studies, world system research	Economic geography
Key concepts	Commodity production as a sequential chain Value creation in chain organization	Commodity production through open-ended and multi-faceted networks Value, power and embeddedness
Major authors	Gary Gereffi, Dieter Ernst, John Humphrey, Hubert Schmitz	Neil Coe, Peter Dicken, Martin Hess, Henry Wai-Chung Yeung
Main strengths	Conceptualization of governance and analysis of organizational power dynamics along global chains	Broad and inclusive network-based approach Explicit geographical orientation
Main weaknesses	Linear and dualistic analyses Under-playing of the role of the state Passive conception of labour Neglect of geography	Neglect of natural environment, labour process and consumption Tendency to gloss over the often unequal nature of the relations between regions and GPNs
Key case study industries	Clothing, footwear, electronics, coffee, horticulture	Automobiles, clothing, telecommunications, retail, computers

Source: Adapted and extended form Hess and Yeung, 2006, p.1194.

intensive, requiring huge amounts of research and development and investment, such as aircraft, motor vehicles and computers. The chain is coordinated and controlled by large manufacturers that rely on conventional top-down hierarchical links with a range of component suppliers. The level of outsourcing in such chains tended to be moderate, although this has increased in recent years, reflecting wider trends. **Buyer-driven commodity chains**, by contrast, are coordinated by large, branded retailers that concentrate on design, sales, marketing and finance, while actual production is outsourced to suppliers in developed countries. This type of arrangement is characteristic of labour-intensive consumer goods industries such

Table 10.2 Characteristics of producer-driven and buyer-driven chains

	Form of economic governance	
	Producer-driven	Buyer-driven
Controlling type of capital	Industrial	Commercial
Capital/technology intensity	High	Low
Labour characteristics	Skilled/high wage	Unskilled/low wage
Controlling firm	Manufacturer	Retailer
Production integration	Vertical/bureaucratic	Horizontal/networking
Control	Internalized/hierarchical	Externalized/market
Contracting/outsourcing	Moderate and increasing	High
Suppliers provide	Components	Finished goods
Examples	Automobiles, computers, aircraft, electrical machinery	Clothing, footwear, toys, consumer electronics

Source: Coe *et al.*, 2007, p.102.

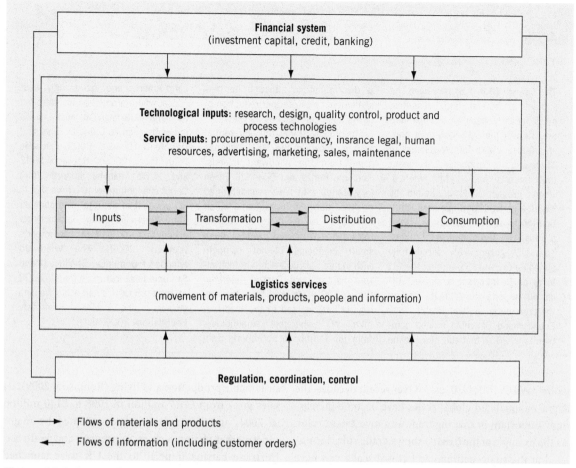

Figure 10.1 A generic production network
Source: Dicken, 2007, Figure 1.4c.

as clothing, furniture and toys. Examples of well-known firms that operate through buyer-driven chains include Wal-Mart, Ikea, Nike, Gap and Adidas. While Gereffi's distinction has proved highly influential, it has become increasingly apparent that reality is often more complex, with a number of industries developing more open, relational forms of governance characterized by close but more equal relationships between firms (Coe *et al.*, 2007, pp.102–4). In response, Humphrey and Schmitz (2002) distinguished between networks and quasi-hierarchical relationships in global commodity chains (Box 10.3). In the former case, firms cooperate in a more information-intensive relationship, with lead firms encouraging knowledge transfer and supplier development, while, in the latter case, lead firms exert more direct control over the supply chain.

Another key concern of the GCCs approach is the institutional contexts in which chains operate, referring to the formal rules, regulations and standards, and informal norms and practices that govern different industries and locations in the world economy. Formal rules and regulations include things such as trade policies, employment legislation, tax regimes, environmental regulations, etc., while informal norms refer to the less tangible ways of working and cultures associated with particular firms, industries and places (Coe *et al.*, 2007, p.104). Commodity chains will typically cross-cut several different institutional contexts, posing considerable challenges to firms in negotiating different rules and regulations. Formal rules are structured by different spatial scales of regulation, from the regional to the national, supra-national (EU and NAFTA) and

Box 10.1

The banana trade war

The banana (Box 1.6) has been the subject of several trade disputes, most recently between Europe and the US over the EU's system of preferential trade with former colonies in Africa and the Caribbean. This meant that imports from Latin America, where production is controlled by large American agri-business multinationals such as Chiquita, faced a range of taxes and restrictions. The US government successfully appealed to the WTO, forcing the EU to agree to introduce a single tariff on all imports from 2006, ending the preferential system. Despite this, Ecuador and Colombia lodged complaints in 2007 that the new regime

is discriminatory, although no new agreement on the further reduction of tariffs has been reached as yet.

Historically, under the preferential system, most of the UK's bananas came from the Windward Islands, Jamaica, Belize and Surinam. These long-standing ties were undermined by the new tariff system, which threatened to decimate the economies of the Windward Islands, where small producers cannot compete with large 10,000-acre industrialized plantations in Central American (Vidal, 1999). As predicted, the new regime has resulted in the displacement of higher-cost smallholders from the Caribbean by imports from

Latin America and, increasingly, West Africa, where production is controlled by large US agri-business multinationals such as Chiquita, Dole and Del Monte (Doward, 2009; Fairtrade Foundation, 2009). Between 1992 and 2007, banana imports from Caribbean countries fell from 70 to 30 per cent of the total with cheaper Latin American 'dollar' bananas now accounting for half of UK imports (Doward, 2009). The Windward Islands of Dominica, St Vincent and St Lucia have lost more than 20,000 of their 25,000 small-scale banana growers since 1992 (Fairtrade Foundation, 2009, p.2).

global (WTO, IMF, G20, etc.). Over recent decades, the supra-national and global scales have become increasingly important in shaping trade and investment rules. As the example of the banana shows, trade rules formulated at the supra-national and global scales can have profound effects on production located in particular places (Box 10.1).

Questions of standards represent another focus of interest in commodity chain research. These are often related to various forms of **ethical consumption**, which aim to improve working or environmental conditions at different points along the chain (section 12.4.1) (Hughes *et al.*, 2008, pp.350–1). As well as being driven by consumers, such standards can also be introduced by firms themselves, NGOs or regulatory bodies like the WTO. As this suggests, standards are highly variable, focusing on different dimensions of chains such as health and safety, labour conditions, environmental impact, price levels or quality assurance, and taking different forms such as labels, codes of conduct or technical standards (ibid.). For example, the UK Fairtrade Foundation, founded in 1992 by a group of charities and NGOs, focuses on guaranteeing economic returns for producers, i.e. a price that covers the costs of

sustainable production and living (Robinson, 2009). Its sales grew from £16.7 million in 1998 to £140 million in 2004, and Fairtrade has been adopted by major supermarket chains such as Sainsbury's and Waitrose. Fairtrade banana imports to the UK were launched in 2000 and they now account for one in every four bananas sold in the UK, providing something of a lifeline for the surviving banana farmers in the Windward Islands (Fairtrade Foundation, 2009, p.10).

The geography of commodity chains has been understood as being either concentrated in a limited number of locations or dispersed across a range of sites. Under globalization the trend is towards increased complexity and dispersal with developments in transport, communication and process technologies allowing the rapid and efficient transfer of goods, services and information between countries. As the example of the laptop industry shows, the geography of commodity production and distribution and consumption is complex and shifting (Box 10.2). Commodity chains are organized through corporate networks based on subcontracting and strategic alliances with other firms, coupled with the use of advanced information and communication technologies. This creates a flexibility that allows

Box 10.2

The laptop PC production network

The laptop production network is highly complex, based upon the bringing together of hundreds of different components into the finished product (Coe, 2008, p.319). Historically, large, integrated companies such as IBM, Hewlett Packard (HP) and Siemens controlled the industry, undertaking the key functions of product innovation, manufacturing and customer relations internally. Since the advent of the PC,

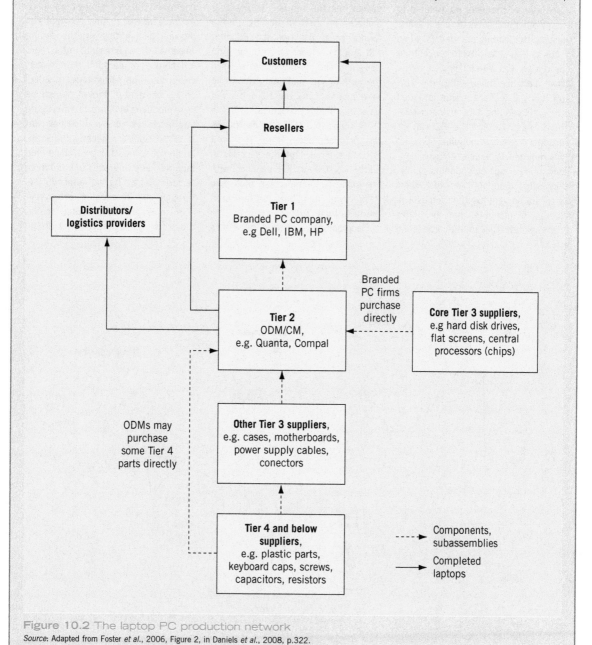

Figure 10.2 The laptop PC production network
Source: Adapted from Foster *et al*., 2006, Figure 2, in Daniels *et al*., 2008, p.322.

Box 10.2 (continued)

however, a far more complex 'tiered' network has developed in which most companies specialize in a particular activity, such as making chips, keyboards or hard disk drives (Figure 10.2) (ibid., p.320). The trend of increased out-sourcing is mirrored in many other industries in recent years, leaving the leading branded PC firms such as Dell or Toshiba to concentrate on design and marketing/sales rather than direct manufacturing. This reflects the fact that value is added through branding and customer relations. As a result, 'the PC can be described as a "modular" product whereby ten to fifteen relatively self-contained sub-components (e.g. keyboard, monitor, hard disk drive etc.) are brought together and assembled, an attribute that facilitates the disintegration of the production network into separate firms' (ibid., p.320).

The vast majority of laptop assembly operations are subcontracted to contract manufacturers, and a kind of subcontractor known as original design manufacturers (ODMs) that are also involved in design. In the former case, the branded PC firm will employ a team of on-the-ground designers and developers to oversee the work of the manufacturers, while in the latter case the PC firms will pass on a product specification to the ODM (ibid., p.321). Toshiba follows the first approach while Dell and HP are examples of the second. In addition, smaller PC vendors with the capacity to undertake their own design will just buy 'off-the-shelf' products from ODMs.

Geographically, most of the world's laptops are produced in Taiwan, which accounted for 72 per cent of world production in 2005, compared with 27 per cent in 1995 (ibid., p.325). Within Taiwan, production is concentrated in the capital, Taipei, and the nearby high-tech region of Hsinchu. At the same time, this geography is dynamic and shifting with the leading Taiwanese firms engaging in the rapid 'offshoring' of production in recent years, to China particularly. In 2001 only 5 per cent of Taiwanese laptops were produced in China, but this figure had increased hugely to 80 per cent in 2004 (ibid.), reflecting the availability of low-cost labour, land and facilities, allowing firms to build huge factories, which further reduced costs through the harnessing of greater economies of scale. Laptop manufacturing is spatially concentrated

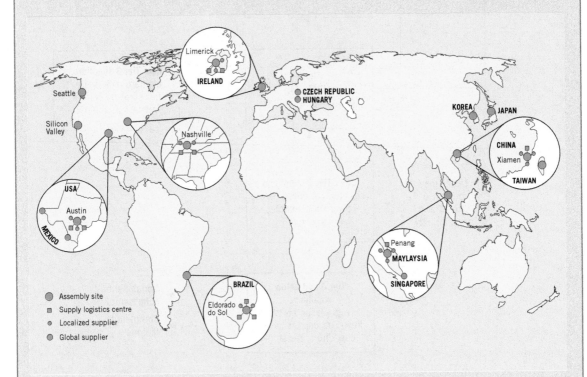

Figure 10.3 Dell's global production network, 2001
Source: Adapted from Fields, 2004, Map 6.3, in Daniels *et al.*, 2008, p.327.

Box 10.2 (continued)

within China in the Shanghai region, with the entire supply chain moved here in many cases. The PC industry generally, however, has a more dispersed geography. For example, Dell operates through six regional production clusters that house global and local suppliers and logistics centres (Figure 10.3) (ibid.).

'lead' firms to switch production between different suppliers and countries without out having to incur the costs associated with moving their own production facilities (Coe *et al.*, 2007, p.98). Such locational flexibility fosters inter-place competition, as existing places within the chain compete to attract new investments from the lead firm controlling the chain.

Processes of **competitive upgrading** have attracted much attention within GCC research. Upgrading can be defined as the improvement of a unit's position in the chain or network, involving increased competiveness through the capture of additional value. Four types of upgrading have been identified: process upgrading based on more efficient production; product upgrading (the production of sophisticated high-value goods or services); functional upgrading which involves moving up the chain by performing additional functions; and intersectoral upgrading whereby a firm uses its existing competences to move into new sectors (Gereffi *et al.*, 2001; Humphrey and Schmitz, 2001). Depending upon the structure and configuration of the underlying relationships, insertion into global commodity chains can provide actors in producing countries with opportunities for upgrading, particularly by offering them access to both lead firms, and buyers and advanced production techniques and skills, potentially enabling them to capture more value (Box 10.3) (Gwynne, 2006). At the same time, of course, lead firms and buyers in the core consuming countries exercise considerable power through the establishment of exacting quality standards and the adoption of global sourcing strategies.

Box 10.3

Competitive upgrading in global commodity chains: the case of Chilean wine producers

The Chilean wine industry provides an interesting case study of how global commodity chains impinge on the 'host' economies in which suppliers are located, illustrating how such chains are grounded in particular places. The Chilean wine industry has experienced rapid export growth over the past couple of decades, with wine exports rising from US$35.4 million in 1989 to almost US$1 billion in 2006 (Gwynne, 2008, p.98). Three types of Chilean wine firm can be identified (Figure 10.4): wine company groups, newly established firms and old established wine firms that are trying to restructure to supply global markets. Only Chile's largest wine company, Concha y Toro has set up its own distribution company in the UK. Unlike the country's fruit sector, the Chilean wine industry remains largely domestically-owned and relatively unconcentrated, although the Concha y Toro group accounts for around 26 per cent of wine exports (ibid., p.104).

The wine commodity chain corresponds to neither the producer-driven nor buyer-driven models. Rather than direct control by large, integrated producers or by buyers who specify detailed designs for suppliers, as occurs in the clothing industry, the wine commodity chain tends to be driven by consumers, and the supermarkets and retail outlets that service them (Gwynne, 2006). In the wine industry, buyers lack direct access to the technology of the suppliers in wine-producing countries, operating instead through the drawing up of requirements that wine exporting firms must meet. As Figure 10.4 indicates, the UK market is highly concentrated, with the supermarket chains accounting for 72 per cent of offtrade wine sales. UK supermarkets

Box 10.3 (continued)

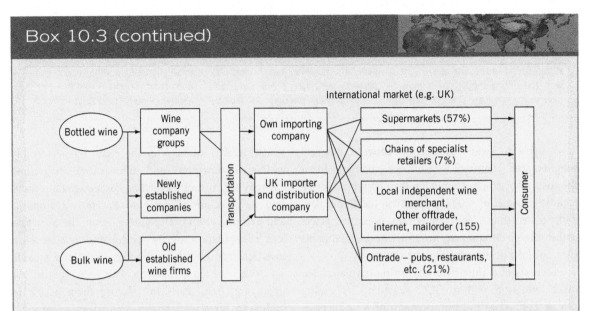

Figure 10.4 Chilean wine: commodity chain downstream to UK market
Source: Gwynne, 2008, p.100.

have stressed the importance of wine sales for the attraction of higher-spending customers in recent years, and have sought to increase the range and quality of the wine they sell. This has created a growing overseas market which Chilean producers have targeted, resulting in rapid export growth. Ultimately, however, the power of supermarkets in the UK wine market has created highly competitive pricing conditions, allowing them to appropriate much of the profit within the chain. Thus the UK wine market has offered Chilean producers rapid export growth but low profitability. In the case of one medium-sized winery, Luis Felipe Edwards, for example, the UK market provided 45 per cent of export volume, but only 30 per cent of export value (Gwynne, 2008, p.1010).

Growing demand in key global markets such as the UK has encouraged product and process upgrading among Chilean wine producers. For many firms, product upgrading involved an initial focus on inserting themselves in international markets through the production of competitive basic and varietal wines (good quality, low prices) before developing new higher-quality wines and increasing their prices. At the same time, Chilean producers drew upon Australian techniques and imported European machinery and equipment during the 1990s. The role of so-called 'flying winemakers', renowned experts (often linked to influential wine critics) who act as consultants and advisers, has also been important. Another key means of upgrading has been increased vertical integration, as wine firms purchase both large tracts of land for vine planting and established vineyards, allowing them to control yields more closely and improve the quality of the raw material.

In overall terms, Gwynne (2008) argues that the global commodity chains in which Chilean wine producers have inserted themselves are closer to networks than quasi-hierarchical relationship (Humphrey and Schmitz, 2002), due to the opportunities they offer for upgrading and the modest bargaining power that Chilean firms have been able to exert. In many respects, however, this story seems fairly unusual as one of the few examples of meaningful upgrading and knowledge transfer benefiting firms from developing countries within agricultural or agro-industrial commodity chains. At the same time, Chilean producers remain highly dependent on the major UK supermarkets which have been highly successful in 'the scramble for value within the chain' (ibid., p.106).

While highly influential in fostering research on the organizational dynamics and governance of global industries, the GCCs approach has also been the subject of some criticism from economic geographers (Table 10.1) (Henderson *et al.*, 2002; Smith *et al.*, 2002). First, it tends to view global industries in linear and dualistic terms, seeing chains in 'vertical' terms as a step-wise sequence of activities and relying on simple distinctions such as that between buyer-driven and producer-driven chains. This means that the GCC approach tends to privilege production and governance over issues of consumption, knowledge and the process of mutual interaction between end-consumers and producers within particular industries (Hughes, 2000). Second, the approach tends to neglect the active role of national states in economic governance and regulation, incorporating them in only a very limited fashion as part of the institutional context. Third, labour processes have also tended to receive rather scant attention, with labour viewed as rather secondary and passive in contrast with the main focus on corporate processes. Fourth, there is little real consideration of regional development issues, reflecting a very basic notion of geography in terms of the role of macro-regional institutional contexts such as the EU. In response, economic geographers have sought to develop a more geographically sensitive and conceptually sophisticated GPN approach in recent years.

Reflect

➤ How useful do think that the distinction between buyer-driven and producer-driven commodity chains is for examining the organization and governance of global industries?

10.3 Global production networks

Global production networks (GPNs) have become a key focus of research in economic geography and related fields in recent years (Coe *et al.*, 2004, 2008; Henderson *et al.*, 2002; Hess and Yeung, 2006). The GPN approach provides a broad relational framework for the study of economic globalization that aims to 'incorporate all kinds of networks relationship' and 'encompass all relevant sets of actors and relationships' (Coe *et al.*, 2008, p.2). As such, it offers an open and geographically sensitive perspective that goes beyond the more restricted and linear frameworks offered by the related concepts of global commodity chains and global value chains (Table 10.1) (Henderson *et al.*, 2002). According to Sturgeon:

A chain maps the vertical sequence of events leading to the delivery, consumption and maintenance of goods and services ... while a network highlights the nature and extent of the inter-firm relationships that bind sets of firms into larger economic groupings.

(Sturgeon, 2001, p.10)

The GPN approach is particularly associated with the work of a so-called 'Manchester school' of economic geographers (Bathelt, 2006). GPN research is, by definition, global in its scope and reach, incorporating all the main nodes within a particular network, although Europe and Asia seem to feature particularly strongly (Coe *et al.*, 2004; Liu and Dicken, 2006; Weller, 2006).

The GPN approach directs attention towards networks of firms, spanning a similar range of activities to GCCs, including R&D, design, production, supplier relations, marketing and sales. It stresses the increased organizational and geographical complexity of such networks under globalization:

The global production network ... is a conceptual framework that is capable of grasping the global, regional and local economic and social dimensions of the processes involved in many forms of economic globalization. Production networks – the nexus of interconnected functions and operations through which goods and services are produced, distributed and consumed – have become both organizationally more complex and also increasingly global in their geographic extent ... At the same time ... firm-centred production networks are deeply influenced by the concrete socio-political contexts within which they are embedded.

(Henderson *et al.*, 2002, p.446)

GPN research is concerned with the distribution of corporate power in networks, particularly in terms of the relations between lead firms and states, and between such firms and their suppliers. It focuses attention on processes of value creation, transfer and capture (see below), paying greater attention to the role of labour, although the main emphasis is on firm and state actors. As the last sentence in the above quotation indicates, the GPN concept also highlights, in a rather similar fashion to GCCs, the importance of particular institutional contexts in shaping firm strategy, referring not only to formal organizations such as the state, business organizations and trade unions, but also to informal rules and norms. While there are important conceptual differences between GPNs and GCCs, with the former having developed out of a critique of the latter (Henderson *et al.*, 2002), this should not blind us to the underlying continuities in terms of the focus on the organization and governance of global industries (Bair, 2008).

The GPN approach is based on three conceptual categories. First, **value**, which attempts to incorporate both Marxist notions of surplus value (see Box 2.3) and mainstream economic definitions of economic rent (Henderson *et al.*, 2002, p.448). It refers to the economic return (profit) or rent generated by the production of commodities for sale, involving the conversion of labour power into actual labour through the labour process (ibid.). Firms may create value through: the control of particular product and process technologies; the development of certain organizational and management capabilities; the harnessing of inter-firm relationships; and the prominence of brand-names in key markets. In the face of competition from other firms for market share, there is a need to also enhance value through processes of technological transfer and the upgrading of skills and production capabilities. All this raises the issue of value capture in terms of which actors and locations in the network are able to appropriate and retain value, highlighting questions of ownership and control. Key aspects of this include: firm ownership (private–public, domestic–foreign, etc.); government policy in terms of property rights, ownership structures and the repatriation of profits; and systems of corporate governance in different national contexts (stakeholder–shareholder) (ibid.).

The second conceptual category of interest is **power** (section 1.4.4), defined primarily as a practice in terms of the capacity to exercise power (Allen, 2003; Dicken *et al.*, 2001). More specifically, Henderson *et al.* (2002) identify three main forms of power. First, corporate power in relation to a lead firm's control of key resources, information, knowledge, skills and brands within a production network. Second, institutional power, which is exercised by national and local states (Box 10.4), supra-national bodies such as the EU, the 'Bretton Woods institutions' (World Bank, IMF and WTO), various UN agencies (especially the ILO) and

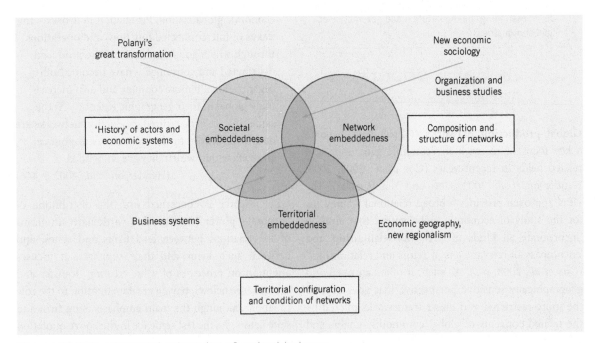

Figure 10.5 Fundamental categories of embeddedness
Source: Hess, 2004, p.178.

Box 10.4

Obligated embeddedness: MNCs in China

Contrary to the hyper-globalist myth of the powerless state in the face of global capital flows, national states remain key actors in the global economy, continuing to control access to their domestic markets. Whereas MNCs seek to maximize their locational flexibility, favouring low levels of political regulation, states aim to capture value within GPNs by trying to embed a MNC's

activities as fully as possible in the local or national economy (Liu and Dicken, 2006). It is the territorial dimension of embeddedness that is invoked here, referring to the effort to fix investment in a particular location and the economic benefits and spinoffs that are derived from this for the host economy in question. Liu and Dicken go on to distinguish between active and obligated forms

of embeddedness as two ideal types. The former refers to cases where local assets are widely available and MNCs source and incorporate such assets on an autonomous basis, while the latter reflects cases when such assets are not freely available and a particular state can control access to them. In this second set of circumstances, states will have much greater bargaining power, requiring

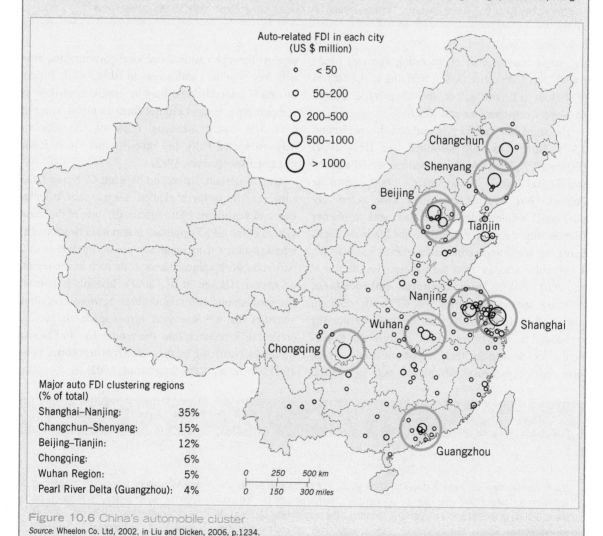

Figure 10.6 China's automobile cluster
Source: Wheelon Co. Ltd, 2002, in Liu and Dicken, 2006, p.1234.

Box 10.4 (continued)

MNCs that wish to invest in their domestic markets to comply with certain criteria.

China became the fastest growing automobile market in the world, emerging as the third largest overall after the US and Japan (ibid., p.1229). The industry is concentrated in six main regional clusters within China (Figure 10.6). Its growth followed the selection of automobiles as a strategic growth sector by the Chinese government in the late 1980s. The basic bargaining strategy

adopted by the state has been one of trading market access for capital and technology transfer from MNCs. Due to their desire to invest in this huge and rapidly-growing markets, foreign investors have complied. All the major players in the global automobile industry had invested in China by 2003 (ibid), and all these investments took the form of joint ventures, as required by the state. This form of obligated embeddedness reflects the unique bargaining power of the Chinese state in controlling access

to the worlds' largest and fastest-growing consumer market, allowing it to reverse the usual scenario of MNCs playing off states against one another into one in which a state could play off different MNCs against each other (ibid., p.1245). Other states are in a much weaker position since they do not control access to such a unique and significant asset as access to the Chinese market, though larger states or groups of states such as the EU will have more bargaining powers over MNCs than smaller ones.

the major international credit rating agencies (ibid., p.450). Third, collective power, referring to the actions of various collective actors, including trade unions, employers' organizations and NGOs.

The third key category is embeddedness (section 2.5.3). Three forms are identified (see Hess, 2004) (Figure 10.5). First, societal embeddedness in a broad sociological sense (Granovetter, 1985), invoking Polanyi (Box 1.7) to emphasize how actors are positioned within wider institutional and regulatory frameworks. Second, network embeddedness highlights the social and economic relationships in which a particular actor or firm participates (Henderson *et al.*, 2002, p.453). Third, **territorial embeddedness**, referring to the 'anchoring' of GPNs in different places (Box 10.4). GPNs can become territorially embedded because of lead firms' historic ties to particular locations, often their regions of origin, which may provide particular advantages such as political

support through national and local governments, links with key suppliers and access to labour skills (ibid.). But such embeddedness may be eroded over time as competitive pressures prompt firms to invest in other, often less costly, locations, reflecting the tensions between spatial fixity and mobility that are endemic to capitalism (Harvey, 1982).

One important distinction between GPN and GCC concerns the former's explicit geographical orientation and sensitivity. More specifically, one of the main virtues of the GPN's approach is that its network methodology allows it to avoid any pre-occupation with a particular geographical site or scale such as the region or nation (Dicken *et al.*, 2001). Instead, it stresses the connections and relationships between economic activities that are 'stretched' across space, supporting comparative research into the production and labour processes occurring in these different territories (Box 10.5) (Coe *et al.*, 2008; Smith *et al.*, 2002).

Box 10.5

The global production network of BMW

BMW, the German car manufacturer, provides an interesting example of the evolution of a GPN and its organization across space (Figure 10.7)

(Coe *et al.*, 2004). In one sense, the example clearly reflects the characteristic of the automobile sector such as a high level of capital-intensity

and technology-intensity, a globalized form of organization, a reliance on 'lean production', and the use of just-in-time strategies, whereby

Box 10.5 (continued)

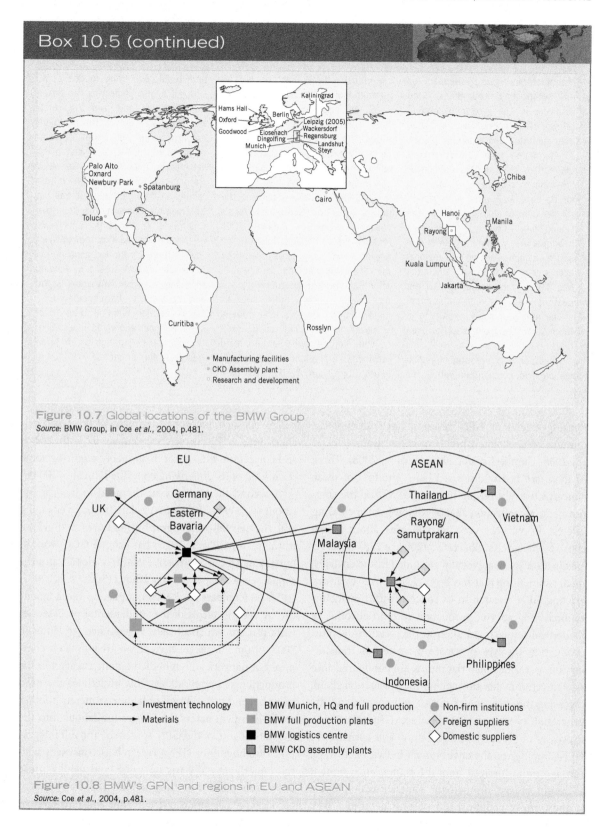

Figure 10.7 Global locations of the BMW Group
Source: BMW Group, in Coe *et al.*, 2004, p.481.

Figure 10.8 BMW's GPN and regions in EU and ASEAN
Source: Coe *et al.*, 2004, p.481.

Box 10.5 (continued)

manufacturers source suppliers as and when they need them, often resulting in the geographical clustering of suppliers around large manufacturing and assembly plants. At the same time, BMW remains distinctive as a niche, upmarket, low-volume producer. It retains a strong base in Bavaria, with around 47 per cent of shares owned by the Quandt family (ibid., p.477). With its headquarters in Munich, BMW has had a major impact on the regional economy of east Bavaria since the late 1960s, accounting for 10 per cent of manufacturing employment in the region, through investment in three plants and one supplier park (ibid., p.478).

In response to pressures of globalization and competition within the car industry, BMW has adopted an internationalization strategy in recent years. In addition to ventures in North and South America and South Africa, it has identified entry to the rapidly growing East Asian market as a strategic priority. Like other car manufacturers, it has invested in Thailand, which has become the prime focus of South-east Asia's nascent automotive industry. This reflects how Thailand has offered foreign investors a fairly liberalized market environment due to the lack of a national car programme compared with nearby counties such as Malaysia and Indonesia. The industry is based to the south of Bangkok in the Rayong and Samutprakarn provinces of Thailand's eastern seaboard (ibid., p.479) (Figures 10.7

and 10.8). According to Coe *et al.* (2004), it also comprises two dozen manufacturers and over 700 first-tier suppliers. BMW's Rayong plant employs about 250 people, assembling nearly 400 cars a year, using vehicle kits imported from Germany. Rayong is BMW's only wholly owned facility in Asia, although it has set up assembly plants in other countries (Figure 10.7). While the small size of the Rayong plant is indicative of BMW's relatively limited contribution to regional development in Rayong thus far, the company intends the Rayong plant to be upgraded to a full production plant as a result of supra-national economic integration, through the negotiation of an Asian free trade agreement.

Recent reviews of GPN research have identified a number of unexplored points of connections (Coe *et al.*, 2008; Cumbers *et al.*, 2008; Hudson, 2008). Three of these are briefly discussed here. The first of these concerns the relationship between GPNs and the environment, a connection that is obviously of increasing interest in view of growing concerns about global climate change. As Coe *et al.* (2008, p.278) observe, 'production is unequivocally grounded in the environment'; something that has been glossed over by much existing GPN research in its concern with the organizational geography of production within networks. In theoretical terms, in addition to the creation of value, production involves processes of materials transformation as various material inputs are configured into new forms of matter throughout the production chain, involving the exchange of matter and energy, and the generation of various waste products (Hudson, 2008). Processes of production, distribution and exchange in GPNs impinge on the environment in two main ways: through inputs derived from the natural environment as resources and through the generation of outputs in the form of pollution and waste (Coe *et al.*, 2008,

p.279). As such, incorporating material flows and balances into GPNs research promises to both enrich and complicate the approach, arguably requiring some reworking of its underlying concepts (Hudson, 2008).

A second theme concerns the need to incorporate labour (Coe *et al.*, 2008), particularly through engaging with the **new labour geography** literature (section 7.5). In the main, GPNs research has echoed GCC work in portraying labour as a passive victim of globalization as capital seeks cheap sources of labour (Smith *et al.*, 2002, p.95). As Cumbers *et al.* (2008) argue, however, labour can be seen in terms of the fundamental processes of work through which all value is generated (cf. Hudson, 2008). From this perspective, GPNs are ultimately composed of networks of embodied labour engaged in the production of commodities. This underlines the need to go beyond the addition of labour as merely another set of non-firm actors, by integrating labour into the theoretical core of the GPN approach. The full integration of labour into GPNs research is consistent with the underlying emphasis on value, and would infuse a dynamic current of class conflict and struggle into accounts of corporate restructuring and outsourcing.

A key issue here concerns the influence of the relationship between place-bound labour and geographically extensive GPNs, as mediated by state institutions at different scales, on the geographies of production.

A third set of connections concerns **consumption**, which has been neglected by GPNs research, reflecting its concern with production. Again, Coe *et al.*, (2008) emphasize the need to respond to the challenge of incorporating consumption research, broadening the scope of GPNS research. Based on a study of ethical trading initiatives among US and UK corporate retailers, Hughes *et al.* (2008) argue for greater incorporation of consumption and retail into GPNs, demonstrating how consumption can shape the governance of GPNs through ethical campaigning and notions of business responsibility. As this research shows, NGOs and social movements can shape the evolution of GPNs, not least through the development of standards and codes of conduct relating to issues such as fair trade, labour rights and environmental sustainability (Barrientos and Smith, 2007). As such, there is also a need to take questions of consumption more seriously.

> ### Reflect
>
> ➤ Compare and contrast the GCC and GPN approaches. Do you think that the differences between them outweigh the similarities?

10.4 Globalizing regional development

The geographical orientation of GPNs research underpins its distinctive contribution to the recent rethinking of regional development processes. Compared with the 'new regionalism' of the 1990s, which was preoccupied with social and institutional conditions within regions (see section 2.5.3), the GPNs approach signals a new interest in extra-regional relations or the 'outside' of regions, while seeking to retain certain 'new regionalist' insights, particularly the notion of regional institutions 'holding down the global' (Amin and Thrift, 1994). The concern with extra-regional relations is

somewhat reminiscent of the Marxist political economy approaches of the 1970s and 1980s (section 2.4), although these relations are conceptualized in a rather different way. The key point is that regions are not isolated and separate, but closely bound up with the operation of global networks, echoing Massey's **global 'sense of place'** (section 1.2.3) and the concept of the **'relational region'** (section 2.5.4). As Dicken *et al.* (2001, p.97) argue, this is 'a mutually constitutive process; while networks are embedded within territories, territories are, at the same time, embedded into networks'.

In particular, Coe *et al.* (2004) aim to 'globalize' regional development through their focus on the link with GPNs, viewing 'the region' as 'a porous territorial formation whose notional boundaries are straddled by a broad range of network connections' (Coe *et al.*, 2004, p.469). Regional development is defined as 'a dynamic outcome of the complex interaction between territorialized relational networks and global production networks within the context of changing governance structures' (ibid.). Regional assets in the form of specific kinds of knowledge, skills and expertise provide an important resource for regional development, but must be harnessed by regional institutions to 'complement the strategic needs of trans-local actors situated within global production networks,' (ibid., p.470). The promotion of local endogenous development based on such assets will not be sufficient to secure increased prosperity. From this perspective, regional development is a product of the **'strategic coupling'** between global production networks and such regional assets.

According to Yeung (2009. p.213), '... in the context of urban and regional development, strategic coupling refers to the dynamic processes through which actors in cities and/or regions coordinate, mediate, and arbitrate strategic interests between local actors and their counterparts in the global economy'. The role of regional institutions is to attract and retain investment by shaping and moulding regional assets to fit the needs of lead firms in GPN (Figure 10.9). Such coupling processes result in cooperation and the construction of a 'temporary coalition' between groups of actors – the managers of MNC affiliates and regional government officials – who might not otherwise work together in the pursuit of a common objective (ibid.).

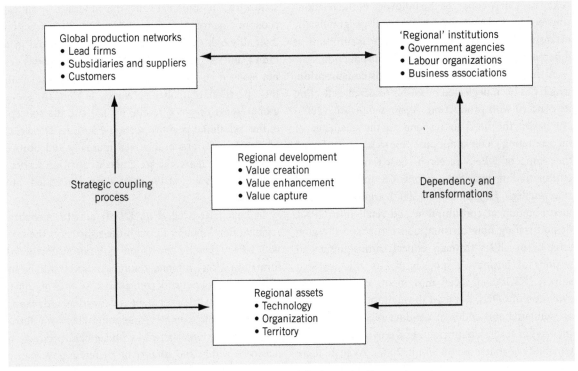

Figure 10.9 The strategic coupling of regions and global production networks
Source: Coe et al., 2004, p.470.

This conceptualization can be seen to overlap substantially with an extensive literature on the role of regional institutions in attracting inward investment (see MacKinnon and Phelps, 2001).

The task for regions is to create, capture and enhance value within GPNs, with regional institutions playing a key role in these processes (Box 10.6) (Coe *et al.*, 2004). Value creation involves the creation of supporting conditions for growth by regional institutions through training and education programmes, the promotion of firm start-ups and the provision of venture capital through private-sector investors. The further enhancement of value can occur through knowledge and technological transfer, industrial upgrading, the provision of more advanced infrastructure and the development of specialized skills (Box 10.6). Value capture refers to scenarios in which regions retain the profits generated by economic activities within them, reflecting, in part, the extent to which key firms are locally owned and embedded in the regional economy. In practice, of course, regions are often involved in a

variety of structural couplings with GPNs in several industries, and each of these will generate different degrees of value creation, enhancement and creations.

In the BMW case, initial value creation in east Bavaria in the late 1960s and 1970s gave way to value enhancement in the 1980s and 1990s, through processes of technological upgrading and the creation of a pool of highly skilled labour (Figure 10.8). Significant value capture was ensured by the local ownership and control of BMW and by elaborate negotiations between management, government and labour at the regional and national scales, with, for instance, the regional branch of the national union, IG metal, agreeing to implement a work-shift system that has become a model for the German car industry (Coe *et al.*, 2004, p.478). In Rayong, Thailand, some value creation occurred through the investment in an assembly plant, supported by the national government, which has set up programmes to develop the skills of the workforce and the capabilities of local suppliers (ibid.). There is a lack of real upgrading, however, reflecting limited

Box 10.6

Value creation, enhancement and capture in Silicon Glen, Central Scotland

West Central Scotland was one of the key crucibles of industrial capitalism in the mid- to late nineteenth century, with 20 per cent of world's shipping tonnage being built on the River Clyde in 1914 (Devine, 1999, p.250). In common with many other 'old' industrial regions, west-central Scotland experienced a long decline over the course of the twentieth century, culminating in severe deindustrialization in the 1970s and 1980s. In

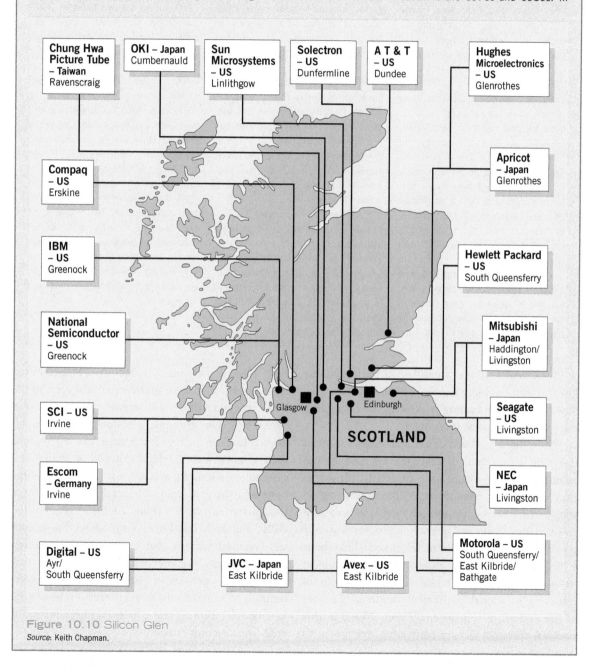

Figure 10.10 Silicon Glen
Source: Keith Chapman.

Box 10.6 (continued)

response, the state sought to attract new industries to Scotland in order to provide replacement employment, resulting in growing investment in electronics, initially from US and European firms, before a new influx from Japan and East Asia in the 1980s and 1990s. This reflected the access that Scotland offered to the European market, a pool of available labour and government incentives (Sutherland, 1995).

By the early 1980s, the term 'Silicon Glen' was being used to describe the Scottish electronics industry, invoking the success of Silicon Valley in California (Box 4.4). The cluster is focused on Central Scotland, but its geography is relatively dispersed in nature, stretching in a south-westerly line from Dundee to Ayrshire, with a particular concentration in West Lothian and Lanarkshire, between the cities of Glasgow and Edinburgh (Figure 10.10). By 2000, electronics had become a very significant sector of the Scottish economy, accounting for over 40,000 direct jobs (and

supporting another estimated 29,000 indirectly) and almost half of Scotland's manufactured exports by value (Scottish Government, 2003, 2004). This reflects substantial value creation through the attraction of investment, infrastructure provision and workforce training with the Scottish Development Agency and its successor, Scottish Enterprise, playing a key role in linking regional assets to the strategic needs of focal firms within GPNs (see Coe et al., 2004). Considerable value enhancement occurred in the 1990s through additional innovation and skills development, reflecting an increasing emphasis on research and development and efforts to move into higher value-added activities.

Such value enhancement has not, however, translated into significant value capture, reflecting a continuing lack of local ownership and control. Despite talk of greater 'embeddedness' and quality flagship investment (see Phelps et al., 1998, 2003), the concept of the branch plant economy (section 6.6.1), prevalent in

the 1970s and 1980s, remains relevant. This was demonstrated by the impact of the sharp downturn in the global electronics sector in the early 2000s, resulting in the Scottish electronics industry contracting by 46 per cent in output terms between 2000 and 2005, compared with 27 per cent for the UK industry, reflecting a continuing specialization in lower value-added activities (Ashcroft, 2006, p.6). Closures included Motorola's mobile phone plant in Bathgate in April 2001, with the loss of 3,000 jobs, despite the visit of the then Scottish Government economic development ministers to Chicago to lobby the Motorola head office. Competition from eastern Europe and East Asia, where costs are lower, has compounded the problems of the Scottish industry with several examples of companies which have closed down their Scottish operations and moved jobs to eastern Europe, particularly the Czech Republic and Hungary.

---

skills and organizational capacities. At the same time, linkages with the wider regional economy remain very limited and value capture is reduced by continuing dependence on outside investment.

While there has been considerable discussion of upgrading, processes of downgrading and capital abandonment or closure have attracted little attention from GPN researchers. Without wanting to suggest that corporate restructuring is a zero-sum game, reinvestment and upgrading processes usually operate in a competitive and selective fashion, benefiting certain locations, while others may be subject to downgrading or rationalization, sometimes involving 'locational tournaments' between branches of the same firm (Phelps and Fuller, 2000). In this context, the emphasis on cultivating linkages with

MNCs can increase the vulnerability of less-favoured regions, exposing them to increased leakage and the threat of relocation to emergent, lower-cost sites (Box 10.6). For example, a historic reliance upon inward investment-based strategies has left regions such as north-east England exposed to 'malignant' external relations structured by financial institutions and the privileging of shareholder value over local material interests (Pike, 2006). In general, the GPNs approach has little to say about the fate of less-favoured regions that are not well connected to global networks, beyond the implication that development agencies in such regions should concentrate on initial forms of value creation involving, for example, training and education programmes, in order to attract investment.

In general, the often unequal nature of the relations between TNCs and local residents and communities – a central theme of earlier political economy approaches to regional development (Bluestone and Harrison, 1982; Clark, 1989) – has been underplayed by the GPNs approach. A key issue here is how the greater mobility of TNCs involves periodic disinvestment in certain locations, linked to reinvestment in others. Regions are viewed largely as nodes within corporate networks, offering a rather circumscribed view of place, which underplays its continuing significance in struggles over livelihoods and social reproduction (Bebbington, 2000; Routledge and Cumbers, 2009). Such struggles may become implicated in GPNs through, for instance, local protests against processes of corporate disinvestment and restructuring (see Pike, 2006). For example, the global drinks corporation Diageo's announced in July 2009 that it would be closing its whisky bottling plant in Kilmarnock in south-west Scotland, ending its historic association with the town (McDonald, 2009). This decision triggered a local campaign against the closure – reflecting the importance of the plant as a source of relatively stable, well-paid work in a deindustrialized town – which gained support from the devolved Scottish government, but failed to save the plant in the face of Diageo's wider corporate objectives.

In summary, our assessment of the strengths and weaknesses of the GPNs approach can be related to the earlier criticisms of GCC research, asking to what extent the former has progressed beyond the identified limitations of the latter (Table 10.1). GPNs research has certainly moved beyond the linear and dualistic character of GCC analyses through the development of a more open-ended and multi-faceted networks-based approach, incorporating a range of actors, relations, sites and scales. Its integrative nature is a key strength, allowing it to potentially address the unexplored areas of the environment, labour and consumption, although this also carries a danger of a loss of analytical focus in GPNs research if it becomes overly broad and inclusive, attempting to incorporate all relevant factors and actors. The second main advance of GPNs work over GCC concerns its explicitly geographical orientation, allowing it to make a major contribution to the rethinking of regional development in relational terms

(Coe *et al.*, 2004). There is, however, a need to extend this contribution by examining the often unequal form of the structural coupling processes that occur between GPNs and regional assets. At the same time, GPNs research has made rather less progress on the other two counts of examining the active role of national states and of incorporating labour processes (Coe *et al.*, 2008).

Reflect

> How useful do you think that the concept of 'strategic coupling' is for examining the relationships between regions and GPNs?

10.5 Summary

This chapter has highlighted the complex geographies of commodity production, distribution and consumption, addressing the tendency towards commodity fetishism whereby individual consumers are concerned with only the price and appearance of the goods that they buy. From this perspective, the commodity can be regarded as a bundle of social relations, encapsulating a complex set of connections between different people and places (Watts, 2005, p.530). The commodity chain, defined as 'a network of labour and production processes whose end result is a finished commodity' (Hopkins and Wallerstein, 1986, p.159), has proved a particularly influential approach in thinking about these connections amongst geographers and other social scientists. Yet, at the same time, it has been criticized for its restricted, linear conception of chains as a fixed sequence of activities performed by different firms and for favouring simplistic, dualistic categories such as that between buyer-driven and producer-driven commodity chains. Its conception of geography, moreover, remains rather crude and impoverished, underplaying the spatial complexity and dynamism of commodity production and neglecting questions of regional development (Henderson *et al.*, 2002).

In response, a group of economic geographers has developed the GPNs approach, which has the following characteristics. First, GPNs research emphasizes the importance of production in the global economy, focusing attention on the social relations between various actors in networks and identifying value as a key category. Second, GPNs researchers have developed a sophisticated and integrated framework which incorporates a range of actors, not just firms and state agencies. Third, the concept of a GPN provides a genuinely multi-scalar framework for research, replacing the local–global binary of the 1990s with a network approach based on following the key actors across various sites and scales. Fourth, the GPNs approach has made a major contribution to the rethinking of regional development in relational terms, successfully overcoming some of the limitations of new regionalist research in the 1990s (see MacKinnon et al., 2002). From this perspective, the 'regional' is a product of the 'strategic coupling' between regional assets and GPNs, emphasizing how regions are incorporated into networks and viewing the role of regional institutions as one of matching regional assets to the needs of TNCs. While certainly capturing some of the dynamics of regional development in a global economy, this approach tends to underplay the often unequal nature of the relationships between regions and MNCs, implying that greater connectivity though 'strategic coupling' is the key to the economic development of less favoured regions.

Exercise

Select a major global industry and examine its organization across space. Which major 'lead' or coordinating firms are involved? What other types of actor are involved in the chain? How is it governed and coordinated? To what extent can it be conceptualized as buyer-driven or producer-driven? What is the geography of the network? What forms of structural coupling can be identified in key regions? Is there evidence of ethical trading initiatives or consumer-led campaigns? What are the relative strengths and weaknesses of the GCC and GPN approaches as frameworks for studying the industry and its geography?

Recommended reading

Coe, N. (2008) 'The geographies of global production networks', in Daniels, P.W., Bradshaw, M., Shaw, D. and Sidaway, J. (eds) *An Introduction to Human Geography: Issues for the 21st Century*, Harlow: Pearson, pp.315–38.

A very useful introduction to the GPNs approach, which defines production networks and examines their operation across a range of industry contexts, providing some good case studies.

Coe, N., Hess, M., Yeung, H.W., Dicken, P. and Henderson, J. (2004) 'Globalising regional development: a global production networks perspective'. *Transactions of the Institute of British Geographers*, 29: 464–84.

An important paper which applies the GPN approach to the issue of regional development, using it to recast regions in relational terms and focusing attention on process of structural coupling between GPNs and regional assets.

Gereffi, G. (1994) 'The organization of buyer-driven commodity chains: how US retailers shape overseas production networks', in Gereffi, G. and Korzeniewicz, M. (eds) *Commodity Chains and Global Capitalism*, Westport, CN: Greenwood Press, pp.95–122.

A key founding statement of the GCC approach from one of its leading figures, offering a conceptualizing of the key relationships and outlining how buyer-driven chains operate with reference to the clothing industry.

Gwynne, R. (2008) 'UK retail concentration, Chilean wine producers and value chains.' *Geographical Journal*, 174: 97–108.

An excellent case study of a global commodity chain in action. Gwynne shows how the growth of the UK wine market has provided increased opportunities for Chilean wine producers, although the power of the UK supermarkets allows them to appropriate value within the chain.

Henderson, J., Dicken, P., Hess, M., Coe, N. and Yeung, H.W.C. (2002) 'Global production networks and economic development.' *Review of International Political Economy*, 9: 436–64.

A key article in which the so-called 'Manchester school' of economic geographers develops its approach to GPNs research, comparing and contrasting this with the more established GCC approach.

Hughes, A. (2006) 'Learning to trade ethically: knowledgeable capitalism, retailers and contested commodity chains.' *Geoforum*, 37: 1008–20.

A detailed examination of UK food and clothing retailers' efforts to develop ethical trading programmes in response to political pressures, focusing on the processes of organizational learning through which these programmes are implemented within the corporation.

Useful websites

http://fairtrade.org.uk

The website of the UK's Fairtrade Foundation, one of the most prominent fair-trade initiatives, which aims to improve conditions for developing country producers in global commodity chains.

http://www.globalvaluechains.org/

This website offers a range of valuable material on global commodity chains and value chains.

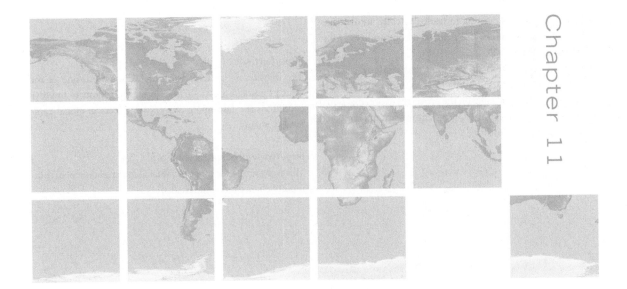

Chapter 11

Knowledge, creativity and regional development

Topics covered in this chapter

- ➤ The recent emphasis on knowledge and innovation as key drivers of economic development.
- ➤ Contemporary theories of the spatial agglomeration (concentration) of economic activity adopted by economic geographers and others.
- ➤ The role of industrial clusters in supporting local and regional growth.
- ➤ The concept of the creative classes and its adoption by urban and regional policy-makers.
- ➤ The significance of wider global linkages for the development of clusters and learning regions.

Chapter map

The aim of this chapter is to assess the role of knowledge and skills (or talent) as key factors shaping regional development. In the introduction, we highlight a renewed interest in regional economic development under globalization, focusing attention on the importance of building and attracting knowledge and skills. Section 11.2 examines the relationship between knowledge and the agglomeration of economic activity, and focuses on Michael Porter's influential theory of industrial clusters. The chapter then turns to discuss Richard Florida's concept of the creative classes, which was widely adopted by urban and regional policy-makers in the early-to-mid 2000s.

11.1 Introduction

A resurgence of interest in localities and regions as economic units has been apparent within economic geography since the late 1980s. At first sight, this 'new regionalism' (see section 2.5.3) seems paradoxical, given the prevailing emphasis on globalization as perhaps *the* political and economic force of the last 15 or so years. As we argue in this book, however, globalization is an uneven process, leading to the concentration of economic activity in particular places and creating increasingly close linkages between the local/regional and global scales of activity. National economic coherence has been undermined since the 1970s, as states have lost control over increasingly globalized flows of investment. The abandonment of Keynesian policies of demand management and full employment has exposed regions to the effects of international competition (section 5.5.4). This has focused attention on the need for regional-level action if regions are to be able to shape their own development prospects in a climate of rapid technological change and increased capital mobility (Amin and Thrift, 1994).

The revival of the region as a focus of interest has breathed new life into the topic of **spatial agglomeration** or concentration, referring to the tendency for industries to cluster in particular places (section 4.3). This has been an issue of recurring interest for economic geographers and spatial economists since the late nineteenth century. Agglomeration can be contrasted with the opposite process of spatial dispersal, where firms move out of existing centres into other, often less developed, regions. The balance between concentration and dispersal process will vary across different economic sectors, with some exhibiting a highly concentrated pattern, while others are more evenly dispersed across the economic landscape. The two processes operate at different geographical scales, for example the global, regional and local, creating distinctive patterns of uneven development (section 1.2.2). The balance between agglomeration and dispersal also changes over time, shaped by the development of technology and organizational structures (for example, the growth of MNCs) in particular.

'New regionalist' policy has placed particular emphasis on the role of knowledge and skills as the key factors in economic development. Recent theories of agglomeration emphasize the role of knowledge 'spillovers' between firms, involving firms working together to solve problems and develop new products and services. As we emphasize in the second half of the chapter, knowledge and skills are increasingly seen as embodied in individual people as workers, entrepreneurs or developers. In line with advances in transport and communication technologies, skilled people are viewed as increasingly mobile, and development policies should focus on attracting skilled or talented people rather than companies. As such, skill or talent is viewed as a flow not a stock (Florida, 2002). In order to remain competitive and attract the right people, places need be promoted as dynamic, 'cool' and interesting, making them appear as exciting places to live or do business in.

11.2 Knowledge, innovation and agglomeration

Since the early 1990s, the growing importance of knowledge in supporting and driving economic growth has been emphasized by academics, business commentators and policy-makers in developed countries (Castells, 1996; Leadbetter, 1999). According to the Scandinavian economist Lundvall, capitalism has entered a new stage in which 'knowledge is the most important resource and learning the most important process' (Lundvall, 1994). This has become something of an overaching 'meta-narrative' or story underpinning economic development policy in the context of globalization, stressing the need for developed countries to follow a 'high road' strategy of specializing in advanced, high-value activities, in response to competition from lower-wage countries in Asia and eastern Europe (with cost-based competition defined as the 'low road').

Knowledge can be defined as a framework or structure in which information is stored, processed and understood (Howells, 2002, p.872). Existing stocks of knowledge shape how people respond to particular events and developments; a process which itself generates new knowledge, enhancing or transforming previous understandings. As this suggests, knowledge is distinct from information, referring to the broader frameworks of meaning through which information or data about real-world events and trends are processed and understood. As Nonaka *et al.* (2001, p.15) – a group of influential management theorists – put it, 'Information becomes knowledge when it is interpreted by individuals and given a context and anchored in the beliefs and commitments of individuals'.

A distinction is often drawn between codified and tacit forms of knowledge (ibid.). **Codified** or **explicit knowledge** refers to formal, systematic knowledge that can be conveyed in written form through, for example, programmes or operating manuals. **Tacit knowledge**, on the other hand, refers to direct experience and expertise, which is not communicable through written documents. It is a form of practical 'know-how' embodied in the skills and work practices of individuals or organizations. Traditionally, in industries such as construction, practical skills were acquired through apprenticeships where new entrants learned on the job by shadowing and assisting established tradesmen.

Computer packages provide a good example of the distinction between codified and tacit knowledge. The operating manual or programme informs the users about the operations and capabilities of the package, telling them what procedures to follow, but it is only by gaining actual experience of operating the package (tacit knowledge) that they become proficient users of it. As we shall see, much of the economic geography literature on learning and innovation is based on the assumption that codified knowledge has become increasingly global in organization and reach, while

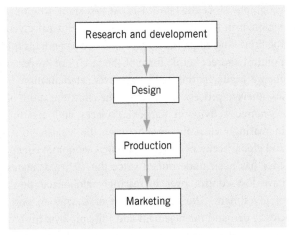

Figure 11.1 The linear model of innovation
Source: D. MacKinnon.

tacit knowledge remains local, relying on geographical proximity to foster communication and interaction between firms in specialist industrial clusters (Maskell and Malmberg, 1999).

11.2.1 Changing approaches to innovation

Innovation is a key theme of knowledge-based economic development. It can be defined as the creation of new products and services or the modification of existing ones to gain a competitive advantage in the market. The commercial exploitation of ideas is crucial, distinguishing innovation from invention.

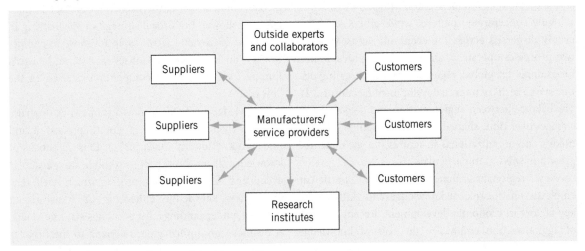

Figure 11.2 The interactive model of innovation
Source: D. MacKinnon.

Box 11.1

The development of Apple's iPod

Apple's iPod rapidly became the most popular and fashionable digital music player in the marketplace, following its launch in October 2001 (Figure 11.3). It is based on the recombination of several existing components, involving collaboration between a number of different companies. Rather than stemming from a revolutionary new invention, the iPod began with a sense in early 2001 that Apple could develop a better product than any of its rivals within the emerging market for MP3 players (Hardagon, 2005). The original idea came from an independent contractor, Tony Fadell, who was hired by Apple to develop the product. The platform design came from Portal Player and the operating system from Pixo (both Silicon Valley start-ups) while the hard disk was developed in collaboration with Toshiba and the lithium battery was obtained from Sony (Nambisan, 2005). By integrating these diverse components during an intensive eight-month design period, Apple was able to produce the most portable, user-friendly and fashionable digital music player, which combines small size and ease of use with a large storage capacity (holding over a 1,000 songs) thanks to the hard drive developed by Toshiba. Sales grew rapidly, equalling Apple's computer sales in two years and reaching 14 million in the three months up to 31 December 2005, as the iPod became a 'must-have' Christmas gift for many people (*The Economist*, 2006).

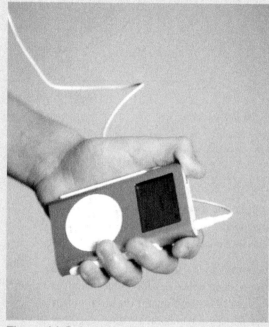

Figure 11.3 Apple's iPod
Source: © Dana Hoff/Beatworks/Corbis.

The traditional **linear model of innovation** was focused on large corporations, breaking it down into a series of well-defined stages running from the research laboratory to the production line, marketing department and retail outlet (Figure 11.1). It emphasizes formal research and development based on advanced scientists and engineers operating separately from other divisions of the company. In recent years, this has been replaced by an **interactive approach**, viewing innovation as a circular process based on cooperation and collaboration between manufacturers or services providers, users (customers), suppliers, research institutes, development agencies, etc. (Figure 11.2) (Box 11.2). The metaphor of the firm as a laboratory draws attention to the experimental nature of innovation, which is often based on trial and error, involving the adoption of existing practices and the trying out of new combinations (Cooke and Morgan, 1998, pp.47–53).

Firms are key agents of learning and innovation in the economy, as emphasized by the competence or resource-based theory of the firm derived from the economist Edith Penrose (section 3.2.2). From this perspective, the firm is a repository of knowledge. By its nature, learning is an evolutionary process, based on developing, processing and absorbing knowledge. Probably the best-known model of knowledge

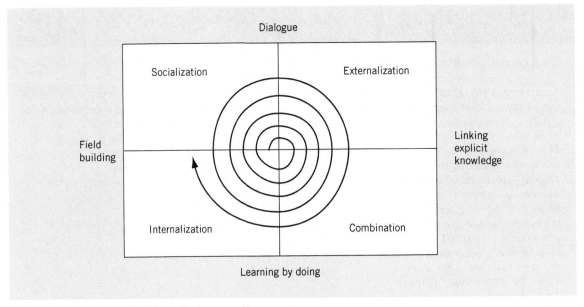

Figure 11.4 The spiral of knowledge creation

Source: The Knowledge-creating Company: How Japanese Companies Create the Dynamics of Innovation, (Nonaka, I. and Takeuchi, H., 1995). By permission of Oxford University Press, Inc.

generation within the firm is the four-stage **SECI approach** developed by Japanese management theorists (Figure 11.4) (Nonaka and Takeuchi, 1995).

➤ Socialization involves the articulation and exchange of tacit knowledge, requiring face-to-face interaction between experts within a firm. The key problem here is that of tapping into tacit knowledge that already exists within the firm, as expressed in the famous statement by Lew Platt, the former Chief Executive of Hewlett Packard, 'if only HP *knew* what HP knows, we would be three times as profitable' (quoted in Morgan, 2004, p.7, emphasis in original). Such knowledge is 'sticky', making it difficult to harness and move to where it is most required within the firm. Efforts to develop a new product need to overcome this problem. The first stage is often to create a 'field' of interaction, whereby experts from different departments get together for rounds of structured discussion and brainstorming (Bathelt *et al.*, 2004, p.35). When Honda, for example, decided to develop a new car in 1978, under the slogan 'Let's gamble', a product development team of young engineers and designers was formed for discussing what an ideal car should look like, in a series of brainstorming sessions (Nonaka and Takeuchi, 1995, pp.11–12).

➤ Externalization is based on the transformation of this tacit knowledge into codified form. It becomes formal and systematic, allowing it be shared with others. The work of the Honda team was shaped by the metaphor of 'automobile evolution', viewing the car as an organism and seeking to identify its ultimate form (Nonaka and Takeuchi, 1995, p.65). This fed into the concept of a car that was simultaneously tall and short, offering maximum comfort and room for the passenger, while taking up least space on the road. The team's concept of the 'tall boy' eventually led to the Honda City, the company's distinctive urban car of the 1980s.

➤ Combination refers to the combination of different bodies of codified knowledge into more complex and integrated systems. This involves the reconfiguration of existing knowledge through sorting, adding, combining and categorizing it. The Honda City design team's revolutionary concept of the 'tall boy' fed back into Honda's senior management sense that a radical change in style was required, helping to launch a new generation of Japanese cars.

> Finally, internalization is the process by which firms embody codified knowledge in the skills of workers and the routines and work practices of the firm, turning it back into tacit knowledge. Members of the Honda City team internalized the knowledge generated through the development of the new product, enabling them to play a leading role in other R&D projects. A key challenge in this internalization stage is to successfully contain the new knowledge within the firm, ensuring that it does leak out to competitors.

A key factor in determining a firm's success in innovation is its '**absorptive capacity**', referring to the ability to recognize, assimilate and exploit knowledge, derived from either internal or external sources (Cooke and Morgan, 1998, p.16). The absorption of knowledge, in turn, depends upon the existence of a common corporate culture and language, meaning that everybody shares the same broad outlook and sense of the company's overall purpose and objectives. In the absence of a common language and outlook, it becomes far more difficult to exploit tacit knowledge without it leaking out to competitors (Box 11.2).

11.2.2. Agglomeration and clusters

A key question for economic geography concerns the role of geographical proximity in facilitating learning and innovation within and between firms. The writings of Alfred Marshall, the renowned Cambridge economist, in the late nineteenth and early twentieth centuries are a key influence on contemporary theories of spatial agglomeration. Marshall's ideas were drawn from his observation of specialized industrial districts in Britain such as the Sheffield steel industry, emphasizing their distinctive 'industrial atmosphere'. More specifically, traditional explanations of the spatial agglomeration of industries, derived from Marshall, emphasize three main factors (compare Box 4.2):

> The growth of various intermediate and subsidiary industries which provide specialized inputs. This refers to the development of close linkages between manufacturers and suppliers of particular components and services with their co-location serving to reduce transportation costs.

> The development of a pool of skilled labour as workers acquire the skills required by local industry.

Box 11.2

'Fumbling the future': Xerox, Apple and the personal computer

A group of scientists working at Xerox's Palo Alto Research Centre in Silicon Valley discovered key elements of the personal computer (PC) in the 1970s. This included not only the processing equipment that sits beneath the desk, but also the screen desktop of icons, folder and menus that make up systems such as Windows, Macintosh and the world wide web.

Yet the development of the PC did not benefit the firm that initially created the technology. This was because of divisions within Xerox between scientists based in Palo Alto and development engineers in Dallas, Texas and the company management in Stanford, California. Not only were the different groups separated geographically, there was also an absence of the common language and outlook required to exploit emerging technologies commercially. The engineers found the scientists naïve and unrealistic, while the scientists viewed employees in other divisions as 'toner heads' who were interested only in photocopiers (Brown and Duguid, 2000, p.151).

Such internal divisions meant that the knowledge embodied in the emerging computer technology leaked out of Xerox, to be exploited by another company, Apple, based nearby in Cupertino. In what has become a well-known story in business folklore, Steve Jobs, one of the founders of Apple, visited the Xerox plant in 1979. After Jobs recognized the potential of the technology being developed there – something that the management of Xerox had failed to do – Apple copied and exploited it, licensing some aspects and replicating others. Thus, 'the knowledge that stuck within Xerox leaked readily through its front door' (ibid., p.151). In this way, the company 'fumbled the future', losing out to one of its main competitors.

Thus, workers can readily find suitable employment and employers can find skilled labour locally, reducing the search costs for both parties.

➤ The establishment of information and knowledge spill-overs between firms and workers that are located close together (Malmberg and Maskell, 2002). Proximity allows economic actors to share specialized knowledge and learning from one another on the basis of the trust and understanding derived from repeated face-to-face contact.

These three factors generate **agglomeration economies**, which can be defined as cost advantages that accrue to individual firms because of their location within a cluster of industrial growth (Knox *et al.*, 2003, p.242). These advantages are sometimes also known as external economies because they stem from circumstances beyond a firm's own practices, reflecting broader features of the local environment. Agglomeration economies be divided into **localization economies** such as those listed above, stemming from the concentration of firms in the *same* industry, and **urbanization economies**, derived from the concentration of firms in *different* industries in large urban areas.

11.2.3 Agglomeration and learning

Since the 1980s, Marshall's work has been rediscovered by a new generation of economic geographers and economists. The focus of economic geography research on agglomeration has generally moved from the traditional Marshallian notion of cost reduction, through agglomeration economies, to an emphasis on the dynamic benefits of clustering in facilitating innovation and learning processes (Malmberg and Maskell, 2002). The third of Marshall's location factors has assumed particular importance in this respect, in terms of how geographical proximity between firms allows firms to create and share information and knowledge, stimulating processes of adaptation, learning and innovation. By contrast, the 'new geographical economics' of mainstream economists such as Krugman (see Box 2.2) has emphasized the more tangible factors of specialized inputs and labour markets, sparking concern

about its continued practical relevance in developed economies where the sources of regional specialization seem to be increasingly rooted in information and knowledge spillovers (Krugman, 2010).

One of the most influential recent models of spatial agglomeration or concentration is the Harvard business economist, Michael Porter's, theory of business **clusters**. Cluster policies have been adopted by government agencies in a range of different countries, reflecting how Porter's theory has been successfully promoted by his Monitor consultancy. Porter was originally concerned with the external conditions that support a firm's competitiveness at the national scale. More recently, he has focused on the role of sub-national clusters in enhancing innovation and productivity. For Porter, clusters are defined as:

> geographical concentrations of interconnected companies, specialised suppliers, service providers, firms in related industries, and associated institutions (for example universities, standards agencies and trade associations) that compete but also co-operate.
>
> (Porter, 1998, p.197)

There are two key elements in this definition (Martin and Sunley, 2003, p.10). First, the firms in the cluster must be linked in some way, for example by the supply of specialized inputs or services. Second, a cluster is defined by geographical concentration, with proximity creating a commonality of interest between firms and encouraging frequent interaction (Box 11.3).

Porter argues that geographical concentration encourages processes of interaction within the 'competitive diamond' by increasing the productivity of constituent firms or industries, stimulating higher rates of innovation and encouraging high rates of business formation. The **diamond model** is based on the contention that clusters enhance competitiveness and productivity by fostering the interaction between four sets of factors (Figure 11.5). Demand conditions refer to the tendency for successful clusters to generally serve global markets with 'leading edge' local customers – firms that sell to global markets – playing a key role in encouraging innovation among suppliers.

Second, supporting and related industries refer to firms that supply inputs or simply industries located in

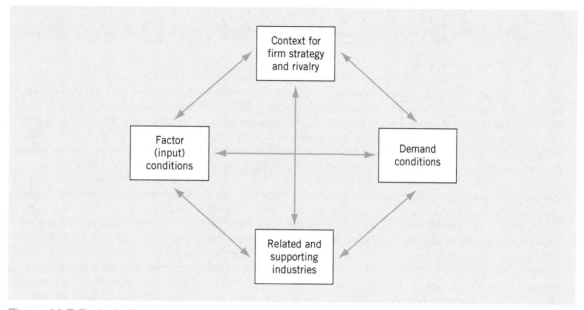

Figure 11.5 Porter's diamond model

Source: 'Locations, clusters and company strategy', in *The Oxford Handbook of Economic Geography*, edited by Clark, G.L. and Gertler, M. (Porter, M.E., 2000). By permission of Oxford University Press.

the same area. Close links between customers and suppliers enable complex communication and interaction to occur, acting as key channels of learning and knowledge transfer, while a concentration of other industries creates the critical mass to support advanced skills, training and infrastructure (Box 11.3).

The third dimension of the model relates to factor conditions, referring to the main factors of production – land, capital, labour and knowledge. The availability of capital to fund investment and growth is important, alongside the presence of a skilled workforce. Such skills need to be continually updated and enhanced through training and education programmes. In addition, clusters need to support innovation, based on close links between universities, research institutes and leading firms, ensuring that scientific research is adequately funded and commercialized.

The final element in the model is firm strategy, structure and rivalry. Maintaining a substantial number and range of firms is crucial, requiring a high rate of new firm formation through mechanisms such as corporate spin-off, which is a key mechanism for this. At the same time, rivalry between co-located firms encourages investment and innovation in order to keep up and remain competitive. Proximity enables

firms to monitor the activities of local rivals, providing them with access to information and making it difficult to ignore new developments. Apple's poaching of PC technology from Xerox provides an excellent example of this effect (Box 11.2).

Recently, however, the emphasis on local linkages in clusters research has been questioned, with critics suggesting that proximity is not only a spatial phenomenon but can also take social and organizational forms (Gertler, 2003). According to advocates of relational thinking, 'close' long-distance relationships are also possible both within and between firms (Allen, 2000). At the same time, studies of well-known clusters have highlighted the importance of extra-local linkages alongside more localized relationships (Box 11.3). The concept of 'global pipelines', developed by a group of Scandinavian researchers, builds on these insights (Bathelt *et al.*, 2004). The key claim is that, in addition to engaging in processes of localized learning within a cluster, firms seek to build channels of communication or pipelines, with selected partners outside the cluster (Figure 11.7). Such strategic partnerships offer access to knowledge and assets not available locally, although their number and scope are limited by the cost and time involved in building them. Successful establishment of

Box 11.3

The Hollywood film production cluster

The US motion picture or film industry has become synonymous with the district of Hollywood in Los Angeles, southern California. Hollywood represents a distinctive geographical phenomenon, consisting of a dense network of film production companies and service providers (Scott, 2002). Most of the industry is clustered in a relatively small area centred on Hollywood itself, stretching from Burbank in the east through to Beverley Hills and Santa Monica in the west (Figure 11.6) (ibid., p.965). At the same time as Hollywood has become increasingly

globalized, production has remained concentrated in southern California (Currah, 2007). This presents something of a paradox, at least for those who believe that electronic communications herald the 'end of geography' (O'Brien, 1992).

Hollywood has historically been dominated by the seven 'majors': the large corporate studios such as Paramount, Twentieth Century Fox and Warner who have controlled film production and distribution. In the 'old' system, these majors directly controlled most aspects of film production, distribution and exhibition,

owning their own theatre (cinema) chains across the US (Storper and Christopherson, 1987). This system began to break up in the 1950s and 1960s, following the Paramount antitrust decision by the US courts in 1948 and the invention of television in the 1950s. The former forced the majors to get rid of their extensive cinema chains, while the latter shrunk the audience for films. The result was increased competition, and the creation of a climate of uncertainty and instability. The majors responded by divesting themselves of many activities, outsourcing these to

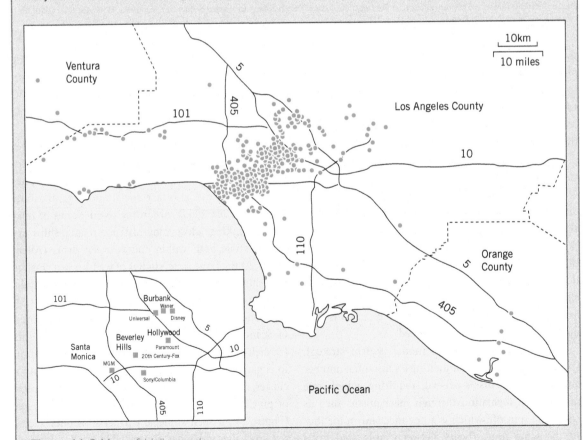

Figure 11.6 Map of Hollywood

Source: 'A new map of Hollywood: the production and distribution of American motion pictures', in *Regional Studies*, 36. Taylor & Francis Ltd (Scott, A.J., 2002). http://tandf.co.uk/journals.

Box 11.3 (continued)

small suppliers and becoming 'nerve centres' of production, exercising overall coordination and control. The number of small firms providing specialized services and inputs, such as script writing or film editing, grew markedly. At the same time, the majors have maintained their economic power, particularly in terms of financing, deal-making and distribution, even in the face of the internet, which allows consumers to access material directly (Currah, 2007).

The Hollywood film production complex is characterized by four key features (Scott, 2002, p.965), which explain the pattern of geographical concentration:

➤ A series of overlapping production networks encompassing majors, independents and providers of specialized services. The frequency, complexity and volume of transactions (contacts) between firms and individuals fosters a

need for face-to-face communication (Storper and Christopherson, 1987, p.112).

➤ The local labour market supplies a large number of individuals with the requisite skills and experiences, constantly replenished by the arrival of new talent from the rest of North America and the world.

➤ The institutional environment consists of many organizations and associations which represent the interest of firms, workers and government agencies.

➤ The broader regional environment – produced in part through the geographical concentration of the industry there over time – provides a range of crucial resources, ranging from a cinematic tradition to background landscapes and proximity to other cultural industries.

The industry has experienced

substantial growth in recent decades with employment in motion-picture production and services expanding by 194 per cent between 1980 and 1997, maintaining its high level of geographical concentration. Some dispersal to satellite locations has occurred in the search for locations and reduced costs, particularly for film-shooting, mainly to Canada. The majors have largely viewed the rise of the internet as a threat, emphasizing the scope for 'piracy' through direct downloading, and have sought to control its implications through the deployment of 'digital rights management' technologies (Currah, 2007). Supported by substantial lobbying and public relations campaigns, these efforts to create a 'closed' system of innovation on a global scale have met with some success, although the majors have been unable to prevent the parallel growth of an 'open' online system of innovation.

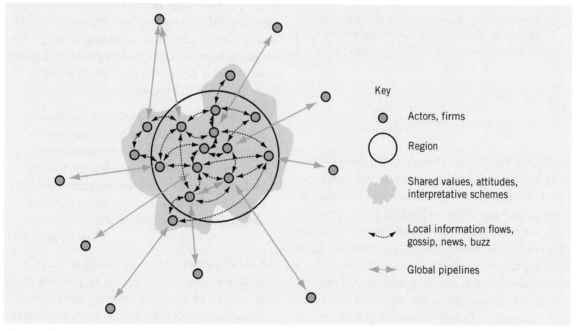

Figure 11.7 Local buzz and global pipelines
Source: Bathelt *et al.*, 2004, p.46.

global pipelines requires firms to develop a shared organizational context, which enables them to learn and solve problems together.

At the same time, such relationships complement and enhance local linkages, rather than being a substitute for them (ibid.). While a firm's embeddedness in a cluster provides automatic and routine access to a range of information and knowledge, pipelines provide access to more specialized forms of knowledge that are not locally available. This specialized knowledge may relate to the development of new technologies or new market opportunities. Bathelt *et al.* (2004) suggest that wider links are particularly important during the early stages of cluster formation, providing access to markets and knowledge before critical mass is achieved locally. Maintaining such links as clusters mature is also seen as important to avoid introversion as local linkages become too close and rigid, leading to 'lock-in' as firms fail to respond to change. In certain circumstances, a conflict is likely to arise between the need to (a) protect and maintain local knowledge and (b) access and plug into non-local networks.

Reflect

> What measures do you think that firms should adopt to stimulate learning, particularly in terms of encouraging communication between departments?

11.3 The rise of the creative classes?

The *Rise of the Creative Classes* was the title of a 2002 book by Richard Florida, an economic geographer now based at the University of Toronto, which became an international best-seller and public policy phenomenon. As a result, Florida himself became something of a celebrity academic, able to command five-figure speaking fees across North America. The basic thesis of the book is that urban prosperity has become increasingly dependent on the attraction and retention of the so-called **creative classes**, highly skilled and educated workers who have distinct lifestyle preferences. As such, while reinforcing the attachment to

the concepts of knowledge-based development and competitiveness (see below), Florida's arguments are associated with an increased emphasis on the attraction of skilled individuals as the basis for successful agglomeration, in contrast with the concern with firm-based dynamics that underpinned the clusters approach. According to one critic, Florida's notion 'that cities must become trendy, happening places in order to compete in the twenty-first century economy is sweeping urban America' (Malanga, 2004, p.1). In a similar fashion to Michael Porter, Florida and his colleagues have been highly successful in promoting their ideas through a consultancy arm – the Creative Classes Group. According to one star-struck local official, getting 'Richard Florida['s Group] to teach you how to create an environment for growth is like getting da Vinci to teach you how to draw' (see http://www.creativeclass.com/).

Florida's basic argument is that a new age of creativity has dawned in recent years, representing a distinctive phase of capitalist development, analogous to the industrial and organizational ages of the nineteenth and twentieth centuries (2002, pp.56–66). From this perspective, the driving forces of economic development are rooted in the human capacity to generate ideas and knowledge rather than in technology or organizational structures *per se*. The basis of Florida's creative class is economic, consisting of people who add economic value through their creativity, engaging in work 'whose function is to create meaningful new forms' (2002, p.68). As such:

> the super-creative core of this new class includes scientists and engineers, university professors, poets and novelists, artists, entertainers, actors, designers, and architects, as well as the thought leadership of modern society: nonfiction writers, editors, cultural figures, think-tank researchers, analysts and other opinion-makers.
>
> (Florida, 2005, p.34)

This super-creative core is estimated to account for 15 million workers in the US, or 12 per cent of the workforce. In addition, the creative classes consist of creative professionals in high-tech sectors, business and finance, law, healthcare and management, who engage in complex problem-solving that involves independent judgement, requiring a high level of education or human capital (Florida, 2002, p.69). As a whole, Florida argues, the creative class has grown markedly to incorporate 30 per cent of the US workforce, increasing tenfold since 1900 (ibid., p.74). At the same time, the 'service class' – made up from low-wage, low-autonomy service workers in sectors such as healthcare, catering, cleaning and administration – has also grown from 5 million in 1900 to 55 million in 1999, while the traditional working classes in manufacturing, transportation, maintenance and construction has fallen from 40 per cent of the workforce in 1950 to around 25 per cent (ibid.). In many respects, Florida's creative class thesis resonates with the American sociologist Daniel Bell's theory of post-industrial society, which was influential in the 1960s and 1970s, arguing that traditional 'blue-collar' manual work in manufacturing would be eclipsed by the rise of new 'white-collar' jobs in advanced services that required high levels of education and knowledge.

The most controversial element of the creative class thesis concerns its implication for economic development, which involve cities and regions competing to attract mobile 'creatives' (Box 11.4). According to Florida, the new economic geography of creativity is based on what he calls the **3 Ts: technology, talent and tolerance**. Technology is defined in terms of the presence of high-tech industry, measured according to the size and concentration of a region's economy in key growth sectors. Talent refers to high levels of human capital, and is measured largely in terms of educational attainment, particularly the percentage of the population with a Bachelor's degree. Tolerance is the most novel of Florida's categories, referring to the openness and diversity of a particular place, measured though various indexes such as the gay index (the

Box 11.4

Scotland's Fresh Talent Initiative

Informed by the writings of Florida and the wider emphasis on competiveness, the devolved Scottish government launched a Fresh Talent Initiative in February 2004. Such 'fresh' or 'foreign' talent schemes have been adopted by a number of states since the 1990s, reflecting the importance attached to the attraction of skilled workers as a key means of ensuring growth and competitiveness with the global economy (see Yeoh and Chang, 2001). Launched against a backdrop of international competition for skilled labour, the Scottish scheme was promoted by the government as a progressive scheme that would help to cultivate a 'global sense of place' in Scotland (section 1.2.3) (Massey, 1994). It attracted particular interest as a distinctive policy initiative adopted in the context of devolution, whereby immigration policy was officially reserved to the UK government. As

such, the introduction of Fresh Talent was negotiated by Scottish ministers with their UK counterparts in a climate in which the UK government was adopting an increasingly restrictive approach. Its genesis reflected specific Scottish concern about population decline, an ageing population and skills gaps in the labour market (Scottish Government, 2006).

The Fresh Talent Initiative has three main aims: to address the problems of a falling population (prior to 2004) and population ageing; to increase the dynamism and cosmopolitanism of Scottish society; and to promote Scotland as an ideal place to live, study, work and do business (ibid.). It was targeted on five main groups: students and people seeking employment; entrepreneurs and the self-employed; Scottish business; and expatriate Scots looking to return. The Initiative was implemented through a range of specific measures,

including: a Relocation Advisory Service; the Fresh Talent Working in Scotland Scheme (FTWiSS), which allows overseas graduates to stay for a further two years in Scotland after the completion of their studies without a work permit; Scottish Networks International; and a renewed emphasis on place marketing and the creation of a distinctive international image. An evaluation of the FTWiSS scheme found that it had been particularly successful in attracting international students to Scotland, although its effects during and after the two years were far less clear, with some graduates experiencing difficulties in finding the right kind of work, while employers' awareness of, and involvement in, the scheme was generally low (Scottish Government, 2008). The Fresh Talent Initiative came to an end in June 2008, when the UK government introduced a new points-based immigration system.

proportion of gay people in the population) and the 'bohemian index' (the percentage of writers, designers, musicians, actors and directors, painters and sculptors, photographers and dancers in the population). More recently, the Creative Class Group has added a fourth T, territorial assets, to provide a more comprehensive framework for building regional prosperity (http://www.creativeclass.com/).

The basic policy agenda is that cities must promote tolerance, viewed as the 'crucial magnet' in Florida's formulation, encouraging openness and diversity, in order to attract talent and generate prosperity. The principal method of the so-called urban creativity industry associated with Florida's ideas is the ranking of cities according to these various indexes and correlation analyses of the relationships between the different measures (Table 11.1) (Figures 11.8 and 11.9). Cities such as San Francisco, Austin and San Diego, already celebrated for their high-tech industries and cultural attractions, emerge as the big-city winners of the urban creativity race (Table 11.1), followed by the likes of Albuquerque, New Mexico, Albany, New York and Tucson, Arizona for the mid-sized cities group and Madison, Wisconsin,

Des Moines, Iowa, and Santa Barbara, California for smaller cities (Peck, 2005, p.747).

According to the economic geographer Jamie Peck (2005), the creative cities script can be seen as a

Table 11.1 The creativity index		
Rank	Region	Score
1	San Francisco	1057
2	Austin	1028
3	Boston	1015
4	San Diego	1015
5	Seattle	1008
6	Raleigh-Durham	996
7	Houston	980
8	Washington, DC	964
9	New York	962
10	Dallas–Fort Worth	960
11	Minneapolis–St Paul	960
12	Los Angeles	942
13	Atlanta	940
14	Denver	940
15	Chicago	935

Source: Florida, 2005, p.157.

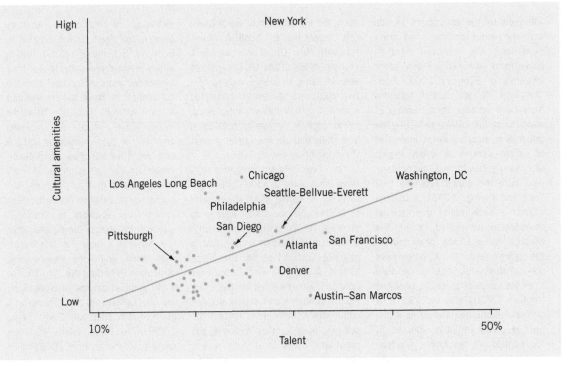

Figure 11.8 Talent and cultural amenities
Source: Florida, 2005, p.98.

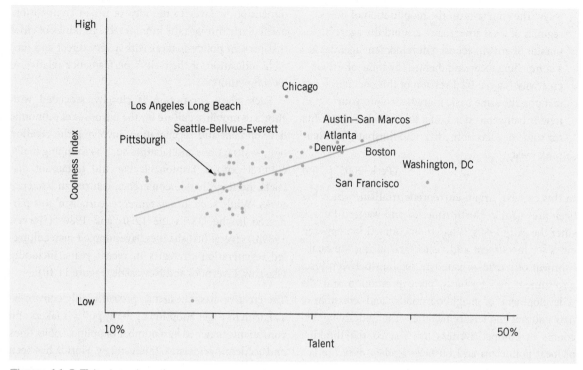

Figure 11.9 Talent and coolness
Source: Florida, 2005, p.98.

'mobilizing discourse', which has prompted policy-makers and business leaders to follow its agenda in order to increase the competitiveness and attractiveness of their cities. Famously, the mayor of Denver ordered multiple copies of *The Rise of the Creative Class* as bedtime reading for his senior staff, while developing a strategy to rebrand the city as a creative centre. At the same time, the Governor of Michigan launched a 'Cool Cities' programme across the state, despite swingeing public spending cuts (Peck, 2005b, p.742). Many other urban development consultancies bought into Florida's ideas on creativity, reinforcing their rapid spread and influence. For example, the UK-based think-tank Demos produced a UK Creativity Index to illustrate the creative potential of the UK's 40 largest cities, measured according to patent applications per head, non-white residents and levels of gay-friendly services. The top ten were Manchester, Leicester and London (equal second), Nottingham, Bristol, Brighton, Birmingham, Coventry, Cardiff and Edinburgh (Nathan, 2005, p.2). Examples of creative class communities initiatives, cited on the Creative Class website, include a Binational Downtown Design Competition in El Paso, Texas, the Tallahassee

Film Festival (showcasing the creative talent of the area) and a 'Third Places' Social Networking Program in Charlotte, North Carolina.

Creating an attractive, 'cool' and 'happening' environment for the creative classes typically involves the redevelopment of neighbourhoods to emphasize the preponderance of authentic historical buildings, converted lofts, walkable streets, coffee shops, arts and live music spaces, generating a sense of buzz and tolerance. As such, it has tended to reinforce and accelerate existing processes of gentrification through which former industrial and working-class inner cities areas have been redeveloped and colonized by middle-class professionals and 'yuppies' (Smith, 1996). More generally, the creative class discourse fits with established modes of urban entrepreneurialism and competition:

the contemporary cult of urban creativity has a clear genealogical history, stretching back at least as far as the entrepreneurial efforts of reindustrialised cities. The script of urban creativity reworks and augments the old methods and arguments of urban entrepreneurialism in politically seductive ways. The emphasis on the mobilisation of new regimes of local governance around the aggressive pursuit of growth-focused development agendas is a compelling recurring theme. The tonic of urban creativity is a remixed version of this cocktail:

just pop the same basic ingredients into your new-urbanist blender, add a slug of Schumpeter-lite for some new-economy fizz, and finish it off with a pink twist.

(Peck, 2005, p.766)

In this context, **urban entrepreneurialism**, which has been prevalent in North America and western Europe since the early 1980s, is based on competition between cities for investment and visitors, requiring the establishment of public–private partnerships between local government and business, place promotion and the redevelopment of neighbourhoods, and creation of new cultural and leisure facilities. The underlying discourse of **competitiveness** has shaped the thinking of local politicians and business leaders over the last couple of decades, helping to explain their receptiveness to Florida's ideas (Box 11.5). In this respect, the creative class approach can be seen as adding a new dimension or layer to the idea of urban competition, particularly through the innumerable rankings of cities that present policy-makers with a very direct and tangible indication of their city's performance relative to its competitors.

Such place promotion is closely associated with efforts to mobilize culture for the purposes of economic development and regeneration, involving the creation of new attractions and facilities such as shopping malls, cultural centres, fashionable bar and restaurant districts, heritage parks, conference centres and science parks. While such 'urban entrepreneurialism' was pioneered in the US in the 1970s and 1980s (Harvey, 1989b), several British cities have adopted such culture-led regeneration strategies in recent years, including Glasgow, Liverpool and Newcastle (Figure 11.10).

The creative class thesis has proved highly controversial, with its eager adoption by many policy-makers and consultants matched by a number of political objections and academic criticisms. Interestingly, Florida has been castigated by both the political right and left for advocating big government-style spending programmes and progressive liberal policies over low taxes and 'family

Figure 11.10 Gateshead Quays regeneration area
Source: D. MacKinnon.

Box 11.5

Competitiveness: a dangerous obsession?

Since the early 1990s, the concept of competitiveness has gained hegemonic status within economic policy debates in developed countries (section 5.5.4) (Bristow, 2010). This means that it has become widely accepted, being regarded as simple common sense. It refers to the notion that cities and regions compete with one another for market share in the global economy, emphasizing the need to promote learning and innovation in order to gain a competitive edge on rival regions. As well as using the notion to develop specific polices such as the identification of clusters, the commercialization of research and the promotion of exports, government bodies have attempted to measure and model competitiveness. 'Benchmarking' studies where regions track and measure their economic performance against other regions have become particularly prominent.

For critics such as the Nobel Prize-winning economist Paul Krugman, the focus on competitiveness is unfortunate since it is fundamentally erroneous, representing a 'dangerous obsession' among policy-makers and commentators. Krugman argues that competitiveness is based on a false analogy between the corporation and

the national or regional economy, involving 'a set of crude misconceptions, presented as if they were sophisticated insights' (1996, p.256). When applied to firms, competitiveness has a clear meaning, referring to their capacity to grow and become profitable in the marketplace, and can be measured in terms of output and profitability (Bristow, 2005, p.287). While regions or nations may compete to attract particular investments, such as a new car factory or government research facility, the idea that they compete for market share in a similar fashion to firms is highly simplistic. In reality, much economic activity remains domestic in nature, particularly in large economies such as the US. Furthermore, international trade is not the zero sum game that it is portrayed as by the competitiveness industry (section 4.2); instead, it generates additional income for both parties, with the demand for imports by country B boosting the export industries of country A and vice-versa. Krugman suggests three reasons for the popularity of competitiveness rhetoric: it is exciting, portraying economic development as a contest between competing territories; it makes economic problems

seem easier to solve, targeting supply-side measures to boost innovation and learning; and it is politically useful, justifying tough policies – for example, welfare cuts or wage restraints – as a necessary response to the pressures of global competition (Coe and Kelly, 2002).

Krugman's critique of the competitiveness agenda is forceful and compelling, offering a necessary rejoinder to the competiveness industry. Yet Krugman's arguments lack much geographical sensibility, providing little sense of the nature of cities or regions as social entities. As a substantial volume of research shows, aspects of the urban and regional environment shape the competitiveness of firms. Advantages derived from the spatial concentration of economic activity in particular places – for example, close links with suppliers, a pool of skilled labour and shared infrastructures – are important in underpinning the success of firms within dynamic clusters. Thus, a modified version of regional competitiveness, which stresses the role of the regional environment in shaping firms' growth, remains useful (Kitson *et al.*, 2004).

values', on the one hand, and for favouring policies that increase social polarization by privileging the needs of educated, high-income workers over the working and service classes, on the other hand (Malanga, 2004; Peck, 2005b). In academic terms, a number of problems has been identified. First, the creative class is a rather fuzzy concept, presenting a number of generalizations that are 'proven' by correlations between particular measures of creativity. Crucially, however, Florida's approach fails to specify the nature of the underlying causal relations between the various dimensions of creativity. While he argues that tolerance attracts talent and technology, it is at least equally plausible to suggest that the direction of causation runs the other way, whereby

centres of high-tech industry attract talent and generate diversity as a by-product of growth, as suggested by the leading position of cities such as San Francisco, Austin, San Diego and Boston (Markusen, 2006). Second, the concept is large and unwieldy, based on the aggregation of a range of distinctive occupations into a single class (Nathan, 2005). Third, Florida fails to really illuminate the workings and meaning of creativity, relying on broad social categories and crude measures such as educational attainment (Box 11.6).

Finally, Florida presents a highly celebratory account of economic change and competition that not only neglects problems of inter-urban inequality and poverty, but is also likely to reinforce them through the policies it advocates. For instance, the redevelopment and gentrification of selected urban neighbourhoods in order to attract talent is likely to result in the further marginalization of the needs of 'non-creative' groups. Here, critics on the left have argued that the most creative cities and regions, as measured by Florida's rankings, also tend to be the most socially polarized and unequal, reflecting how creative strategies have become part of the broader neoliberal policy agenda

of urban competition and entrepreneurialism (Peck, 2005). In recent years, Florida himself has acknowledged this point, providing further statistical evidence of the correlation between his various indicators of creativity and polarization, but attributing this to broader economic shifts which are increasing the returns to higher education and creative ability (Gertler, 2010, p.9).

Box 11.6

Unpacking the creative class: artists in Minneapolis/St Paul

One study which does shed considerable light on the workings of creativity in the economy is Ann Markusen's study of artists in the twin cities of Minneapolis and St Paul in the US Midwest, reflecting the need to break down the very broad and aggregative concept of the creative class into specific creative occupations (Markusen, 2006). Markusen's definition of artists encompasses four sub-groups – writers, musicians, visual artists and creative artists – accounting for 1.4 million jobs in the US. Artists are more likely to be self-employed than

the labour force as whole, seeming to fit the characterization of the creative class as 'footloose' and being prone to choose locations on the basis of place rather than employment. In the 1990s, artists became more concentrated in the three 'superarts' metropolitan areas of Los Angeles, New York and San Francisco, which increased their leads over second-tier arts-specialized metros of Washington DC, Seattle, Boston, Minneapolis–St Paul and San Diego (Table 11.2). Such urban economies both attract and 'homegrow' artists,

with educational institutions and cultural organizations playing a key role. The forces that attract artists to particular cities are complex, but include agglomerations of prospective employers in media, advertising and the arts, in addition to lower costs of living, recreation and environmental amenities, and supportive cultural conventions and activities.

Markusen's study found that artists do make discretionary 'creative class' location choices that are semi-independent of employers. At the same time, the relationship

Table 11.2 Artistic specializations, selected metros

	Location quotient		
	1980	1990	2000
Los Angeles	2.39	2.31	2.99
New York	2.60	2.42	2.52
San Francisco	1.79	1.60	1.82
Washington, DC	1.76	1.63	1.36
Seattle	1.59	1.40	1.33
Boston	1.51	1.49	1.27
Orange County, CA	1.15	1.26	1.18
Minneapolis–St Paul	1.20	1.27	1.16
San Diego	1.24	1.15	1.15
Portland	1.18	1.24	1.09
Atlanta	1.31	1.08	1.08
Chicago	1.03	1.09	1.04
Cleveland	0.82	0.83	0.79

Source: Markusen, 2006, p.1929.

Box 11.6 (continued)

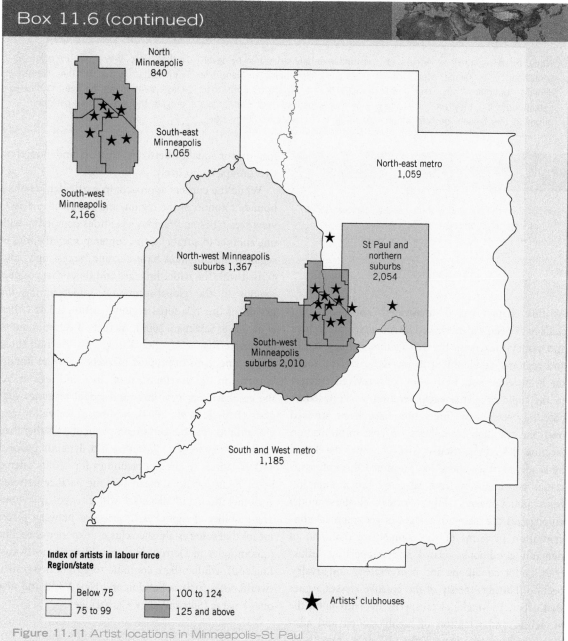

Figure 11.11 Artist locations in Minneapolis–St Paul
Source: Markusen, 2006, p.1930.

between artists and high-tech industry is far from clear, with little evidence of them clustering together in the same cities. Within cities, artists tend to gravitate towards denser, inner-city neighbourhoods, often relatively seedy, transitional ones which provide access to arts schools, performance and exhibition spaces, affordable live/work and studio space, training agencies, artists' centres and amenities such as nightlife and recreational opportunities (Figure 11.11) (Markusen, 2006, p.1930). Artists tend to be politically progressive, voting and campaigning for left and Democratic candidates in elections. In general, they also support more decentralized, neighbourhood-based theatres, galleries and other artist-centred spaces, which they believe to be under-supported, often involving work with racial and minority groups.

Box 11.6 (continued)

While often regarded as agents of gentrification in taking over abandoned buildings in run-down neighbourhoods, they are only one player in this broader process which is fundamentally structured by developers and wider zoning and land use practices. In respect of both their progressive politics and the factors shaping their location within cities, artists have very little in common with other groups of the 'creative class', reinforcing the need to disaggregate this overarching category.

Reflect

➤ Explain the popularity of the creative classes model among urban and regional policy-makers.

11.4 Summary

As this chapter has demonstrated, knowledge and skills have been widely viewed by academic researchers and policy-makers as the key factors shaping urban and regional development in recent years. The focus on knowledge and innovation has revived interest in the topic of spatial agglomeration, as knowledge-based aspects of agglomeration have been stressed over the traditional emphasis on cost minimization (section 4.3). In particular, two concepts developed by academic researchers and promoted through associated consultancies have attracted much attention from policy-makers. First, Porter's clusters model emphasizes the main advantages of geographical concentration in terms of the competitive diamond of demand conditions, related and supporting industries, factor conditions and firm strategy and rivalry. Second, Florida's theory of the creative classes argues that urban and regional prosperity depends upon the attraction and retention of highly skilled and educated workers, who have distinct lifestyle preferences, favouring diverse and tolerant places. While Porter highlights the actions of firms as the basis of urban and regional prosperity and growth, in common with earlier theories of agglomeration, Florida's approach focuses attention on the role of skilled individuals (talent). Florida's writings have proved particularly contentious, provoking criticism from both the political right and left for its 'trendy' liberalism and for

fostering inequality between 'creative' and non-creative' groups respectively.

While the clusters approach, in particular, retains a bounded notion of the region, and the creative classes view sees cities and regions as entities competing with one another to attract 'talent', contemporary theories of regional development have become increasingly concerned with the wider linkages and flows that connect regions to the global economy, underpinning the notion of the 'relational region' (section 2.5.4) (Allen *et al.*, 1998; Saxenian, 2006). As such, it is important to appreciate that wider global linkages – beyond those connecting firms to product markets – play an important role in the development of cities and regions. As the emphasis on local buzz and global pipelines suggests (Bathelt *et al.*, 2004), however, localization and globalization can be seen as complementary rather than contradictory forces. Clustering is a dynamic process, and variations in the opportunities for profits offered by different locations mean that the balance between agglomeration and dispersal will change over time. The mobility of capital as it 'see-saws' between places means that some established clusters experience decline (automobiles in Detroit or shipbuilding in north-east England), while others continue to attract investment (world cities such as London and New York) and new ones emerge (Silicon Valley, Cambridge).

Exercise

Select a long-established cluster of economic activity and review its development over time. Appropriate resources to draw on are academic articles, reports by regional development agencies or local authorities, economic statistics, materials or websites produced by business associations, newspaper articles and media reports.

What were the origins of the cluster? Why did it develop there? What markets did it serve? What were its sources of competitive advantage? Are these consistent with Porter's diamond model? Was it based on a few large firms or a network of smaller firms? What were the key supporting institutions? What was the role of local linkages between firms and institutions? What were the main external linkages (global pipelines)?

Did the cluster remain competitive or experience a process of decline as it matured? How would you explain this outcome? How important were internal conditions (capital, skills, knowledge, infrastructure, attitudes/practices) and external factors (competition, technology, markets)? If the cluster remained competitive, what processes of adaptation were involved? If it experienced decline, why did this occur?

Key reading

Florida, R. (2005) *Cities and the Creative Class*, London: Routledge.

One of the key texts in Florida's creative classes series in which he consolidates and extends the underlying concept and applies this to cities in North America. *Cities and the Creative Class* is particularly instructive in illustrating the methods of the creative class approach through a profusion of tables and charts, which rank cities according to various creative criteria.

Gertler. M.S. (2003) 'A cultural economic geography of production', in Anderson, K., Domosh, M., Pile, S. and Thrift, N.J. (eds) *The Handbook of Cultural Geography*, London: Sage, pp.131–46.

An insightful review of debates about cultures of learning and innovation in economic geography. In particular, Gertler argues than recent research lacks a convincing understanding of the firm as key learning agent. As a result, key questions regarding the importance of local linkages and global networks, respectively in facilitating learning and knowledge transfer, remain unanswered.

Markusen, A. (2006) 'Urban development and the politics of a creative class – evidence from a study of artists.' *Environment and Planning A*, 38: 1921–40.

A very interesting case study of the locational preferences and cultural politics of artists in twin cities of Minneapolis and St Paul in Minnesota. Markusen marshals her research very effectively to question key aspect of the creative class thesis, stressing the need to break it down through the study of specific occupations.

Peck, J. (2005) 'Struggling with the creative classes.' *International Journal of Urban and Regional Research*, 29: 740-70.

An excellent critical assessment of the creative classes approach, which explains its popularity among policy-makers in terms of its fit with existing discourses of urban entreprenurialism and competitiveness.

Porter, M.E. (2000) 'Locations, clusters and company strategy', in Clark, G.L., Feldman, M. and Gertler, M. (eds) *The Oxford Handbook of Economic Geography*, Oxford: Oxford University Press, pp.253–74.

An accessible account of the clusters concept from its founder. Porter defines the concept and explains the key mechanisms of growth within clusters, drawing on a range of examples. The discussion is set within a framework of firm strategy, focusing on the notion of competitiveness that he has done so much to popularize.

Useful Websites

http://www.compete.org/nri/ncric.asp
The US Council of Competitiveness – a partnership between industry, universities and trade unions – website, covering issues of competitiveness and regional innovation.

http://www.creativeclass.com/
The website of the Creative Class Group, featuring Richard Florida as 'author and thought leader'. The site contains background information on the underlying approach, an introduction to services offered by the Creative Class Group, biographical information on members of the groups, summaries of creative class community initiatives and a forum for exchange.

http://www.isc.hbs.edu/
The website of the Institute for Strategy and Competitiveness at Harvard Business School, led by Michael Porter. Contains a wealth of information on clusters, competitiveness and business strategy.

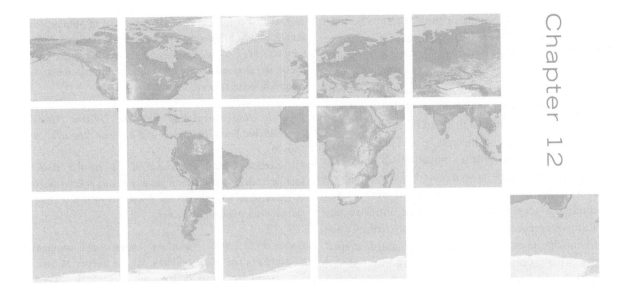

Chapter 12

Alternative economic geographies

Topics covered in this chapter

➤ Capitalism and its alternatives.

➤ Variety, diversity and uneven development.

➤ Alternative economic spaces.

➤ Alternative local economic spaces.

➤ Alternative networks of global trade and exchange.

Chapter map

In this chapter we explore the development of alternative economic practices. After introducing the contexts and purpose of the chapter, we discuss the history of alternative movements to capitalism. This is followed by an exploration of the alternative and diverse forms of economic practice that continue to exist within capitalism in section 12.3. In particular, the writings of Gibson-Graham are explored as an important contribution to the theorizing of alternative economic spaces. At the same time, we assess the role of a number of alternative local institutions such as credit unions and local exchange and trading schemes. In section 12.4, we broaden the scale of analysis to the global through a consideration of emerging norms of fairness and responsibility, apparent, for instance, in the growth of fair trade initiatives.

12.1 Introduction

Given the nature of much media debate and public discourse, it would be easy to assume that capitalism has colonized our 'lifeworld' (Habermas, 1984). One of the hallmarks of the more simplistic readings of globalization is the dominance of an increasingly integrated global economy (Friedman, 2005) based upon capitalist social relations, commodity exchange, intensified competition and the pursuit of profit. Yet such a reading over-simplifies a more complex set of economic and geographical realities. Crucially, as demonstrated by Gibson-Graham (2006), not all economic relations are capitalist, despite the prominence of capitalism. Think, for example, of your own economic activities during a typical day or week, involving things such as working, shopping, paying rent and going for a night out. In what sense are such activities capitalist? They will almost certainly require participation in the formal economy of wage labour and commodity exchange, but is the drive or motivation for this activity economic (maximizing your income and/or reducing your expenditure) or social (for example, building or maintaining relationships or studying for enjoyment or to increase your knowledge)?

While it is easy to take capitalism for granted, given its dominance following the collapse of state socialism in the late 1980s and early 1990s and the apparent naturalness of features such as competition and the pursuit of self-interest (often buttressed by rather crude analogies with biology and Darwin), our starting point in this chapter is the recognition that capitalism is actually a social construction developed by people and institutions at a particular time and in a particular place. Capitalism is a historically-specific mode of production that was invented in western Europe in the sixteenth and seventeenth centuries before spreading to span most of the globe (section 1.3.1). As a human construction, capitalism is open to transformation and change. Indeed, the current dominance of capitalism reflects its periodic success in reinventing itself – often at great social cost – in response to periodic crises and challenges, from the Great Depression of the 1930s to the 'stagflation' of the 1970s. Such crises often create openings or windows for the reconstruction of economic relations. The financial crisis of 2007–08, for instance, prompted debate on the need for the re-regulation and 'rebalancing' of the economy in the US and UK, although by 2010 this seemed to have given way to a kind of reinforced neoliberalism, in the European context at least, through the introduction of austerity programmes to cut public expenditure.

12.2 Capitalism and its alternatives

In recent years, the evidence that inequality, poverty and alienation are features of global capitalism has resulted in a resurgence of interest in alternative forms of economic organization development, linked to the growth of the **counter-globalization movement** (section 1.2.1). This search for more equitable forms of economic and social organization is not new, but has been a feature of capitalism from its very early phases, spawning a wide range of movements and philosophies, from anarchism to communism to 'deep green' environmentalism.

The history of industrial capitalism (since its inception in the late 1700s) is replete with resistance and oppositional movements from those dispossessed by its accumulative thrust (Harvey 2003). What equally characterizes the history of capitalism, right up to the present day, is the proposal of alternatives, both within capitalism (e.g. social democracy) and by those who completely reject it and attempt to create new modes of production outside capitalist social relations (e.g. anarchist, communist, socialist, etc.).

Attempts by wealthy elites to enclose common land in England in the sixteenth and seventeenth centuries, for the purposes of making profits, spawned resistance movements such as the Levellers and the Diggers who were committed to more democratic and egalitarian forms of government and economy. In the nineteenth century, the development of an industrialized capitalism and factory system, and the growth of massive new urban centres to house an emergent industrial working class – or proletariat in the words of Engels and Marx – produced new forms of exploitation and alienation. Out of these conditions grew new social movements, most notably labour unions, but also other individuals or groups committed to radical social reform, concerned with not only improving the lives of workers but also constructing new forms of society, based upon non-capitalist principles of equality and solidarity.

12.2.1 The cooperative movement as an early alternative to capitalism

Cooperatives are economic organizations that are owned and democratically controlled by a group of people for their mutual benefit (O'Sullivan and Sheffrin 2003, p.202). A key feature is that members, who may include managers, employees and other committed individuals, all have an ownership stake in the organization, often receiving a share of the wealth that it generates in the form of a dividend, according to their participation rather than degree of share ownership. Cooperatives have diverse roots going back to the late eighteenth and early nineteenth centuries, but one of the key influences was Robert Owen, a Welsh factory owner who envisaged a society where workers owned and ran their own companies, rather than being exploited by a class of 'bosses'. Owen developed a series of utopian schemes from 1800 onwards, seeking to put his principles into practice (Box 12.1).

Over the past 200 years, Owen's cooperative principles have developed into a global movement, in both the global north and south. By 2010, in the UK alone,

---

Box 12.1

Robert Owen and the cooperative movement

The idea of employees owning and controlling their work as an alternative to the exploitation and alienation apparent under capitalist employment relations is almost as old as industrial capitalism itself. Awareness of the harsh economic and social consequences of early industrial capitalism for the workers led to experiments by social reformers such as Robert Owen in forms of employee ownership and mutualism. As an early socialist, Owen believed firmly in the principles of social cooperation as a replacement for the socially destructive competition unleashed by capitalism. Owen had witnessed the harsh and oppressive realities of the capitalist workplace as a manager of a cotton spinning mill in Manchester. In 1800, he began to put his ideals into practice in the purchase of a cotton mill at New Lanark in Scotland (Figure 12.1). New Lanark combined more humane working conditions with new, efficient methods of production, although it was ultimately a moral order imposed from above, rather than an example of a grassroots initiative. Owen subsequently attempted to establish a number of utopian cooperative communities in Scotland and the US, although they ultimately foundered due to a lack of funds and community support.

Nevertheless, Owen's ideals spawned a new movement, known as Owenism, which believed in the principles of a society premised on 'equal exchange' (Pollard, 1967, p.106) rather than exploitation. These ideas were to influence the cooperative movement, but can also be found in the more recent emergence of a Fair Trade movement. Although cooperative ideas have proved very popular and enduring, the more radical intentions of Owen and others, that a cooperative society should replace a competitive one, have not been realized. In practice, capitalism has been able to exist alongside cooperative movements, often absorbing them – where the latter are in competition with capitalist-run organization – low profits have often meant the collapse or absorption of cooperative movements into the economic mainstream.

Figure 12.1 New Lanark in 1799
Source: RCAHMS Enterprises: © Royal Burgh of Lanark Museum Trust. Licensor www.scran.ac.uk.

Table 12.1 Cooperative market share in selected countries and sectors, 2010

Country	Sector	Proportion controlled by coops
Brazil	Agriculture	40% GDP
Bolivia	Banking	25% of savings
Denmark	Consumer retail	36.4% of market
France	Farming	9 out of 10 farmers in coop
Kenya	Coffee growing	70% of market
Norway	Milk production	99% of total production

Source: International Cooperative Alliance website, at http://www.ica.coop/al-ica/.

9.8 million people were members of cooperatives, in India 236 million, 9.1 million in Argentina, one in three people in Norway and Canada, and one in four in the US (http://www.ica.coop/al-ica/). The UN in 1994 estimated that the livelihoods of 3 billion people had been 'made secure' by the cooperative movement (ibid.). Cooperatives are a worldwide phenomenon and are active in virtually every major sector of the economy, from agriculture, where farmers' cooperatives play a major role in many countries in the Global North and South, to retailing, to financial services and even in house-building and ownership. Indeed, in some countries they dominate particular sectors (Table 12.1).

One of the most successful and frequently cited examples of cooperative operation is the Basque corporation, Mondragon, which operates now as a multinational. Mondragon has 30,000 worker-owners, over 100 separate cooperatives in industrial, service and retail sectors, plus its own bank which is tasked with financing more cooperatives (Gibson-Graham, 2006). Founded in 1941, by a catholic priest, Father Arizmendiarrieta, who was influenced by Robert Owen and early Spanish anarchist cooperatives, its mission was to develop 'democratic economic and social arrangements that might benefit all in the community and give a strong footing for postwar society' (Cheney, 1999, quoted in Gibson-Graham, 2006, p.223).

Cooperatives are formed for different reasons, and not all sign up to the ideals espoused by Owen and Father Arizmendiarrieta. While most have the aim of providing their members with security from the vicissitudes of free markets (whether this is in regulating prices for farmers and consumers, or providing decent wages and employment safeguards for workers), the criticism that is often leveled at them is that they still operate within the wider capitalist economy and are therefore not insulated from the same competitive pressures facing other firms. While they may be able to cushion the effects of capitalism, they cannot work against it.

This is to take a rather deterministic view of capitalism and markets. From a more agency-centred perspective, one might argue that cooperatives – such as the example of Mondragon – evoke different values and alternative perspectives of the economy that provide a challenge to the more rapacious forms of capitalism, associated, for example, with contemporary forms of neoliberalism. They can, in turn, shape the economy towards a different set of ethics and values. Cooperatives can be powerful influences, particularly in situations where they control a large share of the market (Table 12.1), although this does not automatically mean that cooperative actors will pursue more social, as opposed to capitalistic, ends.

12.2.2 Developments in anti-capitalist thought and practice

From a very early stage, there was a basic division between those seeking to reform capitalism in favour of a more socialized model (a social democratic tendency) and those seeking to overthrow capitalism through revolution. This schism emerged first in what was known as the 'Second International' from the 1890s onward (of the international working-class movement, the First International having been founded by Marx and others in the 1860s).

These opposing tendencies were most evident in the largest Marxist inspired movement of the time, the German Social Democratic Party (SDP), which was split between revolutionary theorists such as Rosa Luxembourg and Karl Liebnecht on the one hand, who argued for violent class struggle to overthrow capitalism, and more reformist positions taken by intellectuals such as Karl Kautsky and Eduard Bernstein, on the other, who argued for reform from within.

After the First World War, these tensions split the SDP and the wider international labour movement, shaping the history of class struggle and economic alternatives during the twentieth century. In the Soviet Union (after the Bolshevik Revolution in 1917), eastern Europe (after 1945), China (after 1949), and Cuba (post-1959) successful revolutions tried to install communist-inspired models as alternatives to capitalism, while in other countries, especially in northern Europe and Scandinavia, labour unions and social democratic parties were successful in developing more socially oriented and egalitarian forms of capitalism (at least between 1945 and the mid 1970s). By the late 1980s, the collapse of the Soviet bloc and the opening up of China to global capitalism effectively ended these alternative experiments, although most observers would accept that the more egalitarian and democratic impulses of these movements were thwarted much earlier.

The collapse of communism in the Soviet bloc led to a phase of mainstream theorizing about the triumph of liberal capitalism, based upon market freedoms and individual liberty, over socialism, heralding the end of history (Fukuyama, 1992). Attempts to reform capitalism from within were also dealt a setback in the 1980s, as globalization and the growing dominance of neoliberal ideas in economic policy-making meant that most states retreated from social democracy in the face of increasing competition from the 'Tiger' economies of East Asia. Business elites were successful in projecting their own views onto the public policy agenda with the result that alternative and more egalitarian views of society were in retreat and competitiveness agendas in the ascendant. Indeed, as the US writer Robert McChesney (1999) put it: 'Neoliberalism's loudest message is that there is no alternative to the status quo, and that humanity has reached its highest level.' Originally associated with Margaret Thatcher, British Prime Minister during the 1980s, the phrase, 'there is no alternative' became a handy slogan for those advocating neoliberal free market principles in the context of globalization.

Since the late 1990s, however, growing concerns about increased social inequality at the global and local levels and the environmental destruction caused by capitalism have given rise to a new set of counter-globalization movements (section 1.2.1, Box 1.3).

Table 12.2 The Bamako Accord: a manifesto for an alternative globalization

1. The cancellation of the debt of countries in the global south
2. Implementation of the Tobin Tax on financial speculation
3. The dismantling of tax havens
4. The implementation of basic rights to employment, welfare and a decent pension and equality in this regard for men and women
5. Rejection of free trade and implementation of fair trade and environmentally sound trade principles
6. Guarantees of national sovereignty over agricultural production, rural development and food policy
7. Outlawing of knowledge patenting on living organisms and privatization of common goods, especially water
8. Fight by means of public policies against all kinds of discrimination, sexism, xenophobia, anti-semitism and racism
9. Urgent action to address climate change, including the development of an alternative model for energy efficiency and democratic control of natural resources
10. Dismantling of foreign military bases except those under UN supervision
11. Freedom of information for individuals, the creation of a more democratic media and controls on the operation of major conglomerates
12. Reform and democratization of global institutions incorporating institutions such as the World Bank and IMF under the control of the UN

Source: Derived from Routledge and Cumbers, 2009, p.195.

Through the growth of the World Social Forum (WSF) and various regional and local forums, this movement has developed an important momentum and role in promoting alternative ideas, with its slogan 'Another World is Possible'. At the Mali WSF in Bamako in 2006, the movement promoted a manifesto for an alternative globalization (Table 12.2)

As we noted in Chapter 1, there are considerable divisions within this counter-globalization movement, once again exhibiting the tensions between those who wish to reform the current system of economic globalization and those 'anti-capitalists' who aim to overthrow the current order and develop a new society. Even within the latter, there are divisions between those who want to rebuild from below, through a diversity of local forms, and those that hold to a singular model of communism or socialism (Routledge and Cumbers, 2009, Chapter 7). The Bamako Manifesto itself was the subject of considerable controversy between its supporters who wanted to promote a more global vision and those who maintained the primacy of local autonomy over alternatives.

12.2.3 Variety, diversity and alterity

Although many writers correctly talk of the predatory and competitive forces that drive capitalism in the form of 'primitive accumulation' or 'accumulation by dispossession' (Harvey, 2003), as it seeks to impose itself as a system over other forms of social relations, this tendency is at the same time a rather schizophrenic one. One of the hallmarks of capitalism is its ability to coexist with, and even adapt, other forms of social relations into its operation, resulting in the development of diverse forms of economy emerging in practice. In this regard, we would agree with the institutional economist, Geoff Hodgson, in recognizing the principle of variety that is at work in shaping economic life (Hodgson, 1999).

The literature on varieties or national systems of capitalism provides one conceptual handle on economic variation, demonstrating how sets of national institutions structure economic performance and development (for example, Amable, 2003; Hall and Soskice, 2001). Such theories emphasize the difference between coordinated-market economies (for example,

Germany and Japan) and their 'long-term, structural relationships' with the decentralized and short-termist liberal-market economies such as the US and UK (Peck and Theodore, 2007, p.736). The later varieties of capitalism approach favoured by Hall and Soskice (2001), among others, positions economies on a continuum between coordinated market economies and liberal-market economies in which institutions adapt to complement one another. At the same time, the 'varieties' approach is characterized by some serious limitations in terms of how it conceptualizes institutional variation within capitalism. These include its pre-occupation with the national scale and neglect of regional and local institutions and its tendency to conceive of change as a process of convergence to a pre-defined (neo) liberal market model (Peck and Theodore, 2007).

A rather different take on economic diversity and variety is offered by recent work on alternative social and economic practices in economic geography. Instead of seeking to bracket variety into pre-defined categories, this work seeks to uncover and celebrate diverse economic forms and practices in all their plurality and complexity, drawing theoretical and political inspiration from the writings of Gibson-Graham (section 12.3.1). The principle of alterity emphasizes the 'otherness' of alternative economic practices compared with the capitalist mainstream. Rather than being driven by the prevailing economic logic of profit maximization, they are governed by a range of social logics (for example, meeting local needs, fairness), and often bracketed together under the rubric of 'the social economy' (section 12.3.3). For some commentators, such as Gibson-Graham (2006), existing forms of economic diversity need to be tended and nurtured as seeds of an alternative 'post-capitalist' future. But alternative practices are entangled with the mainstream economy in numerous ways, raising concerns about their possible cooption and dilution (Fuller *et al.*, 2010).

Reflect

> Are there any lessons that the contemporary counter-globalization movement can learn from the earlier cooperative movement?

12.3 Alternative economic spaces

Two of the leading thinkers in debates on emergent 'alternative economic spaces' are the feminist geographers Kathy Gibson and the late Julie Graham who wrote together under the moniker, J-K Gibson-Graham. Their approach is both radical and controversial in regarding the economy not as a single system dominated by capitalist imperatives, but rather as 'a zone of cohabitation and contestation among multiple economic forms' (Gibson-Graham, 2006, p.xxi).

12.3.1 Gibson-Graham and diverse economies

In a series of writings over two decades, notably their two monographs *The End of Capitalism (As We Knew It: A Feminist Critique of Political Economy* (1996) and *Postcapitalist Politics* (2006), Gibson-Graham seek to challenge what they view as the capitalocentric views of the economy held by both mainstream economists and Marxists. Drawing upon feminist, post-colonialist and post-structuralist philosophies, their work seeks to decentre capitalism as the predominant set of processes to highlight the varied, overlapping and competing forms and practices that constitute the economy. Although still framed by a political economy understanding of the world, class is just one of the axes through which economic power relations are played out; others being gender, race, age, caste and even more primitive social forms such as slavery and feudalism.

A key concept is **diverse economies** (section 1.3.1), whereby capitalist economic relations represent 'just one particular set of economic relations situated in a vast sea of economic activity' (ibid. p.70). Most vividly demonstrated in their 'iceberg model' (Figure 12.2), this draws attention to the other forms of activity that exist in the economy beneath the visible surface of wage labour in capitalist enterprises (see Box 12.2). Indeed, it has been estimated by feminist economists that between 30–50 per cent of economic activity globally is non-capitalist in the form of unpaid work in the home or non-market transactions (Gibson-Graham, 2008; Ironmonger, 1996). Offe and Heinze (1992) made

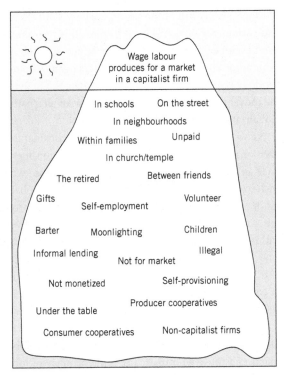

Figure 12.2 The iceberg model
Source: Drawn by Ken Byrne, Gibson-Graham, J-K, 2006, *a Postcapitalist Politics*, Minneapolis and London: University of Minnesota Press, p.70.

a similar point about the DIY economy in showing how much economic activity is taking place outside the capitalist mainstream (Box 12.2).

Gibson-Graham are motivated by a progressive left politics which views a fixation with 'capitalocentric' (Gibson-Graham, 2008, p.72) discourses as a barrier to developing more egalitarian and democratic ways of organizing the economy. Their aim is to produce 'a discourse of economic difference as a contribution to a politics of economic innovation' (ibid., p.5). For them, subjugation under capitalism is as much about language, discourse and identity as it is about the hard material realities of exchange and surplus value. Recognition of the diversity that already exists in the economy, opens us to anti-capitalocentric readings (ibid., p.72) where we can choose between alternative forms of social relations (collective over individualistic, cooperative over competitive) that can produce fairer economic outcomes. Their concept of the **community economy**, where markets are regulated by ethical concerns around fair trade and exchange, and private

Box 12.2

The diverse economies and geographies of childcare

A good example of the diverse forms of social relations which exist is childcare. While often undervalued (in terms of economic rewards), childcare is fundamental to the operation of the economy, when thought of as reproducing the future labour force. Yet, most of the ways that the work of childcare is organized have little to do with market relations or mechanisms (Gibson-Graham, 2006, p.73), although, as an earlier generation of feminists argued, this does not mean that exploitative relations are absent. The intersection of patriarchal and capitalist forms of power meant that, historically, women's work in the sphere of social reproduction (such as childcare and housework) has been excluded from paid employment.

In the contemporary economy, childcare exists in both capitalist and non-capitalist forms (Table 12.3). There are paid childcare workers who work in the home, or in private childcare centres (e.g. after-school clubs, work-based nurseries and kindergartens), but the majority of childcare takes place through other forms of exchange. The most usual is 'nonmarket' forms, through parents and grandparents sharing childcare responsibilities, friends and family offering babysitting and babysitting clubs or cooperatives set up by groups of parents. In many non-western societies, there are also traditions of children being raised by other family members than their parents. A range of alternative non-capitalistic market mechanisms such as childcare trading schemes (as part of wider local non-monetary trading networks) and informal economy payments to babysitters can also be identified. State run and community schemes would also feature in this group where the price for childcare is not driven by the profit motive.

The diverse economy of childcare is, in turn, producing an increasingly diverse geography of places and flows. Traditional forms of childcare centred upon the home and family have given way to a diversity of spaces, including the home, the specialist childcare centre, workplace facilities and state-run nurseries. Focusing upon the space of flows, through which work and social interaction is organized, also draws attention to diverse geographies. Paid childcare in the home often involves the use of a transnational migrant workforce (typically on low pay and often in exploitative conditions) (Gregson and Lowe, 1994; Pratt, 1999). Trans-local and transnational flows of labour might also feature in other forms of childcare; for example, in the use of transnational relations of care within the extended family networks of migrant communities. Other spaces – such as childcare cooperatives and parent support networks – tend to be more locally bounded. State nurseries often have strict territorial boundaries, which in common with school catchment areas can have important knock-on effects in driving up house prices in local economies that are perceived to have better performing childcare services.

Table 12.3 Diverse identities of childcare and their geographies

	Market	Alternative market	Non-market
Forms of relation	➤ Domestic service (e.g. nanny, au-pair, housekeeper) ➤ Care worker in childcare centre ➤ workplace	➤ Childcare trading schemes ➤ Informal economy (e.g. babysitters)	➤ Parents sharing ➤ Grandparents ➤ Friends, neighbours ➤ Extended family ➤ Volunteer, cooperative forms
Spatial dimensions	➤ Home ➤ Workplace ➤ Transnational migrant flows	➤ Domestic home ➤ Community centres ➤ State nursery ➤ Local flows of reciprocal exchange ➤ Bonded labour	➤ Domestic home ➤ Family-friend's home ➤ Cooperative space ➤ Local and non-local family/communal networks

Source: Derived from Gibson-Graham, 2006, p.73.

ownership is replaced by forms of common ownership, represents an alternative to capitalist market economics.

There are important geographical implications in their approach which emphasize openness and difference in economic processes and outcomes, rather than an underlying overdetermined capitalist logic informing the relations between places. For them, the economic landscape is a mosaic of 'specific geographies, histories and ethical practices' (ibid., p.71). Rather than capitalism spreading and colonizing the global in an unproblematic fashion, their approach reinforces the importance of place in shaping economic relations.

While Gibson-Graham's approach is radical and inspiring, it has also proved controversial, as we indicated at the start of this sub-section. Perhaps their most valuable contribution is their theoretical and political recovery and re-valuing of the economic diversity that exists alongside the formal capitalist economy, rendering it visible through the brilliant metaphor of the 'iceberg'. This diversity has been neglected and devalued by mainstream economics and, to a lesser extent, economic geography. What seems more questionable, however, is their characterization of diverse economic formations as 'post-capitalist' when they are often entangled in the capitalist economy, notwithstanding their transformative potential. Moreover, 'the focus on the small-scale and often intimate practices of alternative or diverse economies ... can seem somewhat at odds with the increasing scale and scope of capitalist activity' (Lee, et al., 2010, p.122). Finally, Gibson-Graham's post-structuralist conception of post-capitalist politics, based upon the fostering of diversity through discourse, thought and collective action, seems partial and idealist (Scott, 2006), neglecting the need for alternative movements to challenge the economic, political and institutional foundations of capitalism rather than operating outside of it.

12.3.2 Alternative local economic spaces

Following Gibson-Graham's lead, a body of work by economic geographers has grown up since the mid-1990s interested in alternative spaces of local economic practice (e.g. Gomez and Helmsing, 2008; Leyshon et al., 2003; North, 1998, 2005; Seyfang, 2006). The focus of these studies is on how the exclusion of some places from the capitalist economy results in the creation of **alternative local economic spaces** of exchange and circulation, where relations of economic reciprocity – rather than commodification – are created to provide essential goods and services outside the free market. The philosophy in such local economic spaces runs directly counter to the integrating dynamics of globalization; instead, the emphasis is upon creating endogenous power and capacities that allow individuals to carry out sustainable economic activities at the local level, detached from broader circuits of capital (Gomez and Helmsing, 2008). The idea of decoupling local and regional economies from the broader global economy is not a new idea (e.g. Stohr and Todling, 1978), but interest has grown in response to the pressures and threats of globalization.

A number of alternative local institutional forms has been created (Seyfang, 2006), ranging from local currencies, time banks (where people volunteer their time on a reciprocal basis for activities such as dog walking, DIY, etc.), community or social enterprises, and credit unions. Here, given space constraints, we focus on two: local or community currencies and credit unions.

Local currencies are viewed as among the more radical initiatives in alternative local economics; aimed at 'short-circuiting' the global economy, creating spaces of closure that prevent leakages out of the local economy (Gomez and Helmsing, 2008). Seen as a direct response to increasing globalization and the loss of power of local communities over their economies, local or community currencies in particular are viewed by many radical and green economists as a way of challenging the disciplinary power of national and international monetary regimes. It has been estimated that there are as many as 2,700 local currencies active in 56 countries. They take diverse forms and are known by a range of terms: Local Exchange Trading System (LETS) (see Box 12.3) in the UK, *monnaies parallèles* in France, *moneda sociale* in Italy, for example.

One of the most celebrated examples of local currencies, known as the Red de Trueque (RT), occurred in Argentina in the wake of the 2001–02 financial crisis. Although there was a number of very small-scale local currency schemes operating prior to the

Box 12.3

LETS schemes: lifeboats of uneven development or piloting alternative futures?

A well popularized initiative is the **Local Exchange and Trading System (LETS)**, whereby a group of people, usually in a local area of a city or small town, decide to create their own currency through which they can trade economic activities (Williams *et al.*, 2001). An association is established to administer this currency, which exists in the form of cheques issued every time a transaction takes place between people. The currency is virtual – no money actually changes hands – but provides a record of the amount of work taking place and the surpluses and deficits of individuals. Note that to purchase a service, one does not need to have 'money' beforehand.

The first LETS scheme was established in the UK in 1985 (Williams *et al.*, 2001) and it has been estimated that there are now around 300 LETS schemes nationally with around 22,000 people involved and a total estimated value of around £1.4 million (or around $2 million) (North, 2005; Seyfang, 2006). They range in the numbers involved from 10–15 people doing basic activities such as babysitting and odd-jobs in small communities to up to 300 in bigger towns and cities (North, 2005). As such, they clearly remain a marginal pursuit, often the domain of green and other more ethically conscious groups, outside the economic mainstream, with the suggestion therefore that they do little to challenge existing practices or deal with deeper-seated structural and spatial inequalities.

However, an alternative argument is that LETS are important in constructing new networks of social support – social capital – for disadvantaged groups that enable them to improve their lives. In their extensive survey of LETS schemes in the UK in the early 2000s, Williams *et al.* (2002, p.162) found that 76.2 per cent of participants had benefited from improved social networks on which they could call for help. Moreover, invoking Gibson-Graham's arguments, even if they work alongside mainstream market forces, LETS are important because, in a practical everyday sense, they bring more reciprocal community economy relations into play in challenging dominant capitalist social identities. There is often an uneasy tension between developing 'social capital' to allow individuals and groups to participate in the capitalist economy (Williams *et al.*, 2002) and the more radical intent of providing genuine alternative models of economic and social relations.

crisis, it was this event that marked a critical acceleration in their usage. As the economy plummeted, as a result of the broader financial crisis in Asia and Latin America, with GDP falling by 25 per cent between 1999 and 2002, growing numbers of people were made unemployed or destitute (among traditionally more prosperous middle-class groups as well as the poor). One of the government's desperate policies was to freeze bank accounts in a bid to prevent capital leaving the country. In these circumstances, people turned in desperation to alternative forms of economic exchange. Resources released from the capitalist economy – land, labour and technology – were taken up in what one set of observers terms an 'emergency circuit of low productivity and small-scale production' (Gomez and Helmsing, 2008; p.2495). Out of this situation, alternative currencies and barter networks developed across the country (North, 2005).

The desperate conditions of the mainstream economy meant that many millions of people participated in this alternative economy. A system of credits (*creditos*) developed to allow people to barter goods and services at informal street markets that sprung up across the country. These markets operated through a system of 'nodes', which met at organized times in a range of places, including church halls, disused factories and car parks. North (2005) suggests that the largest of these nodes – in the western city of Mendoza – had over 36,000 members at its height. A wide variety of food and clothing, as well as services such as haircuts, were on offer in these nodes. Nodes were not territorially confined but were overlapping, with people able to travel across cities and regions to exchange goods and services. Without a central coordinating body, the system somehow allowed people to exchange credits from one part of the country to another, as long

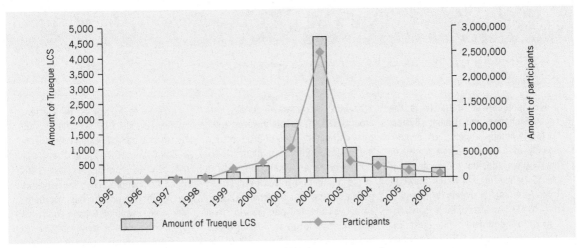

Figure 12.3 Participants and amount of Trueque local currency scheme
Source: Gomez and Helmsing, 2008, Graph 1, p.2496.

as the person they traded with considered their credits to be valid (North, 2005). Although attempts were made to develop alternative currencies, at a national scale, such as the Arbolito, many activists remained avowedly localist, seeing any attempt to 'scale up' monetary regulation as an erosion of the local participatory democracy through which local currencies could be monitored and controlled, and prone to inflationary pressures. Over time, interest in the alternative currency declined (Figure 12.3), due to not only a number of corruption scandals involving those administering the alterative currencies, but also the recovery of the national economy (Gomez and Helmsing, 2008; North, 2005). There are also suggestions that business owners and the Peronist government sought to undermine the currencies, fearful of the longer term effects on the economic mainstream (ibid.).

Credit unions are another form of alternative economic institution (Fuller and Jonas, 2003). They effectively operate as financial cooperatives that are owned and run by their members, who are the only people allowed to deposit and borrow money and take out loans. In order to borrow money, members must usually have deposited some of their own savings; often a ratio of three to one, in terms of loans to savings, is the maximum, helping to ensure the financial sustainability of the unions, in contrast with many mainstream banks implicated in the 2007–08 financial crisis.

Credit unions offer opportunities for hard-pressed individuals to save or borrow money on better terms (particulary much lower interest rates) than mainstream banks or 'loan sharks' of the underground economy, which are frequently the only option for people in poorer communities to access credit. Once again, credit unions usually operate at the local or 'community' level, and may be linked to particular community groups or organizations, sometimes based along ethnic or religious lines. Like LETS schemes, they often represent a response to the withdrawal of 'mainstream' economic actors, in this case financial institutions such as banks and building societies from disadvantaged areas, resulting in **financial exclusion** (Leyshon and Thrift, 1995, see also Box 9.2). Economic geographers have made similar claims about credit unions, as alternative institutions that can strengthen communities and create autonomous economic spaces for local people (Lee, 1999).

The credit union movement originated in Germany in the 1850s, but by 2008 had reached over 53,000 credit unions worldwide in 93 countries with over 185 million members (Table 12.4). Its geography varies dramatically by world region, from North America where its penetration rate is over 40 per cent (number of credit union members divided by the economically active population) to continental western European countries where membership is negligible, although this is probably because many people are already members of

Table 12.4 Credit unions worldwide, 2009

Number of countries involved	97
Number of credit unions	49,330
Number of members	184 million
Financial statistics:	
➤ savings	$1.146 trillion
➤ loans	$912 billion
➤ assets	$1.354 billion
Penetration of credit unions by world region:	
➤ Africa	6.8%
➤ Asia	2.6%
➤ Caribbean	18.9%
➤ Europe	3.6%
➤ Latin America	4.8%
➤ North America	44.6%
➤ Oceania	17.9%

Source: Derived from World Council of Credit Unions, 2009 Statistical Report.

long-established local and regional banking coopera-tives, such as the Sparkasse network in Germany, which fulfils a similar role. Ireland is a notable exception with a 70 per cent penetration rate for credit unions.

12.3.3 The 'social economy' and the limits to alternative local economic practice

Although many schemes set out with the more radical intention of creating alternative models to capitalism, in practice activists find it difficult to maintain their autonomy from the mainstream economy and broader processes of uneven development. LETS schemes and credit unions were incorporated into social economy policy agendas pursued by many centre-left govern-ments in the 1990s and 2000s as a way of dealing with socially excluded regions and communities in North America and western Europe (Giddens, 1998). The **social economy** – composed of organizations such as cooperatives, NGOs and charities that are not part of the private or public sectors of the mainstream economy – was perceived as being a way of delivering local economic and social regeneration in communities left behind by globalization, by governments otherwise committed to neoliberal economic policies (Hudson, 2009). As Amin *et al.* (2003b, p.48) noted, policies were only adopted in those 'localities, communities and neighbourhoods where conventional economic

practice is deemed to be no longer possible, practical or desirable'. Rather than innovative alternatives to main-stream economic practice, social economy strategies tended to lead to the development of a 'second-class economy' (Bowring, 1998, p.106) divorced from the mainstream, and often serving to reinforce inequalities between places.

Even for those schemes determined to maintain their autonomy from wider government programmes and initiatives, activists are inevitably drawn into wider economic structures and processes beyond their control. For example, in their study of credit unions in the UK, Fuller and Jonas (2003, p.70) suggested that, the 'space for community credit unions to retain their financial autonomy and alternative-oppositional iden-tity may be contracting' as these institutions became involved in filling the gaps left by the withdrawal of mainstream financial institutions.

Moreover, small-scale alternatives, which are typically dependent upon a handful of activists and enthusiasts for their functioning and organization, often encounter problems of sustainability and regen-eration over time. For example, the Manchester LETS scheme studied by North (2005) dwindled from 485 members at its height in 1995 to 125 by 2001, with most of the original participants having left as a result of changing personal circumstances – such as having children, becoming unemployed and being unable to afford the participation fees or having less time to commit because of working in regular schemes. As North (2005, p.227) puts it, 'the resistances of everyday life meant that [people] did not maintain membership, so that, rather than being the precursor of a new localised experiment, LETS became episodic'.

As the example of the Argentinian barter economy demonstrates, even where alternative practices get scaled up to the national level through local initiatives coalescing into broader solidarity networks, they are at present more useful as escape valves for people and communities when the mainstream economy enters periodic crises. In this sense, North (2005, p.222) describes alternative economic spaces as 'lifeboats against globalization developed by the marginal in spaces suffering from uneven capitalist development'. The decline of alternative currencies in the wake of the recovery of the Argentinian national economy would appear to bear this out, although it should be noted that alternative forms of economic organizing continue there to a greater degree than other countries.

12.3.4 Alternative corporate geographies

Contrary to the mainstream economic assumption that the actions of large corporations are driven by underlying economic and political imperatives, an alternative approach views corporate decision-making as a more open and contested process that is shaped by competing discourses (O'Neill and Gibson-Graham, 1999; Schoenberger, 1997). Even within privately owned or joint-stock capitalist firms, Gibson-Graham argue that we cannot assume that the pursuit of profit dominates to the exclusion of all other motivations. Instead, there is considerable diversity in practice. Multinational corporations, for example, do not have singular dominant strategies, but are home to competing and sometimes contradictory discourses between different groups (O'Neill and Gibson-Graham, 1999).

From this perspective, corporate strategies can be challenged and even transformed for more ethical

Box 12.4

Fair Pensions and the Tar Sands Campaign

Fair Pensions is a UK-based campaign of shareholder activism aimed at changing corporate behaviour by realizing the latent power tied up in pension schemes:

> Our mission recognises the simple but under-appreciated fact that the trillions of pounds invested in pensions have the power to change our world. The money people save every month for their retirement is the money that pension funds invest in corporations across the world.
>
> (http://www.fairpensions.org.uk/about)

Its aim is to pressurize for corporate strategies that respect the rights of people and the environment in the regions where the large corporations owned by their members operate. Using the concept of Responsible

Investment, it seeks to campaign through shareholder activism. Member organizations include high-profile NGOs such as Amnesty and Greenpeace but also some of the largest unions in the UK, such as public sector union Unison and the Universities and Colleges Union (UCU). Members of the UCU have also been involved in the formation of Ethics for the Universities Superannuation Scheme (USS), a group which has been successful in changing USS towards more ethical investment priorities.

One of the most high-profile recent campaigns has been against the production of oil from tar sands in Alberta, Canada, by BP and Shell (Figure 12.4). The extraction of oil from tar sands generates more than double the greenhouse gases

of conventional oil, leading to the practice being condemned by many environmental groups. The Fair Pensions campaign was successful in generating two shareholder resolutions at the 2010 annual general meetings of the two companies, requesting that the companies publish details of the environmental, social and financial consequences of their actions. Although the resolutions were not passed, they were formative in the amount of shareholder backing which they received from pension funds and other fund holders: one in seven from BP and one in ten from Shell. The campaign has also shed light on broader corporate practices, especially the two oil companies' assumptions and strategies with regard to climate change and future energy policy, which

Box 12.4 (continued)

demonstrated a refusal to prioritize climate change and the future of the planet.

Figure 12.4 Shell's sticky problems
Source: David Simonds.

and progressive ends. Campaigns for corporate social responsibility and ethical trade seek to intervene in broader corporate discourses to challenge and shape executive decision-making for more progressive ends. Although these campaigns have grown in prominence since the late 1990s, their effectiveness in promoting a more progressive globalization is open to debate, with most of the evidence suggesting business as usual in terms of how MNCs function worldwide. In addition, the increasingly complex networks and webs through which corporations operate, using many different suppliers and subcontractors, allows them to render many of their environmental and labour abuses less visible. At the same time, however, the availability of the internet, which enhances activists and campaigners' ability to highlight corporate malpractice, means that MNCs are now being subjected to a greater level of scrutiny than ever before.

Within corporations, alternative discourses are being developed by shareholder groups that are effectively developing networks of common concern between the ordinary members of pension schemes – which have dual identities as workers and shareholders – and workers and communities affected by MNC activities (see Box 12.4). Pension activism operates through intervening in the complex and contradictory spaces of corporations (O'Neill and Gibson-Graham, 1999) challenging existing narratives in pursuit of more social and environmentally conscious forms of development.

Reflect

➤ To what extent would you agree with Gibson-Graham's view of the community economy as an alternative to capitalism?

12.4 Alternative global norms of fairness and responsibility

12.4.1 Fair trade initiatives

A different – but often complementary – approach to the more localist alternatives identified above is provided by movements that seek to intervene in global economic networks to promote fairer and more ethical forms of social relations. Since the 1970s, concerned consumers in the more developed economies of western Europe and North America have campaigned to improve the working and living conditions of farmers and workers who are involved in **global commodity chains**. This has resulted in the growth of movements such as the **Fair Trade** network, the Clean Clothes Campaign – a Netherlands based alliance of trade unions and NGOs, which exists to improve the rights of workers in the clothing and textile sectors – and the Ethical Trade Initiative, a similar UK initiative that also involves some of the major multinational retail chains.

Because many developing countries remain dependent upon one or two commodities for export, producers can be dramatically affected by sudden price fluctuations in global markets over which they have little control. Many farmers in the Global South, for example, are in highly dependent power relations with multinational food or retail chains over which they have virtually no control. In contrast, ethical or fair trade networks seek to develop a different and less exploitative set of relations between producers and consumers in global trade networks. The political aspects of fair trade are evident in the following quote: 'When you buy Fair Trade you are supporting our democracy.' (Guillermo Vargas, Costa Rican farmer, talking to the UK House of Commons 2002, cited in Morgan et al., 2006, p.1).

The Fair Trade Initiative describes itself as follows:

> Our mission is to connect disadvantaged producers and consumers, promote fairer trading conditions and empower producers to combat poverty, strengthen their position and take more control over their lives.

> Our vision and mission will be reflected in the values by which we work as an organization so that we ourselves set an example for the changes we seek in others. Therefore we will work collaboratively and seek to empower those who wish to be partners in our mission. Trust is a crucial factor in our work and we will be mindful of our responsibilities to those who place their trust in us. Embracing transparency and stakeholder participation is an important way that we will be accountable for our work.

> (website of the Fairtrade Labelling Organizations International: http://www.fairtrade.net/our_vision.html)

Thus, at one level it can be seen as quite revolutionary, going beyond Marx's famous critique of the mainstream economy's **commodity fetishism** to reveal the true nature of relations between people and place as a way of developing more equitable connections between consumers and producers.

Making the geographies of production, consumption and circulation more visible is a key theme of much recent writing in economic geography (e.g. Cook et al., 2010; Goodman et al., 2010; Hughes et al., 2008), and there have been detailed criticisms of the limits of fair trade initiatives (see the excellent discussion of these issues in Cook et al., 2010). One criticism is that one set of dependency relations – between multinationals and producers – is replaced by another set, this time between producers and middle-class consumers in the west, who are able to afford to pay higher prices for fair trade products. Other research suggests that there is considerable variation in the practice of fair and ethical trade, often highly dependent upon relations in developed countries between consumers and retailers involved in ethical initiatives (Hughes et al., 2008).

There is no doubt that fair trade has expanded dramatically in recent years (Figure 12.5). In 2009, fair trade products globally accounted for around $3.4 billion, up 15 per cent on 2008, with over 27,000 fair trade products on sale in over 70 countries (Fairtrade Labelling Organizations International, 2010, p.13). As these figures indicate, fair trade products have maintained impressive sales even during the global recession of 2008–09, suggesting that some people choose to consume ethically, even in more difficult economic

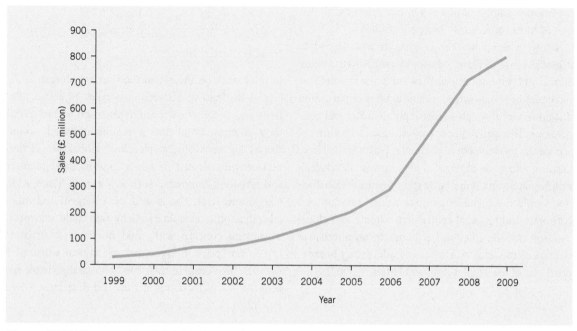

Figure 12.5 The growth of fair trade products in the UK 1998–2009
Source: Fair Trade website http://fairtrade.org.uk.

circumstances. At the same time, doubts are sometimes raised about the real economic significance of these schemes, echoing the criticisms of the alternative local economic initiatives discussed earlier. For example, the above figures can be put into perspective somewhat by comparing them with the global sales of the British retail MNC, Tesco, which amounted to £62.5 billion in 2009–10 (Tesco, 2010, p.8).

12.4.2 The moral economy and geographies of responsibility

The sociologist Andrew Sayer (2000) uses the term **'moral economy'** in advocating an approach that replaces capitalist economic values and the exploitation of one group by another, with the recognition that everyone is entitled to certain basic economic rights (such as decent housing, food and clothing, standard of living, education, etc.), alongside a recognition of economic responsibility to others. Of course, such principles are nothing new; they have been the concern of theorists as diverse as Smith, Marx, Keynes and Polanyi. A global set of economic rights has also

existed in principle since the end of the Second World War and the United Nations' Declaration of Human Rights:

> Everyone has the right to a standard of living adequate for the health and well-being of himself and of his family, including food, clothing, housing and medical care and necessary social services, and the right to security in the event of unemployment, sickness, disability, widowhood, old age or other lack of livelihood in circumstances beyond his control.
>
> (UN Declaration of Human Rights 1948, clause 25)

The practice of contemporary globalization, however, is driven by the pursuit of corporate profitability, sometimes riding roughshod over these lofty ideals. The values espoused by fair trade groups are clearly motivated by alternative ideals that connect with Sayer's moral economy. With regard to the consequences of global economic processes and connections, Doreen Massey (2004, p.7) similarly argues for a **geography of responsibility** in which we recognize how the 'lived reality of our daily lives, invoked so often to buttress the meaningfulness of place, is in fact pretty much dispersed in its sources and its repercussions'.

Drawing upon her earlier insights with regard to a 'global sense of place', Massey's point is that a more ethical and responsible globalization is one in which we recognize how our own economic actions impact upon distant others through increasingly dense global connections. This also requires contesting certain forms of economic development in particular places if they have harmful effects on others. So, for example, developing an alternative and more egalitarian financial globalization would mean challenging the role of London as a corporate and financial centre that is home to a global decision-making elite, and arguing for an alternative politics of the local as a means of addressing broader processes of uneven development (Massey, 2007).

12.4.3 Reconnecting places through responsible geographies

In their work on the global food sector, Morgan *et al.*, (2006) distinguish between two types of social relations and geographies that are practised within global food commodity chains: a regime of 'hard power' related to increasingly 'placeless' global production networks dominated by larger multinational retailers and agri-food companies such as Walmart, Tesco, Aldi, Cargill and Kraft Foods; and, an emergent and more ethical regime, characterized by fair trade networks, a growing concern with food quality and attempts to recover 'place' by reconnecting ethical consumers with the producers of their food by making visible the geographies of food production and distribution (see Box 12.5).

Box 12.5

Making responsible connections through Fair Trade Towns

The fair trade concept was extended through the establishment of Fair Trade Towns, an initiative first started by an Oxfam member, Bruce Crowther. Crowther spent eight years attempting to sell the fair trade concept to cafés and restaurants in his home town of Garstang, a market town in rural Lancashire with a population of 5,000. A key turning point for Crowther was watching a demonstration by local farmers protesting at falling prices of their own dairy products. As he noted: 'When I saw dairy farmers marching down Garstang High Street with banners saying 'we want a fair price for our bottle' I realized that we couldn't just campaign for developing countries.' Crowther recognized that local farmers in a developed country such as the UK were connected to the same relations of exploitation as developing country peasants and farmers in global food commodity chains. Subsequent support from local farmers groups was crucial in making Garstang the first Fair Trade Town in April 2000. The movement has since spread to include 480 towns in the UK and 800 worldwide in 19 countries.

Fair Trade Towns sign up to five key principles:

➤ the local council passes a resolution supporting fair trade, and agrees to serve fair trade products (for example, in meetings, offices and canteens);

➤ a range of fair trade products is available locally (targets vary from country to country);

➤ schools, workplaces, places of worship and community organizations support fair trade and use fair trade products whenever possible;

➤ media coverage and events raise awareness and understanding of fair trade across the community;

➤ a fair trade steering group representing different sectors is formed to coordinate action around the goals and develop them over the years.

There is now a European network of Fair Trade Towns, with support from the EU, which hosts conferences that bring together members to communicate activities and share best practice. The ethos shared by Fair Trade Towns of encouraging their residents to use their everyday local consumption choices to benefit and bring about positive changes in the livelihoods of food producers globally, connects strongly with Massey's concept of a geography of responsibility. (http://wwwfairtrade.org.uk/get_involved/campaigns/fairtrade_towns/default.aspx)

Table 12.5 Mainstream and alternative economic geographies of globalization

	Mainstream global economy	Moral/alternative global economy
Dominating economic values	Competitive advantage, free trade	Solidarity relations, fair trade, consumer quality
Underlying rationalities	Profit maximization	Ethics of care and responsibility
Relations of power	Hard power, hierarchical and coercive	Soft power, collaborative and discursive
Organizational forms	Private ownership (although usually MNC dominated), subcontractors, home workers, child workers	Family firms, SMEs, cooperatives
Geographies	Placeless, encouraging mobility and low cost competition between places	Re-envisaging place through politics of responsibility and reconnection
Economic outcomes	Highly unequal distribution of income with value captured by larger actors (e.g. MNCs, global retailers)	Income more evenly allocated through commodity chains with producers in Global South provided with living incomes

Sources: Derived from Massey, 2005; Morgan *et al.*, 2006; Sayer, 2000.

We think that these principles can be applied more generally to theorizing an alternative geography of the global economy, where the profit maximization rationalities of mainstream or neoliberal capitalism can be juxtaposed against Sayer's moral economy and Massey's geographies of responsibility (see Table 12.5). An alternative economic geography of counter-globalization would emphasize solidarity between places within commodity chains, and fair trade, contrasting with the mainstream espousal of free trade and competitive advantage as providing the optimum economic outcome. It would also favour organizational forms that are likely to evoke alternative ethical and moral values (e.g. family firms, cooperatives) to private firms and multinational corporations driven by narrow monetary values.

The hierarchical and uneven power relations that characterize global commodity chains in the mainstream economy would be replaced by 'softer' and more decentred forms of power that would involve more collaboration and dialogue between actors to organize commodity chains and production networks. Economic actors in commodity chains develop awareness of the varied and entangled geographies of their commodity chain and forge closer and more intimate connections by recognizing how their own actions (as consumer, producers, traders) have impacts on distant others. Creating such an alternative economic geography is an immensely difficult task, of course, but such difficulty should not blind us to the fact that the roots of this alternative formation already exist within capitalism in the diverse forms of economic practice as highlighted in this chapter.

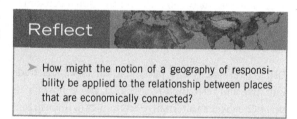

Reflect

➤ How might the notion of a geography of responsibility be applied to the relationship between places that are economically connected?

12.5 Summary

Capitalist economic practices have never become completely dominant over other forms of social relations. The evolution of capitalism as a global economic system has always been accompanied by resistance and the articulation of alternatives. During the 1980s and 1990s, however, globalization, the emergence of neoliberal economic policies, the collapse of concrete alternative systems in the Soviet Union and eastern Europe and the entry of China into the global capitalist economy seemed to herald the end of history for many commentators (Fukuyama, 1994). The onset of new capitalist crises in the 1990s and 2000s has exposed the limits to such thinking, while the counter-globalization movement and more localist forms of experimentation show the increased vitality of alternative economic thinking.

Geographers such as Gibson-Graham have also been at the forefront of critiquing capitalocentric thinking, showing how the geography of the global economy is shaped by non-capitalist relations and practices, both within firms and within the broader economy. Important work on alternative economic spaces has highlighted not only the diversity and variety that exist in the economy but also the political limits of such alternatives. Critiques of the limits to more localist alternatives such as LETS schemes or credit unions emphasize the dangers of incorporation into the mainstream capitalist economy. By contrast, the work of Doreen Massey (2005) is important in highlighting more global visions that develop new spatial connections between places, based upon different values to those of existing global commodity chains. Fair trade initiatives attempt to evoke more ethical relations and a more responsible set of geographies (Massey, 2004) that question uneven development and the spatial imbalances that exist between places in the global economy.

Exercise

Using a case study of a particular fair or ethical trade initiative and referring to Table 12.5, assess how far its practice evokes an alternative economics to mainstream commodity chains. Pay attention to the connections and relations within all aspects of the commodity chain and theorize how far a geography of responsibility is or can be established in the face of broader processes of uneven development. How does the selected initiative seek to monitor and regulate processes of trade and exchange? And, how far is it successful in replacing dependency relations within mainstream networks by new and more egalitarian sets of relations? How are the relations between places reconfigured as a result?

Key reading

Cook, I. *et al.* (2010) 'Geographies of food: "Afters."' *Progress in Human Geography*, online publication at http://www.phg.sagepub.com/content/early/.
A really useful discussion based on an online blog discussion between Cook and others that debates the ethics and politics of alternative food production systems. A very accessible and informative introduction to the issues in a novel format.

Gibson-Graham, J.K. (2006) *Postcapitalist Politics*, Minneapolis: University of Minnesota Press.
An influential and stimulating account of the diversity of economic practices. Gibson-Graham provide an important counterweight to perspectives from mainstream economics and Marxism, emphasizing how the economy is shaped by a diverse range of social relations and cannot be reduced to profit maximization and economic rationality.

Massey, D. (2004) 'Geographies of responsibility.' *Geografisker Annaler B*, 86(1): 5–18.
Massey's 2004 article is a useful summary of her arguments for a geography of responsibility between places, although the ideas are developed further in more theoretical form in her 2005 book *For Space*, and more empirically in her 2007 book *World City*.

Morgan, K., Marsden, T. and Murdoch, J. (2006) *Worlds of Food: Place, Power and Provenance in the Food Chain.* Oxford: Oxford University Press.
Morgan *et al.* present a well grounded set of arguments around the moral economy, using a series of case studies from the food sector to highlight the links between alternative food practices in particular places (e.g. Tuscany, California, Wales) and the broader economic geographies of power and governance within which these places are embedded.

North, P. (2005) 'Scaling alternative economic practices? Some lessons from alternative currencies.' *Transactions of the Institute of British Geographers*, 30(2): 233–5.
A useful summary article on the possibilities and limitations of alternative local economic spaces with a discussion on the prospects for scaling up these alternatives to higher levels. The article also contains comparative case studies from the Global North and South.

Useful websites

http://www.communityeconomies.org/Home
The website of the Community Economies Collective, which developed out of the work of J.K. Gibson-Graham but expanded to incorporate a range of collaboration between academic researchers and community groups in Australia, North America, Europe and South-east Asia.

http://www.ica.coop/al-ica
The website of the International Cooperative Alliance.

http://www.woccu.org
The World Credit Union website.

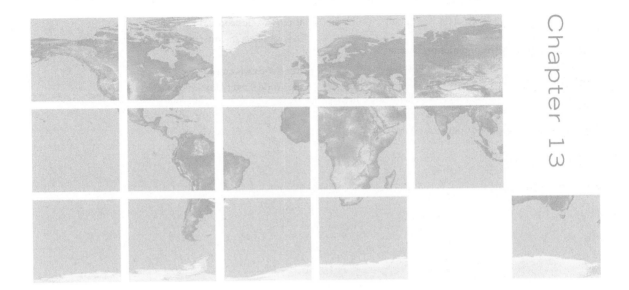

Conclusion

In this book, we have examined the changing geography of the contemporary world economy, focusing particularly on the location of different types of activity, the economic relationships between regions and the economies of particular places. Our approach to these issues is based on a revised political economy approach informed by cultural and institutionalist insights, making it more open to the importance of context, difference and identity. From this perspective, we view the economic geography of the world as the product of a process of interaction between general processes of capitalist development and pre-existing local conditions and practices. In this short concluding chapter, we summarize the key themes of the book, and offer some reflections on globalization and uneven development after the financial and economic crises of 2007–10.

13.1 Summary of key themes

The three interrelated themes running though the book are globalization, uneven development and place. Globalization should be viewed as an on-going process, not a final outcome or 'end state', which is shaped by the actions of a wide range of individuals, organizations, firms and governments (Dicken et al., 1997). The economy has become more globally integrated in terms of hugely increased flows of goods, services, money, information and people. The process of globalization has been facilitated by the development of a new set of 'space-shrinking' transport and communications technologies since the 1960s, resulting in 'time–space compression' as the costs of moving materials, capital and information across space have been dramatically reduced (Harvey, 1989a). MNCs, financial markets and international economic organizations such as the World Bank and IMF can be viewed as key agents of globalization. States should be viewed as active facilitators of globalization, rather than its passive 'victims', through the adoption of neoliberal policies including the abolition of exchange and capital controls, the reduction of trade barriers and the implementation of privatization programmes. At the same time, the project of neoliberal globalization has encountered opposition from a loose coalition of counter-globalization groups and organizations, sparking periodic protests and the beginnings of an alternative model of 'bottom-up' globalization through bodies such as the World Social Forum. Instead of economic integration creating a 'flatter' or more equal world through the reduction of international inequalities (Friedman, 2005), globalization is an uneven process, creating prosperity in some places, while others become increasingly marginalized. On a global scale, this is symbolized by the dramatic growth of East Asia, especially China, and the stagnation of sub-Saharan Africa since 1980. At the same time, levels of inequality within many countries have risen, particularly in rapidly growing economies such as China, India and Russia (Tomaney *et al.*, 2010).

Uneven development represents a basic characteristic of capitalism as a mode of production, reflecting the tendency for capital and labour to move to the areas where they can secure the highest returns (profits and wages respectively). In this way, growth becomes concentrated in core regions though a process of 'cumulative causation' (Myrdal, 1957) (Box 4.2). Surrounding regions get left behind, becoming subordinate peripheries supplying resources and labour to the core. Patterns of uneven development are not static, however, with capital and labour moving between locations in search of higher profits and wages. The process of economic growth in a particular region eventually tends to undermine its own foundations, leading to overheating as the prices of labour, land and housing spiral inexorably and underinvestment in infrastructure creates congestion. Lower costs in other, underdeveloped regions attract capital and labour. This 'see-sawing' of capital between locations creates shifting patterns of uneven development at different geographical scales (Smith, 1984). This process of uneven regional growth and decline is apparent from the decline of so-called 'rustbelt' regions (northern England, the American Midwest) and the rise of 'sunbelt' ones (the south and west of the US, southern Germany).

An important pattern of uneven development to emerge in recent decades is the spatial division of labour, where different parts of the production process become located in different types of region. On a global scale, MNCs moved routine manufacturing and assembly operation in industries such as textiles, footwear and electronics to certain developing countries in the 1960s and 1970s. This relocation process was driven by the availability of large surpluses of low-cost labour in developing countries, and facilitated by the increasing division of labour, in manufacturing and new transport and communication technologies. It created a new international division of labour where routine assembly and manufacturing was increasingly carried out in low-wage countries, while higher-level functions such as strategic management and research and development remained based in developed countries. More recently, the offshoring of IT-enabled services to developing countries such as India has prompted talk of a 'second global shift', creating a new international division of labour in services (Bryson, 2008).

Patterns of uneven development are structured by the conflicting processes of spatial concentration and dispersal. The question of which of these processes is predominant will vary according to a range of factors, including the type of economic activity in question, the level of available technology, the development of the (technical) division of labour and the size and location of the market being served. Information and communication technologies have been a key force in reshaping the economic landscape, facilitating the relocation of both manufacturing and IT-enabled services on a global scale. This does not herald the 'end of geography' (O'Brien, 1992), however, as shown by the increased concentration of corporate headquarters and advanced financial and business services in world cities. This brings us to the crucial observation that high-value activities show a marked tendency towards spatial concentration in large urban areas, while lower-value activities are more susceptible to dispersal to lower-wage locations. We regard this as a key underlying 'rule' or 'stylized fact' in economic geography. Various qualifications can be made, of course, and various exceptions found, but these do not diminish the importance of the basic point.

The role of place represents the third key theme of the book, partly reflecting the effects of wider processes of economic development in creating distinctive forms of production in particular places. These general processes interact with pre-existing local conditions and practices (for example, resource bases, employment patterns, skills, income levels, cultural values, institutional arrangements and political orientations) to create new geographies of production and consumption. Rather than simply reflecting geographical isolation, the distinctiveness or uniqueness of places is actually reproduced through this process of interaction (Johnston, 1984). From this perspective, place can be regarded as key meeting point or node constructed out of wider social relations and connections (Massey, 1994). Such relations and connections span the spheres of production, consumption and circulation as transport and communication networks and financial flows bind places together. The volume and intensity of these spatial flows and connections have increased markedly under globalization, inspiring Massey's effort to develop a global sense of place.

The assumption that globalization is erasing the distinctiveness of place, making distant localities appear increasingly similar, is only true at a very superficial level. In terms of consumption, for instance, the spread of global brands such as McDonald's and Coca-Cola is often cited, together with the profusion of large retail malls and centres. Yet the associated idea of a single global consumer culture is crude and simplistic, masking a more complex process whereby different local cultures have become increasingly mixed and entangled (Crang, 2005). Local sites of consumption remain highly significant, not least in terms of the growth of tourism and 'heritage' where place itself is a central object of consumption. In the sphere of production also, place remains important in the global economy, although many of the specialized industrial regions of the past have disappeared in the face of wider processes of economic restructuring. Particular places continue to be associated with distinctive forms of economic activity. Numerous examples have been cited and discussed through the text, from the densely clustered financial districts of Wall Street and the City

of London, the high-tech centre of Silicon Valley to the post-industrial regions of Wales and nort-east England and the mining districts of the Andes. While some places have prospered under globalization, others have become marginalized and impoverished, reflecting the inherently uneven nature of uneven development under capitalism. The impact of processes of uneven economic development upon particular places cannot be easily predicted or modelled in advance, since this will depend upon a range of contingent factors and relationships. This means that the shaping of the economic landscape is characterized by a basic openness and unpredictability, providing much of what makes economic geography such an interesting subject to study.

13.2 Globalization and uneven development after the crisis

The political economy approach that we have adopted in this book emphasizes the dynamic and unstable nature of capitalism as an economic system. Over the long-term, it has been characterized by periods of growth and stagnation (Figure 3.3), punctuated by moments of crisis. It is this dynamism that underpinned Schumpeter's famous notion of creative destruction which emphasizes how the periodic emergence of new technologies undermines and destroys established industries and skills. This dynamism and instability has been reinforced by the widespread adoption of neoliberalism as the dominant economic policy framework since the late 1970s. Policies of deregulation and liberalization have unleashed renewed market forces and eroded the ability of states to control global flows of capital.

The instability of capitalism has been underlined by the experience of the financial crisis and subsequent global recession since 2007. While triggered by failures in the 'sub-prime' mortgage market in the US, the increasing global integration of the financial system ensured that the crisis soon enveloped the world economy, since banks and financial institutions from a range of countries had become heavily involved in the re-packaging and trading of US mortgage liabilities (section 9.5). Globalization and speculation were fuelled by the deregulation of financial markets in the 1980s and 1990s, which removed the barriers between different segments of the market (e.g. between building societies, retail banks and investment banks) and led to a lack of proper scrutiny and regulation. At the same time, economic growth from the mid-1990s to 2006 was underpinned by growing private debt with households relying on credit to support consumption (Blackburn, 2008). The financial crisis spilled over into an economic crisis through the 'credit crunch', meaning that businesses could not obtain funds for investment and forcing households to reduce their expenditure.

The other source of crisis to affect the global economy in recent years is, of course, environmental. Concerns about climate change have grown markedly, attracting increasing attention from the media and politicians, although momentum seems to have stalled somewhat in the wake of the failed Copenhagen Summit in late 2009. The issue of climate change is shot through with questions of uneven development and justice, as the impact will be felt most severely in developing countries due to their greater dependence on natural resources and ecosystems to meet their everyday material needs. Even in the developed world, however, the presumption that economic development can proceed without any adverse effect on the environment is no longer valid, with resources likely to come under increasing pressure, focusing renewed attention on the physical geography of development (Tomaney et al., 2010) and geopolitical struggles over scarce resources (Klare, 2002). More broadly, the era of cheap energy and low transport costs that underpinned processes of economic globalization may be coming to an end, raising questions about the scope for the re-localization of supply chains (Morgan et al., 2006).

The immediate responses to the economic crisis in 2008–09 appeared to signal a shift away from neoliberalism to something more akin to neo-Keynesianism, with governments in both North America and western Europe lowering interest rates and undertaking fiscal stimulus packages to try to boost economic recovery. For many economists, this action was viewed as successful in averting a 1930s-style depression (Krugman, 2010). By the end of 2009, however, this window for remaking the economy appeared to have closed. Media

and public debate shifted back dramatically from the banking crisis, the perils of free markets and deregulation to a concern with public debt, much of it caused by propping up the failed financial system, which generated a fiscal crisis of the state (Hendrikse and Sidaway, 2010). Countries as diverse as Greece, Iceland and Ireland embarked upon IMF-backed austerity programmes, while at the time of writing (October 2010) the new UK coalition government was about to embark upon the most savage cuts to government spending in living memory. These developments reflect the entrenched institutional power of neoliberalism as a policy discourse that benefits financial and business elites at the expense of other groups in society (Harvey, 2005). As one commentator put it:

> The austerity measures are part of the plan to rescue the banks again. Governments throughout the Eurozone have succumbed to an alliance of banks and large holders of public debt who are desperate to avoid the implications of their foolish lending … The price was austerity across the region in a naked attempt to shift the costs of the crisis on to taxpayers and public sector workers.
>
> (Lapavitsas, 2010)

There is a sense here of Naomi Klein's 'shock doctrine' in operation where business, financial and political elites take advantage of crises to remake societies in pursuit of their neoliberal agenda (Hendrikse and Sidaway, 2010). Meanwhile, the underlying causes of the crisis have not been addressed, while the solution of cutting government spending throughout the OECD countries threatens to further undermine economic recovery. The impacts of all this are likely to further exacerbate spatial inequalities and conditions of uneven development, with the evidence to date suggesting that the social costs of the crisis may be borne by those poorer social groups and less-favoured cities and regions that bore the brunt of previous recessions (Chapter 9), rather than the financiers, policy-makers, regulators and economists who were implicated in the crisis.

The revival of a reinforced neoliberalism to deal with the effects of the economic crisis has an added irony. For the past 30 years, governments across the world have sought to liberalize and deregulate their economies, adopting policies of low inflation, open competition and privatization. Such policies have proved controversial, often becoming associated with cuts in public expenditure and increased levels of inequality. In the context of developing countries particularly, IMF- and World Bank-inspired austerity (structural adjustment) programmes have foisted often inappropriate policies of liberalization and deregulation on heavily indebted economies, with harsh social consequences in terms of, for instance, the reduction of health and education expenditure and the introduction of user charges for basic services. The same economic medicine is now being applied to western Europe, in particular, with the IMF giving its approval to the UK government's austerity package. As such, an austere form of neoliberalism is coming home with a vengeance in a renewed assault on the welfare state regimes that were established in the post-1945 period, while remaining social and state assets face further marketization (Hendrikse and Sidaway, 2010.).

The consequences for the global economy are likely to be a period of stagnation or very slow recovery. Globally, this has been the worst economic recession since the end of the Second World War, with most commentators expecting a deep and long recession because of the 'balance sheet' problems built up by high levels of public and private debt (Dolphin, 2009). As governments and the private sector attempt to reduce debt levels,

overall demand will fall further. From a longer-term Schumpeterian perspective (section 3.2.4), recovery begs the question of where the technological innovation and investment to generate renewed growth will come from. Perhaps it will be from the newly emerging BRIC (Brazil, Russia, India and China) economies which have generally displayed continued growth in recent years, fuelling much speculation about an underlying geopolitical transfer of economic power from the 'old' core of North America and western Europe.

While all this might seem depressing or, at best, frighteningly uncertain and unstable, it is important to remind ourselves that the economy itself remains socially constructed and open to economic alternatives. The resurrection of neoliberalism reflects its power as a discourse articulated by influential actors in government, business and media circles to serve their political and economic interests. While this emphasizes the deep institutional entrenchment of neoliberalism and its ongoing capacity to reinvent itself in response to events (Peck *et al.*, 2010, there is nothing inevitable about its continued rule. Opposition to austerity measures may grow as cuts to state provision really begin to take effect, generating increased interest in the alternative and fairer ways of regulating and organizing economies that we covered in Chapter 12.

Glossary

3 Ts: technology, talent and tolerance. According to Richard Florida's theory of the creative classes, the new economic geography of creativity is based on these so-called 3 Ts. Technology is defined in terms of the presence of high-tech industry, while talent refers to high levels of human capital, and tolerance refers to the openness and diversity of a particular place.

Absorptive capacity. This refers to a firm's ability to recognize, assimilate and exploit knowledge. It seems to rely upon the existence of a common corporate culture and language, meaning that everybody shares the same broad outlook and sense of the company's overall purpose and objectives.

Accumulation function, of the state. This refers to its activities in supporting and promoting economic development within its territory, ensuring that business can accumulate capital for investment and growth.

Agglomeration economies. A set of economic advantages for individual firms that are derived from the concentration of other firms in the same location – such as the availability of skilled labour, proximity to a large number of customers, access to specialist suppliers of services and the provision of an advanced infrastructure.

Aid. Official development assistance offered by developed countries to promote economic development and poverty reduction in developing countries.

Alternative local economic spaces. Local economic initiatives that seek to develop alternative economic identities and practices outside the mainstream capitalist economy.

Back-office functions. Routine clerical and administrative tasks such as the maintenance of office records, payroll and billing, bank checks and insurance claims.

Backwash effects. The adverse effects of growth in a core region on a surrounding region in terms of capital and labour being 'sucked out' of the latter of region into the former, which offers higher profits and wages. In this situation, the virtuous circle of growth in the core is matched by a vicious circle of decline in the periphery, expressed in classic symptoms of underdevelopment such as a lack of capital and depopulation.

Basic needs. An approach to development prominent in the 1960s and 1970s which focused attention on the everyday lives of the poor, reflecting concerns that these were being neglecting by orthodox modernization policies. The concern for identifying and meeting such needs in relation to food, shelter, employment, education, health, etc is shared by the recent work on **grassroots development**.

Branch plant economy. A term coined to describe the way particular regions become dominated by 'branch plants', undertaking basic production activities but controlled from outside. Branch plant economies emerged through economic restructuring processes in the 1960s and 1970s whereby processes of corporate growth at national and international level produced a new spatial division of labour taking over from or displacing local ownership.

Buyer-driven commodity chains. These are coordinated by large, branded retailers which concentrate on design, sales, marketing and finance while actual production is out-sourced to suppliers in developed countries. This type of arrangement is characteristic of labour-intensive consumer goods industries such as clothing, furniture and toys.

Capital accumulation. The process of investment and reinvestment in production to generate higher profits. Accumulation lies at the heart of the capitalist system, representing the basic driving force for economic growth and innovation.

Capitalism. A particular mode of production, dominant since the eighteenth century, based on the private ownership of the means of production – factories, equipment, etc. – and the associated need for the majority of people to sell their labour power to capitalists in exchange for a wage.

Capital switching. The process by which capital is moved between sectors of the economy and regions, in response to changing investment opportunities. In geographical terms, capital is often transferred from regions dominated by declining sectors to 'new industrial spaces' in distant regions offering more attractive conditions for investment.

Central banks. The state institutions that are responsible for controlling the supply of money in a country, usually involving the setting of interest rates and the monitoring of inflation. Although central banks are 'independent' of government in many countries, the dominance of central bank committees by mainstream economists tends to mean that they adopt a very orthodox approach to managing the economy.

Central place theory. Developed by Walter Christaller in the 1930s, this is perhaps the best known of the **German locational theories**. Based on the assumption of economic rationality and the existence of certain geographical conditions such as uniform population distribution across an area, central place theory offers an account of the size and distribution of settlements within an urban system. The need for shop owners to select central locations produces a hexagonal network of central places.

Centralization of finance. The process by which financial markets have become larger and more integrated over time, with local and regional banks giving way to national and, increasingly, global systems.

Circuit of capital. The basic process of producing commodities for profit under capitalism, involving the combination of the means of production – factories, machines, materials, etc. – and labour power to produce a commodity for sale, generating a profit above the initial money outlay. Part of this profit is reinvested back into the production process, which recreates the circuit anew and forms the basis of capital accumulation. In addition to the primary circuit of capital in production, secondary and tertiary circuits are also commonly identified, referring to investment in the built environment and education, health and welfare respectively.

Clusters. A more popular term for agglomeration involving the concentration of economic activities in a particular location. The term has become associated with the work of Michael Porter, a Harvard business economist, in recent years. Porter (1998, p.197) defines clusters as 'geographical concentrations of interconnected companies, specialised suppliers, service providers, firms in related industries, and associated institutions (for example universities, standards agencies and trade associations) that compete but also co-operate'.

Codified knowledge. Also termed explicit knowledge, this refers to formal, systematic knowledge that can be conveyed in written form through, for example, programmes or operating manuals.

Cold War. The political and ideological conflict between western capitalist countries led by the US and a communist block led by the Soviet Union, lasting from the mid-1940s to the late 1980s.

Colonialism. A political and economic system based on territorial empires, involving the direct political control of colonies by the colonial powers.

Commercial geography. A key component of the traditional approach, prominent from the 1880s to 1930s. Closely linked to European imperialism, it provided knowledge about colonial territories in Africa, Asia and Latin America, identifying and mapping key resources, crops, ports and trade routes, relating these to climate and settlement patterns.

Commodity chains. The network of connections involved in the production, circulation and consumption of a commodity, covering the different stages from the supply of materials to final consumption. They typically incorporate a range of actors, for example farmers, sub-contractors, manufacturers, transport operators, distributors, retailers and consumers. Commodity chains have a distinct geography, linking together different stages of economic activity carried out in different places.

Commodity fetishism. The tendency for the geographies of production and distribution that actually generate the goods on sale in western shops to be obscured by the emphasis consumers attach to the physical appearance and price of goods.

Community economy. An attempt to encourage a more locally contained and reciprocal economic system in opposition to the profit-centred global economy. Examples of community economy initiatives are LETs schemes and credit unions.

Comparative advantage. A key principle of international trade theory, classically expressed by David Ricardo in

1817. It states that a country should specialize in exporting goods which it can produce more cheaply than other goods and import goods which are more expensive to produce domestically. Through specialization, both countries gain by focusing on the goods in which they have a comparative or relative cost advantage, and importing those in which they have a comparative disadvantage.

Competence or resource-based theory of the firm. Derived from the work of the economist Edith Penrose (1959), the competence perspective views firms as bundles of assets and competencies that have been built up over time. Competencies can be seen as particular sets of skills, practices and forms of knowledge.

Competition state. The idea that a new type of state is emerging in developed countries, the key task of which is the promotion of innovation and economic competitiveness rather than the provision of social welfare services.

Competitive advantage. The dynamic advantages derived from the active creation of technology, human skills, economies of scale, etc. by firms. This can be contrasted with the rather static and naturalistic notion of **comparative advantage**, which relates efficiency to pre-existing endowments of the main factors of production. Competitive advantage helps to account for patterns of trade and regional specialization at a more detailed level.

Competitiveness. A key concept underpinning economic development policy since the early 1990s. It refers to the underlying strength of the economy in terms of its capacity to compete with other countries, based on the assumption that nations and regions compete for global market share in a similar fashion to firms. Key aspects of competitiveness include levels of innovation, enterprise and workforce skills and government is charged with fostering these capacities.

Competitive upgrading. The improvement of an actor's or firm's position in a commodity chain or production network, involving the capture of additional value. Four types of upgrading haven been identified: process upgrading based on more efficient production, product upgrading (the production of sophisticated high-value goods or services), functional upgrading and intersectoral upgrading whereby a firm uses its existing competences to move into new sectors.

Conditionality. The requirement for developing countries to meet strict conditions set by the World Bank and International Monetary Fund in return for providing grants and loans. In particular, the Bank and IMF have required countries to undertake a set of economic reforms, generally known as **structural adjustment programmes**, based on the principles of the **Washington Consensus**.

Consumer culture. A key feature of contemporary society, which is increasingly organized around the logic of individual choice in the marketplace, with shopping and consumption an increasingly central part of people's lives and identities. Modern consumer culture developed through the targeting of middle-class women, in particular, during the nineteenth century, with the establishment of large department stores in cities playing a key role.

Consumption. The processes involved in the sale, purchase and use of commodities by individuals.

Cooperatives. Mutual (as opposed to private) forms of ownership of businesses or other organizations (e.g. housing associations) set up to act collectively in the best interests of particular groups (e.g. employees, consumers, tenants) and usually attempting to pursue more ethical and social goals alongside economic ones.

Counter-globalization movement. The movement that has arisen in recent years to oppose the inequalities brought about by the neoliberal or free market agenda of economic globalization. It espouses a more open, participative 'bottom-up' model of globalization, evident in initiatives such the World Social Forum The movement is significant in bringing together diverse social groups (e.g. farmers, trade unionists, environmentalist) from both the Global North and South.

Creative classes. A group of highly skilled and educated workers that has distinct lifestyle preferences, according to Richard Florida and his theory of the creative classes. They consist of people who add economic value through their creativity, incorporating a super-creative core (scientists and engineers, university professors, poets and novelists, artists, entertainers, actors, designers and architects) and a range of creative professionals who engage in complex problem-solving that involves independent judgement, requiring a high level of education or human capital. Florida argues that cities and regions should actively seek to attract the creative classes which are drawn to open and diverse places.

Creative destruction. A term coined by the Austrian economist Joseph Schumpeter to capture the dynamism of capitalism in terms of innovation and the development of new technologies. As new products and technologies are developed and adopted (creation), they often render existing industries obsolete, unable to compete on the basis of quality or price (destruction).

Cultural turn. An important development in human geography and the social sciences from the late 1980s, involving a shift of focus from economic to cultural questions.

Cumulative causation. A model of the process of uneven regional development derived from the work of the Swedish economist Gunnar Myrdal. This explains the spatial concentration of industry in terms of a spiral of self-reinforcing advantages that build up in an area and the adverse effect this has on other regions, creating a core-periphery pattern.

Debt crisis. The problem of large-scale indebtedness facing many developing countries since the 1980s, undermining development efforts and threatening the viability of the world financial system at times. The origins of the debt crisis lie in the interactions between three sets of factors: the borrowing of large sums by developing countries from northern banks and institutions in the 1970s; rising interest rates in the late 1970s and early 1980s; and reduced commodity prices and thereby export earnings since the early 1980s.

Deindustrialization. A decline in the importance of manufacturing industry as a sector of the economy, expressed in terms of its share of employment or output. Generally associated with the growth of service industries and the closure of older heavy industries, such as coal, steel and shipbuilding, deindustrialization has been a common experience across developed countries since the 1960s.

Dependency theories. The most prominent set of **structuralist theories**, particularly associated with the radical economist Andre Gunder Frank. According to Frank, the metropolitan core exploits its 'satellites', extracting profits (surplus) for investment elsewhere. **Colonialism** was a key force here, creating unequal economic relations which were then perpetuated by the more informal imperialism characteristic of the post-war period.

Deregulation. The reduction of the rules and laws under which business operates, a key component of **neoliberalism**.

Derivatives. New financial instruments, defined as 'contracts that specify rights/obligations based on (and hence derived from) the performance of some other currency, commodity or service used to manage risk and volatility in global markets' (Pollard, 2005, p.347). Key forms of derivatives include futures, swaps and options.

Deskilling of labour. The removal of more rewarding aspects of work, such as design, planning and variation, often associated with an increased division of labour. The subdivision and fragmentation of the labour process increases employers' control of production, reducing workers to small cogs in the system and making them vulnerable to replacement by machines, which often perform standardized tasks more efficiently. This is what Marx meant when he spoke of the alienation of labour under capitalism, with industrial workers playing an increasingly minute role within tightly controlled factories, becoming divorced from the products of their labour.

Development. In the context of economic policy, the term conveys a sense of positive change over time, making a particular country or region more prosperous and advanced. As an economic and social policy, development has been directed at those 'underdeveloped areas' of the world that require economic growth and modernization.

Developmental state. A particular type of state which is heavily involved in the promotion of economic development. The term is derived from the experience of East Asian countries such as Japan, South Korea and Singapore, where the activities of the state were crucial in fostering export-led expansion. Weiss's (2000, p.23) definition of the developmental state emphasizes three key criteria: the aim of increasing production and closing the economic gap with the industrialized countries; the establishment of a strong government department to coordinate and promote industrial development; and close cooperative ties with business.

Dialectics. A way of thinking which sees change as driven by the tensions between opposing forces, usually in the form of thesis–antithesis–synthesis. Capitalist society in particular is characterized by continual change and flux, driven by the search for profits in the face of competition. Dialectics is originally derived from the eighteenth-century German idealist philosopher Hegel, before being adapted by Marx and subsequent Marxists such as David Harvey.

Diamond model. Michael Porter's representation of how business clusters actually operate, enhancing competitiveness and productivity by fostering the interaction between four sets of factors: demand conditions; supporting and related industries; factor conditions; and firm strategy, structure and rivalry.

Discourses. A key term derived from poststructuralist philosophy which refers to networks of linked concepts, statements and practices that produce distinct bodies of knowledge. Crucially, meaning is generated through particular discourses which, instead of simply reflecting an underlying reality, actively create it.

Diverse economies. A term that signifies an emphasis on the different forms of economic activity and organization that co-exist with capitalism, such as domestic work, cooperatives and gift-buying.

Division of labour. A key principle of industrial society that has technical, social and geographical dimensions. The technical division of labour can be defined as the process of dividing production into a large number of highly specialized parts, so that each worker concentrates on a single task rather than trying to cover several. Under-industrialization, as Adam Smith argued, and an increased division of labour results in huge rises in productivity.

Doha Round of trade negotiations. The latest round of discussion launched in Qatar in late 2001, under the auspices of the World Trade Organization (WTO) which replaced GATT in 1995.

Economic geography. A major branch of human geography which addresses questions about the location and distribution of economic activity, the role of uneven geographical development, and processes of local and regional economic development.

Economics. An important neighbouring discipline to economic geography, which views the economy as governed by market forces that basically operate in the same fashion everywhere, irrespective of time and space. In contrast with the diversity and open-endedness of economic geography, economics is a formal theoretical discipline based on modelling and quantification.

Economies of scale. The tendency for firms' costs for each unit of output to fall when production is carried out on a large scale, reflecting greater efficiency. Industrialization led to huge economies of scale through the establishment of large factories employing sophisticated machinery and an elaborate division of labour.

Employment. The act of selling your labour to work for an employer in the formal economy, usually paid in the form of a weekly wage or a monthly salary. Paid employment as the norm is a form of work distinctive to capitalist society.

Equity finance. Funds raised by investors buying a stake or share in a firm. For large, publicly quoted firms, this occurs through the issuing of shares on the stock market.

Ethical consumption. Forms of consumption which aim to improve conditions for workers and to enhance environmental sustainability within commodity chains. They usually involve buying guaranteed products, identified by labels such as Fair Trade (see **Fair or ethical trade**).

Eurocentric/Eurocentrism. A preoccupation with European or western experiences and practices, reflecting an assumption that western values and methods are always superior to those found in developing countries. This has been widely criticized as arrogant and condescending, particularly by post-colonial writers.

Export-oriented industrialization (EOI). Often regarded as the opposite strategy to i**mport-substitution industrialization (ISI)**, EOI involves countries producing goods and services for selling in external markets. It is compatible with traditional notions of free trade and comparative advantage in contrast with ISI which involves high levels of protection and state intervention.

Export platform. Production enclaves in parts of the less-developed world set up by MNCs (or their suppliers) from the developed world to supply markets in advanced economies. These are typically disconnected from the local or national economy in which they are situated and often have low levels of regulation and labour standards, prompting claims of exploitation, especially of women workers.

Export processing zones. Similar to export platforms but with the added incentive for foreign firms of offering tax and investment incentives to locate there. Typically these will include up to 100 per cent rebates on local taxation, the provision of all infrastructure and the relaxation of the usual rules governing foreign ownership (Dicken, 2003a). Some of the most notorious of these zones have been set up in Mexico along the border with the US and have led to the establishment of branch plants known as *maquiladoras* (literally translated as 'assemblers').

Factors of production. The different elements that are bought together to produce particular goods and services: capital, labour, land and knowledge.

Fair or ethical trade. The emergence of an alternative system of international trade in commodities and products such as clothes that links concerned consumers (primarily in richer countries) with farmers and producers in the Global South. Such alternative trade networks aim at giving producers a fair and stable price that allows decent living standards and is not subject to global market fluctuations or cost pressures from dominant multinationals. 'Fair Trade is a trading partnership, based on dialogue, transparency and respect, which seeks greater equity in international trade' (FINE, 2001, cited in Moore, 2004, pp.73–4).

Fictitious commodity. A material or product that takes on the appearance of a commodity, but cannot be regarded as a 'proper' one. Land, labour and money were originally identified as 'fictitious commodities' by the economic sociologist Karl Polanyi in his analysis of nineteenth-century industrialization. Such commodities are not directly produced for sale on the market and often require state intervention to balance supply and demand.

Financial crises. The term refers to occasions when the assets of financial institutions and actors lose a large proportion of their value and markets crash. They do not always have a large negative impact upon other parts of the economy, depending upon their nature and the ability of governments to respond, but at various points in time (e.g. the 1930s and 2007–08) they can result in severe global economic recessions.

Financial exclusion. Defined as 'these processes that serve to prevent certain social groups and individuals from gaining access to the financial system' (Leyshon and Thrift, 1995, p.314). It is based on income, compounding the difficulties facing disadvantaged individuals and groups.

Firm, the A legal entity involved in the production of goods and services, owned by individual capitalists or, more commonly, a range of shareholders. The standard organizational form of capital.

Fiscal crisis of the state. A situation in which the state lacks the resources to meet its financial obligations. The term was coined to describe the financial difficulties facing many national and local governments in the developed world in the 1970s as the post-war era of economic growth gave way to recession. This meant that tax receipts fell while welfare expenditure continued to grow, fuelled by rising unemployment. New York City became technically bankrupt in 1975, for example, while the UK was forced to seek an emergency loan from the IMF in 1976.

Flexible labour markets. An employment regime in which wages, conditions and worker attitudes become more responsive to the pressures of competition and the needs of business, requiring workers to accept varying pay rates and hours of work while learning new skills and undertaking new tasks. The creation of flexible labour markets has been a key goal of neoliberal policy since the 1980s.

Flexible production. A new form of production regarded in the 1980s and early 1990s as the successor to **Fordism**, based on the rise of new 'sunrise' industries such as advanced electronics, computers, financial and business services, and biotechnology. In the sphere of production, flexibility was associated with the widespread use of information and communications technologies which enabled processes and equipment to be continually reprogrammed and reset. In the sphere of consumption, it was rooted in a new emphasis on niche markets and customization compared with the standardized mass markets of the post-war period.

Fordism. A regime of accumulation, dominant from the 1940s to the 1970s, based on a crucial link between mass production and mass consumption, provided by rising wages for workers and increased productivity in the workplace. This regime of accumulation was supported by a Fordist–Keynesian mode of regulation where the state adopted highly interventionist policies of demand management, full employment, welfare provision, trade union recognition and national collective bargaining. Fordism is named after the American car manufacturer, Henry Ford, who pioneered the introduction of mass-production techniques. As such, the term is also used in a narrower sense to refer to an **intensified labour process**, based on a highly elaborate division of labour and the introduction of moving assembly lines.

Geographical expansion. The spatial expansion of the economy outwards from original core areas to encompass new territories, an inherent feature of capitalism since the sixteenth century, reflected in the development of an integrated world economy from an initial core in western Europe. Geographical expansion has been driven by the profit-seeking and competition between firms, generating a search for new markets, new sources of raw materials and new supplies of labour. It gained new momentum as the industrial revolution took off, culminating in the 'age of empire' between 1875 and 1914 (Hobsbawm, 1987). The shift towards globalization since the 1970s can be seen as the latest chapter in this ongoing story.

Geographies of responsibility. A phrase associated with the geographer Doreen Massey whereby people living in one location recognize the inter-connections between events and relations in their local economy and distant others as a result of globalization processes and take responsibility for them.

German locational theory. A body of spatial economic theory from the nineteenth and early twentieth centuries, based on the work of German theorists such as Von Thunen, Weber, Christaller and Losch, which developed models of the economic landscape derived from the assumptions of **neo-classical economic theory**.

Global commodity chains approach. An approach to the study of commodity chains that emerged out of world systems research in historical sociology and political economy. Much of this literature focused on the issue of governance, referring to how chains were organized, controlled and coordinated by 'lead' firms.

Global consumer culture. A phrase emphasizing the creation of a single global market, centred upon brands such as McDonald's, Coca-Cola and Nike. For many commentators, this is erasing the distinctiveness of local places and cultures, heralding the 'end of geography'. As a number of studies have shown, however, this cultural homogenization argument is highly simplistic.

Global production networks (GPNs). An approach developed by economic geographers to the study of economic globalization, incorporating a wide range of actors and relationships, in contrast with the more restrictive **global commodity chains approach**. GPN research has made an important contribution to the rethinking of regional development by developing a more globalized perspective which emphasizes the process of **strategic coupling** between regional assets and GPNs.

Global sense of place. An attempt to rethink place in an era of globalization, associated with the geographer Doreen Massey. This approach rejects the idea of place as isolated and bounded, viewing it as a meeting place, a kind of node where wider social relations and connections come together.

Globalization. A process of economic integration on a global scale, creating increasingly close connections between people and firms located in different places. Manifested in terms of increased flows of goods, services, money, information and people across national and continental borders.

Globalization from below. A new model of 'bottom-up' globalization which emerged out of the increasing linkages between local social movements, facilitated by modern communications and the internet. Networks such as People's Global Action and the World Social Forum exemplify this, providing an alternative to the dominant neoliberal model of 'globalization from above'.

Governance. A term used to refer to the growth of new ways of governing societies in place of the traditional notion of government, incorporating special purpose agencies, business interests and voluntary organizations alongside government. Geographers are particularly interested in the emergence of new forms of local governance since the 1980s.

Host regions. Destination regions for foreign direct investment (FDI).

Household or domestic labour. Work undertaken within households as opposed to in workplaces and usually associated with the performance of tasks associated with social reproduction (i.e. looking after children, performing work for the family). Under capitalism, such work is usually unpaid and predominantly performed by women.

Human Development Index (HDI). A widely known composite measure of development established by the United Nations Development Programme (UNDP), and published annually since 1990 in its *Human Development Report*. The HDI 'measures the overall achievement of a country in three basic dimensions of human development – longevity, knowledge and a decent standard of living' (UNDP, 2001, p.14).

Hyper-mobility of capital. A phrase increasingly used by geographers and economists in the context of globalization and, more specifically, the deregulation and unleashing of financial flows across borders since the 1970s and the growth of electronic forms of money. It refers to the dramatic and potentially crisis-inducing effects that stem from the ability of financial actors to instantaneously move vast amounts of money and finance from one part of the globe to another at the touch of a button. It is also viewed as having a strong effect on local and national actors by requiring them to develop policies that serve business interests in order to maintain the confidence of markets and avoid outflows of capital.

Import-substitution industrialization. Involves a country attempting to produce for itself goods that were formerly imported. Newly created **infant industries** are protected from outside competition through the erection of high tariff barriers, allowing the country's economy to be diversified and dependence on foreign technology and capital reduced.

Industrial districts. Specialized industrial areas based on networks of small firms and craftsmen. Associated with nineteenth-century regions such as the Sheffield cutlery district until the 1980s, when a number of scholars drew attention to the revival of industrial districts in central and north-eastern Italy.

Industrial society. Used usually to define the period of mass industrialization in western (North American and western European) countries, from around 1750 through to the 1960s, when society went through major social and economic changes, not least of which was the transition from a rural and feudal way of life to an urban and capitalist one. Increasingly used as a term to distinguish it from the current post-industrial society (post-1970).

Industrial working class. The class of workers that emerged with the development of modern industry from 1750 onwards, who as Marx noted, were distinguished by the fact that their sole means of 'earning a living' was employment or waged labour. As an increasingly large and concentrated group of workers, with considerable collective strength through their trade unions, the industrial working class were viewed as being at the forefront of struggles for social change in the first half of the twentieth century. With the deindustrialization of many societies in the Global North, this class is perceived as in terminal decline and increasingly marginal to mainstream capitalism, although this is a very western-centric perspective. A new industrial working class is growing in many parts of the Global South (particularly in China) with the growth of manufacturing industry.

Infant industries. New industries with a particular country at an early state of development that are regarded as strategically important by the state, which seeks to foster their development. This involves protection from outside competition until such industries are strong enough to compete in global markets (Brohmann, 1996, pp.115–16).

Innovation. The creation of new products and services or the modification of existing ones to gain a competitive advantage in the market. The commercial exploitation of ideas is crucial, distinguishing innovation from invention.

Institutional foundations of markets. The wider social rules, norms and practices upon which the workings of markets depend, allowing transactions to take place and contracts to be respected.

Institutions. Broad social and organizational conventions, practices and rules that shape economic life. Key institutions include firms, markets, the monetary system, the state and trade unions.

Intensification of the labour process. An increase in the volume and speed of work within a factory or office, often associated with the introduction of new technology and organizational techniques. **Fordism** involved an intensification of the labour process, based on the introduction of mechanized assembly lines.

Interactive approach to innovation. A more recent perspective which views innovation as a circular process based on cooperation and collaboration between manufacturers or services providers, users (customers), suppliers, research institutes, development agencies, etc.

International division of labour. A pattern of geographical development which involves different countries specializing in different types of economic activity. The classic 'old' international division of labour of the nineteenth century, associated with colonialism, involved the developed countries of Europe and North America in producing manufactured goods while the underdeveloped world specialized in the production of raw materials and foodstuffs. This can be contrasted with the **new international division of labour** of the 1970s and 1980s.

International policy transfer. The transfer of policies between states, involving initiatives and experiments that have been introduced in one country being adopted in another.

International strategic alliances. The growing number of collaborations between firms which are competitors in the same markets. They are particularly pronounced in sectors that require high levels of research and product development and therefore have high start-up costs (e.g. computers, biotechnology). Their growth highlights the vulnerability of MNCs to changes in global markets and the desire to minimize uncertainty.

Just-in-time (JIT) production. A form of production organization prevalent in Japanese companies whereby firms maintain minimal stocks of supplies and components by maintaining a close relationship with suppliers. JIT systems were viewed by many commentators as being more flexible and efficient than Fordist production methods, providing Japanese car manufacturers with an important source of competitive advantage over US and European rivals during the late 1970s and early 1980s.

Keynesian economic policies. A set of economic policies, widely adopted by developed countries between the 1940s and the 1970s, derived from the economic theories of John Maynard Keynes. Keynesianism involved the state taking an active role in managing the aggregate level of demand for goods and services in the economy. This involved stimulating demand in downturns by increased government expenditure – supporting public works and employment schemes, for example, road building projects – or reducing taxes (Box 6.2). In periods of economic growth, the state should dampen down demand by increasing taxes or reducing expenditure, thus preventing inflation.

Knowledge-based economy. The idea that a new type of economy has developed since the early 1990s in which knowledge has become the key resource and learning the key process for firms and individuals.

Kondratiev cycles. Named after the Soviet economist Kondratiev who first identified such cycles in the 1920s, the term refers to long waves of economic development, based on distinctive systems of technology. Five Kondratiev cycles are usually distinguished since the late eighteenth century. Each cycle consists of two distinct phases: one of growth (A) and one of stagnation (B).

Learning regions. A concept of the region as a nexus of innovation and learning processes, rooted in the local concentration of **tacit knowledge** and formal and informal ties and relationships between firms.

Legitimation function of the state. The range of activities undertaken by the state to maintain social order, ensuring that the capitalist system and the associated social order is regarded as legitimate and 'natural' by the majority of citizens.

Liberalization. The opening up of protected sectors of the economy to competition, a key component of **neoliberalism**.

Linear model of innovation. The traditional understanding of innovation as a series of well-defined stages running from the research laboratory to the production line, marketing department and retail outlet. The linear approach is focused on large corporations, emphasizing formal research and development involving scientists and engineers operating separately from other divisions of the company.

Local exchange and trading schemes. Usually locally based initiatives where groups of individuals come together to establish their own economic micro-system for exchanging their labour and services among themselves on a reciprocal basis (e.g. exchanging haircuts for DIY for childcare). More often than not, this will also involve the setting up of a micro-currency to facilitate transactions, but can only function as long as the system is broadly in balance with users consuming as many services as they themselves produce.

Localization economies. A particular type of **agglomeration economy**, stemming from the concentration of firms in the *same* industry.

Local labour markets. A term used to signify the importance of the local scale in the operation of most forms of employment, whereby the majority of the population live and work within relatively small spatially delineated areas. The term 'travel-to-work-area' has been developed to statistically define local labour markets, identifying geographical boundaries that correspond to the limits within which the majority of a population of an area live and work. A major contribution of economic geography (e.g. Peck, 1996) is to highlight the way that labour markets vary widely in their processes and practices, both at the local and national scales.

Marxism. A set of social and economic theories derived from the writings of Karl Marx (1818–83) Marxism adopts a materialist view of society, stressing the importance of underlying social relations and forces over ideas. The economy is structured by a capitalist **mode of production** defined by the antagonistic social relations between the capitalist and working classes. The exploitation of workers forms the basis of capitalist profit, but this contradictory relationship will also ensure that capitalism is ultimately overthrown by socialism.

Mass consumption. A form of consumption based on the purchase of standard consumer durables such as automobiles, fridges and washing machines by large numbers of households. This was a key dimension of post-war **Fordism**, linked to mass production and rising wages for workers.

Millennium development goals. A set of objectives for development agreed by the United Nations in 2000, providing a focus for the new anti-poverty agenda. The millennium development goals list specific targets to be achieved by 2015 against which progress can be measured and monitored on an annual basis. They include the eradication of extreme poverty and hunger, the achievement of universal primary education and the promotion of gender equality.

Modernization school. The dominant approach to development in 1950s and 1960s. The basic idea is that developing countries undergo a linear process of transformation (modernization), analogous to the changes experienced by developed countries in the nineteenth century following the industrial revolution. Economic growth is paramount, generating increased income and employment opportunities which, it was assumed, would 'trickle down' to the poorest groups in society. This thinking was famously expressed in Rostow's **stages of growth** model.

Modes of production. Economic and social systems by which societies are organized, determining how resources are utilized, work is organized and wealth is distributed. Economic historians have identified a number of modes of production, principally subsistence, slavery, feudalism, capital and socialism.

Modes of regulation. A concept developed by the French regulationist school of political economy that emphasizes the role that wider processes of social regulation play in stabilizing and sustaining capitalism. These wider processes of regulation find expression in specific institutional arrangements which mediate and manage the underlying contradictions of the capitalist system, enabling renewed growth to occur. Modes of regulation are focused on five key aspects of capitalism, in particular: labour and the wage relation, forms of competition and business organization, the monetary system, the state and the international regime. The post-war period of economic growth was based upon a Fordist–Keynesian mode of regulation (see **Fordism**).

Moral economy. A term that has its origins in pre-capitalist forms of social relations where, in certain religions (notably Christianity and Islam), there were strong moral (and legal) pressures against greed and usury (e.g. charging interest on loans is still illegal in some Muslim countries). Marxist historian, E.P. Thompson, famously refers to the 'moral economy' of the crowd in relation to riots by the poor in response to rising food prices due to periods of shortage in the eighteenth century. The term has more recently been used in relation to the attempt to develop ethical or fair trading systems for global commodities, replacing pure capitalist values with more ethical ones.

Movimento Sem Terra (MST). Based in Brazil, this is a mass social movement of an estimated 1.5 million members. Founded in 1984, it is made up of mainly landless labourers and peasants from rural areas, many of whom have been dispossessed and displaced by agricultural reforms, mechanization and land clearance. The Movimento's strategy involves targeting large unused private estates and illegally squatting and occupying the land.

Multinational corporations (MNCs). Companies that conduct operations in a number of countries, allowing them to access different markets and take advantage of geographical differences in conditions of production, such as the skills and costs of labour. MNCs have been key agents of **globalization** since the 1970s.

Nation, the A group or community of people who feel themselves to be distinctive, on the basis of a shared historical experience and cultural identity, which may be expressed in terms of ethnicity, language or religion.

National collective bargaining. A system in which representatives of employers, trade unions and government got together to agree pay rates and awards at a national level, often on an annual basis. This form of bargaining was a key feature of **Fordism** between the 1940s and 1970s, with labour gaining higher wages in exchange for increased productivity.

Neo-classical economic theory. Mainstream economic theory on economics based on the assumption that people and firms act in a rational and self-interested manner, continually weighing up alternatives on basis of cost and benefits. The market is viewed as essentially self-regulating, tending towards a state of equilibrium or balance through the role of the price mechanism in mediating between the forces of demand and supply.

Neoliberal approach/neoliberalism. A political and economic philosophy and approach to economic policy that seeks to reduce state intervention and embrace the free market, stressing the virtues of enterprise, competition and individual self-reliance.

Neo-Fordism. A phase of reinforced Fordism in the 1960s and 1970s as mass production technologies became increasingly routine and standardized.

New economic geography. A label applied, somewhat loosely, to describe the new culturally and institutionally informed research that has grown since the early 1990s, characterized by a focus on the links between economic action and social and cultural practices in different places.

Confusingly, the same term is also used by spatial economists to describe their research, a body of work that is better defined as the **'new geographical economics'**.

New geographical economics. An approach to economic geography developed by the economist Paul Krugman and others, involving the application of mathematical modelling techniques to analyse issues of industrial location. The new geographical economics addresses questions such as why, and under what conditions, do industries concentrate? It applies the methods of mainstream economics, devising models based on a number of simplifying assumptions.

New industrial spaces. Areas distinct from the old industrial cores that became centres of flexible production from the 1970s onwards. These areas offered attractive environments and a high quality of life for managerial and professional workers. Three different kinds of 'new industrial spaces' have been identified in Europe and North America: craft-based industrial districts such as central and north-eastern Italy, centres of high-technology industries such as Silicon Valley in California, and clusters of advanced financial and producer services in world cities.

New international division of labour. A term that refers to the process by which MNCs based in western countries have shifted low-status assembly and processing operations to developing countries where costs are much lower. It is a form of the **spatial division of labour**, operating at the global scale, facilitated by the increasing division of labour in large MNCs, new transport and communications technologies and the creation of a pool of available labour in developing countries.

Newly industrializing countries (NICs). Formerly underdeveloped countries that experienced rapid economic development and significant foreign direct investment between 1960 and 1990, including Asian economies such as Hong Kong, Korea, Singapore and Taiwan, but also including Brazil and Mexico.

New labour geography. A growing branch of economic geography (e.g. Herod, 2001) that is concerned with how workers and their trade unions help to influence the changing geography of the economy. A central concern of the 'new labour geography' is to counteract accounts dominated by business and state actors, stressing the agency of workers and their representatives.

New model of local and regional development. The prevailing approach to economic development since the 1980s has stressed the need to facilitate growth and enhance the competitiveness of the regional economy. In contrast with traditional regional policy, this new model is more 'bottom-up' in nature, focusing on the need to develop local skills and stimulate enterprise with regions.

New regionalism. A collective label used to describe a body of research in economic geography that stresses the renewed importance of 'the region' as a scale of economic organization under late capitalism.

New trade theory. An approach developed by the economist Paul Krugman and others since the 1970s, which recognizes that comparative advantage does not simply reflect pre-existing endowments of the **factors of production**. Rather, it is actively created by firms through the development of technology, skills, economies of scale, etc.

Non-government organizations (NGOs). Organizations, often of a voluntary or charitable nature, which make up the so-called 'third sector' belonging to neither the private nor public sectors.

Offshoring. The relocation of economic activities from developed countries to low-wage economies. A characteristic of manufacturing since the 1960s, creating the **new international division of labour**, it has become a feature of service operations in recent years, e.g., the move of some call centres to developing countries such as India.

Open political economy approach. A revised political economy framework informed by the culturally and institutionally informed approaches to economic geography that have been developed in recent years. This revised approach is not limited by its adherence to fixed categories, evolving in line with capitalism as its object of analysis.

Overproduction. The tendency for the volume of output to grow more rapidly than market demand. This is viewed as inherent to a capitalism system based on decentralized decision-making by many Marxist and institutionalist economists.

Partnership. The growing tendency for different organizations to work together in order to address particular problems and to coordinate services, involving the development of common objectives and the sharing of resources.

Path dependence. A key evolutionary idea adopted by economic geographers, referring to how past decisions and experiences shape the economic landscape, particularly in terms of structuring and informing economic actors' responses to wider processes of economic change.

Place. A particular area, usually occupied, to which a group of people have become attached, endowing it with meaning and significance. Often associated with notions of family, home and community.

Places of consumption. The particular sites at which goods and services are bought and consumed, including the department store, the mall, the street, the market and the home, as well as a host of more inconspicuous sites of consumption (for example, charity shops and car boot sales).

Political economy. A broad perspective on economic life which analyses the economy within its wider social and political context, focusing on production and the distribution of wealth between different sections of the population as well as the exchange of commodities through the market.

Positivism. A philosophy of science which states that the goal of science is to generate explanatory laws which explain and predict events and patterns in the real world.

Post-Fordism. A term widely used in the 1980s and 1990s to describe the growth of new production methods. In contrast to the standardized mass production techniques of Fordism, post-Fordism is defined by flexibility in face of more fragmented and customized patterns of market demand (see **'flexible production'** and **'flexible labour markets'**).

Post-Fordist consumption. This is defined by flexibility, with markets becoming fragmented into distinct segments and niches since the 1970s. Accordingly, consumption is defined individually rather than collectively with choice and identity becoming increasingly important as individual consumers regard the purchase and consumption of commodities as expressions of their lifestyles and aspirations.

Post-industrial society. A concept which emphasizes that services have become the dominant sector of the economy in developed countries, taking over from manufacturing. The theory of post-industrial society was developed by the American sociologist Daniel Bell in the late 1960s and early 1970s, highlighting the increased importance of white-collar employment, the role of knowledge and information as the key resources in the economy, and the liberation of individual workers from routine manual labour.

Postmodernism. Defined as 'a movement in philosophy, the arts and social sciences characterised by scepticism towards the grand claims and grand theory of the modern era, and their privileged vantage point, stressing in its place openness to a range of voices in social enquiry, artistic experimentation and political empowerment' (Ley, 1994, p.466).

Poverty reduction strategies (PSRs). The development of strategies by the governments of developing countries to address poverty, based on consultation with the World Bank and IMF, NGOs and local communities.

Power. The ability or capacity to take decisions that involve or affect other people. While neglected by mainstream economists, alternative approaches such as political economy tend to stress the importance of power in shaping the operation of the economy, particularly in terms of the (social) relations between different economic actors.

Privatization. The policy of transferring state-owned enterprises into private ownership, a key component of **neoliberal** reform programmes since the early 1980s.

Producer-driven commodity chains. These are coordinated and controlled by large manufacturers that rely on conventional top-down hierarchical links with a range of component suppliers. They are typical of industries which are highly capital and technology intensive, requiring huge amounts of research and development and investment, such as aircraft, motor vehicles and computers.

Qualitative state. A conception of the state as a dynamic process rather than a fixed 'thing' or object. The term, coined by the Australian economic geographer Philip O'Neill (1997), reflects a general shift of emphasis from a concern with quantitative aspects of state intervention to an interest in its qualitative characteristics.

Radical geography. A new kind of geography, developed in the US in the late 1960s in response to the perceived irrelevance of much mainstream geography in the face of pressing social problems such as inner city poverty, racism and inequality. Initially finding expression in a reformist liberalism, radical geography embraced Marxism in the 1970s.

Redlining. This is defined as cases 'where goods and services are made unavailable, or made available only on less than favourable terms, to people because of where they live' (Squires, quoted in Pollard, 1996, p.1209). Typically, low-income inner city areas are regarded as high risk by financial institutions and are 'redlined' or marked out on a map. As a result, people living within these areas are either denied access to credit and insurance services or, more commonly, charged higher premiums.

Regime of accumulation. Another key regulationist term, referring to a relatively stable form of economic organization that structures particular periods of capitalist development, creating a balance between production and consumption. Regimes of accumulation are supported by particular **modes of regulation**.

Regional division of labour. A form of the **spatial division of labour** operating at the macro-regional scale. It generally involves the location of high-value activities in economically advanced 'core' areas and low-values ones in less advanced peripheries. A good example is East and Southeast Asia, where Hong Kong and Singapore in particular have focused on high-level managerial and professional activities while production is carried out in neighbouring low-wage countries such as China, Indonesia, Thailand and the Philippines.

Regional geography. This represents another key element of the traditional approach, structuring human geography as a whole between the 1920s and 1950s. It was defined as a project of 'areal differentiation' which describes and interprets the variable character of the Earth's surface, expressed through the identification of distinct regions (Hartshorne, 1939, p.21).

Regional inequalities. Differences in income and wealth between regions. These can be seen as products reflecting processes of **uneven development**.

Regional policy. A set of programmes and measures established by governments to promote regional growth and development. As a key instrument of 'spatial **Keynesianism**', conventional regional policy involved governments inducing companies to locate factories and offices in depressed regions by offering grants and financial incentives. At the same time, development in core regions such as south-east England and Paris was restricted. Classical regional policy reached its peak in the 1960s and 1970s, helping to reduce the income gap between rich and poor regions in Europe.

Regional sectoral specialization. A pattern of industrial location during the nineteenth century whereby particular regions become specialized in certain sectors of industry. Characteristically, all the main stages of production from resource extraction to final manufacture were carried out within the same region.

Relational approach. A philosophical approach that sees space as constructed out of wider social relations and networks. Relational thinking is defined by three basic ideas about space: it should be seen as the product of interrelations, rather than something which simply exists in an *a priori* fashion; it emphasizes the possibilities of difference and multiplicity; it is always changing and becoming rather than static or fixed.

Relational region. This is based on the application of the **relational approach** to regions, re-defining them as open and discontinuous spaces, according to the wider social relations in which they are situated. In particular, the concept of the relational region emphasizes the wider networks and flows that connect regions under globalization.

Relative immobility of labour. The tendency for labour to remain tied to particular places, reflecting its dependence on family and community for reproduction. This is often contrasted with the geographical mobility of capital.

Reproduction of labour. The daily processes of feeding, clothing, sheltering and socializing, which support and sustain labour, processes that rely on family, friends and the local community, occurring outside the market.

Resource wars. Conflicts over the use of local economic resources such as land, forests and water between communities, which rely on them to meet their material needs on a day-to-day basis, and states or private interests that often want to exploit them for economic gain, threatening the basis of local livelihoods.

Resurgence of local and regional levels of government. The increased prominence of local and regional organizations in economic development policy since the early 1980s.

Scale. The different geographical levels of human activity: local, regional, national, supra-national and global.

Scientific management or Taylorism. An approach to industrial organization, associated with the Fordist mass production system, named after its founder and principal advocate, F.W. Taylor. Taylorism involved the reorganization of work according to rational principles designed to maximize productivity, based on an increased division of labour, enhanced coordination and control by management and the close monitoring and analysis of work performance and organization.

SECI approach. A theory of **innovation** developed by Japanese management theorists, drawing on the practices employed by Japanese corporations. Four stages are identified: socialization, involving the articulation and exchange of **tacit knowledge**; externalization, based on the transformation of this tacit knowledge into codified form; the combination of different bodies of **codified knowledge** into more complex and integrated systems; and internalization, by which firms embody codified knowledge in the skills of workers and the routines and work practices of the firm, turning it back into tacit knowledge. The process is viewed as a spiral, based on continuous interaction between tacit and codified knowledge.

Second global shift. The movement of service activities to developing countries such as India, where labour costs are much lower than in developed countries. This represents a new stage in the evolution of the international division of labour, following the original global shift based on the relocation of manufacturing activities in the 1970s and 1980s.

Shrinking of space. A reduction in the effects of geographical distance through the development of new transport and communications technologies, which effectively reduces the time and costs of moving goods, services, capital and information.

Social division of labour. The vast array of specialized jobs that people perform in society.

Social economy. A broad term that has developed to characterize economic activities that are not viewed as part of the capitalist mainstream, but are involved in providing activities to meet social needs rather than produce a profit. A wide range of activities is frequently and somewhat problematically included in this category, from privatized welfare services to more radical attempts to develop autonomous activity in opposition to capitalism (e.g. LET schemes).

Social movements. Movements and groups that have been established to protest against extreme poverty and to defend local livelihoods in face of their commodification or appropriation by the state and corporations in developing countries.

Social networks. The informal social ties between individuals working in different firms, providing a channel for the sharing of information and ideas. The role of such networks has often been cited in accounts of the rise of new industrial spaces such as Silicon Valley.

Social relations. The sets of relationships between different groups of economic actors. The relations between employers and workers have attracted particular attention, but other relations include those between producers and consumers, manufacturers and suppliers, supervisors and ordinary workers and government agencies and firms.

Space. An area of the Earth's surface, such as that between two particular points or locations. Often thought of in terms of the distance and time it takes to travel or communicate between two points.

Spatial analysis. An approach to economic geography that became influential in the 1960s and 1970s as part of the so-called 'quantitative revolution' in geography. Spatial analysis in economic geography involves the use of statistical and mathematical methods to analyse problems of industrial location, distance and movement.

Spatial agglomeration. The tendency for industries to cluster in particular places, underpinned by the operation of **agglomeration economies**.

Spatial dispersal. The opposite process to agglomeration, where industries or firms move out of existing centres of production into new regions.

Spatial division of labour. A concept developed by Doreen Massey (1984) to explain how an increasing division of labour within large corporations produced new spatial patterns. Companies were locating the higher-order functions in cities and regions where there are large pools of highly educated and well qualified workers, with lower order functions such as assembly locating increasingly in those regions and places where costs (especially wage rates) are lowest.

Spatial Keynesianism. The application of Keynesian economic principles to spatial issues, based on the redistribution of resources from rich and poor regions in order to close the gap in income between them.

Spatial fix. The establishment of relatively stable geographical arrangements that facilitate the expansion of the capitalist economy for a certain period of time. Examples include imperialism during the nineteenth century, **Fordism** in the post-war period and **globalization** since the 1980s, which has involved the **deindustrialization** of many established centres of production in the 'rustbelts' of North America and western Europe and the growth of new industry in 'sunbelt' regions and the newly industrializing countries of East Asia.

Specialized industrial regions. Regions that became dependent on a particular set of industries, providing the basis of the local economy. Regional specialization was a product of industrialization in the nineteenth and early twentieth centuries, leading to a profusion of mining villages, steel towns and shipbuilding districts in Europe and North America. Many of these regions have experienced severe deindustrialization since the 1960s, although some have diversified successfully in recent years.

Spread effects. A contrasting set of effects that allows surrounding regions to benefit from increased growth in the core region. One important mechanism here is increased demand in the core region for food, consumer goods and other products, creating opportunities for firms in peripheral regions to supply this growing market. At the same time, rising costs of land, labour and capital in the core region, together with associated problems such as congestion, can push investment out into surrounding regions.

Stages of economic growth. An influential model of economic development produced by the US economist Walt Rostow in the late 1950s which identified distinct stages of growth. The process of 'take-off' is crucial in bringing about the transition from traditional society to the final stage of 'high mass consumption'.

Stagflation. A term coined to describe the unprecedented combination of economic stagnation and rising inflation that occurred in the 1970s. Generally, high unemployment and inflation had been regarded as incompatible by economists, with the former reflecting stagnation or recession and the latter regarded as an expression of overly rapid growth or 'over-heating'.

State, the. A set of public institutions that exercise authority over a particular territory, including the government, parliament, civil service, judiciary, police, security services and local authorities.

Strategic coupling. The dynamic processes of interaction between **global production networks** (GPNs) and regional assets which underpins regional development. Regional institutions play a key role in attracting and retaining inward investment by shaping and moulding regional assets to fit the needs of lead firms in GPNs.

Structural adjustment programmes (SAPs). An economic reform package developed by the IMF and World Bank in the 1980s and 1990s as part of the **Washington Consensus**. SAPs have been adopted by a large number of developing countries in exchange for financial assistance. They encompass a range of measures requiring countries to open up to trade and investment and to reduce public expenditure.

Structuralist theories. A set of theories of development which explained global inequalities in terms of the structure of the world economy, particularly the relationship between developed countries and developing countries. They emphasized the mode of incorporation of developing countries into the world economy, viewing this as a key source of exploitation. The structuralist approach was particularly associated with a group of theorists (such as Raul Prebisch) and activists based in Latin America in the 1950s and 1960s.

Structured coherence. A term introduced by David Harvey to describe the social, economic and political relations that develop in association with particular forms of production in specific places (Harvey, 1982). It has been particularly associated with the main centres of heavy industry in developed countries during the late nineteenth and twentieth centuries, which became known as working-class regions with strong socialist and trade union traditions.

Sub-prime housing market. A particular segment of the housing market made up from high-risk borrowers (usually from poorer groups) whose levels of income make it unlikely that they will be able to meet mortgage payments where market conditions deteriorate.

Subsistence agriculture. The production of food and crops predominantly for use by the household itself, rather than for sale as a commodity. Subsistence agriculture still characterizes much of the Global South.

Sunk costs. The costs of investment not directly recoverable if a firm were to pull out of a particular location. A neglected topic in economic geography but potentially important in restricting the mobility of firms.

Supply-side of the economy. This can be is defined in terms of the quality of the main **factors of production** such as labour (training, skills), capital (emphasizing enterprise and innovation) and land in terms of sites and infrastructure for investors. Improving these supply-side factors is the central focus of the new model of economic development described above, seen as vital to the **competitiveness** of the regional economy.

Supranational tier of government. This term refers to the increased prominence of supranational agencies and organizations operating above the level of the national state. Examples include the European Union, World Trade Organization, International Monetary Fund, World Bank and United Nations.

Tacit knowledge. In contrast with **codified knowledge**, this refers to direct experience and expertise, which is not communicable through written documents. It is a form of practical 'know-how' embodied in the skills and work practices of individuals or organizations.

Territorial embeddedness. A concept that emphasizes how certain forms of economic activity are grounded or rooted in particular places. Economic geographers have adapted the concept from economic sociology, where it is used in a broad societal sense to refer to the idea that economic action is grounded in social relations. A third form of network embeddedness is also identified to highlight the social and economic relationships in which particular economic actors or firms participate.

Time–Space compression. A term that refers to the effects of information and communication technologies in reducing the time and costs of transmitting information and money across space. This reduces the 'friction of distance' which geographers have traditionally stressed.

Third way. A term used to refer to the policies of centre-left leaders such as Blair, Clinton and Schroder in the late 1990s, which aimed to find a new path between the conflicting extremes of free market capitalism and state socialism. Marrying the efficiencies of markets to a revived social democratic notion of social justice is the key notion. The term has been widely criticized as vague and meaningless.

Third World. A collective label for the 'underdeveloped areas' in Africa, Asia and Latin America widely used between the 1960s and 1990s. The term is a product of the **Cold War**, with the 'Third World' distinguished from the 'First World' of western democracies and the 'Second World' of communist states in the USSR and eastern Europe.

Trade unions. Collective organization representing workers who grew in strength in the late nineteenth and early twentieth centuries, and in many countries were associated with the formation of parliamentary Labour Parties.

Traditional approach to economic geography. This was factual and descriptive in nature, focusing on the compilation of information about economic conditions and resources in particular regions. It was the dominant approach from the late nineteenth century to the 1950s.

Trans-nationality index. A measure compiled by the United Nations, which gauges the international extent of multinational corporations' operations by calculating the average of three ratios: foreign assets to total assets, foreign sales to total sales and foreign employees to total employees.

Uneven development, Neil Smith's theory of. This important contribution to Marxist geography explains uneven development in terms of the movement of capital between locations. Capital is attracted to areas that offer high profits for investors, with their resultant development leaving other areas behind. Over time, however, development leads to rising costs in core areas, prompting capital to move to other, less developed regions where costs are lower. It is this 'see-sawing' of capital between regions, driven by the need to maintain profit levels, that creates patterns of uneven development.

Urban entrepreneurialism. A new focus on economic development and regeneration that became a key part of urban policy from the early 1980s. It is often contrasted with the managerialism of the post-war decades, which was primarily concerned with managing the delivery of welfare services to local residents. The entrepreneurial approach saw cities focus on the need to generate growth and employment, seeking to attract new investment and funds from outside and generate new business and income from within.

Urbanization economies. A second type of **agglomeration economy**, derived from the concentration of firms in *different* industries in large urban areas.

Uruguay Round of trade negotiations. The last of eight rounds of trade negotiations held between the late 1940s and the mid-1990s under the umbrella of the General Agreement on Trade and Tariffs (GATT). Running from 1986–1994, the Uruguay Round was the most ambitious and wide-ranging of the GATT agreements, incorporating not only agriculture, textiles and clothing but also services for the first time.

Value. The economic return (profit) or rent generated by the production of commodities for sale, involving the conversion of labour power into actual labour through the labour process. Firms may create value through: the control of particular product and process technologies; the development of certain organizational and management capabilities; the harnessing of inter-firm relationships; and the prominence of brand names in key markets.

Venture capital. A form of private equity finance provided by outside investors to new or growing firms that are generally not quoted on the stock exchange. Such investment tends to be high risk, focusing on firms with a high growth potential and attracting investors who aim to make high returns by selling their stake at a later date.

Washington Consensus. A set of economic policies, based upon **neoliberal** economic principles, adopted and implemented by the US Treasury and World Bank and IMF, all based in Washington DC. Key elements include reducing public expenditure, economic **liberalization, privatization,** and the promotion of exports and foreign direct investment. The role of the IMF and World Bank in imposing such policies on highly indebted poor countries has generated substantial controversy.

Welfare state. A particular type of state constructed between the 1930s and 1970s in developed countries, based on the establishment of elaborate welfare systems to spread the benefits of growth to all sections of the population and to offer social protection to its citizens against the vagaries of the market, including unemployment and ill health.

Work. The basic physical tasks needed to reproduce daily life. These would include hunting, growing or finding food, finding or making shelter, making clothes, looking after and raising children, etc.

Workfare. A system introduced in the US, which requires people to work in exchange for welfare benefits and payments.

References

Adler, G. and Webster, E. (1999) 'The labour movement: radical reform and the transition to democracy in South Africa', in Munck, R. and Waterman, P. (eds) *Labour Worldwide in the Era of Globalization*, Basingstoke: Macmillan, pp.133–57.

Aglietta, M. (1979) *A Theory of Capitalist Regulation: The US Experience*, London: New Left Books.

Albert M. (1993) *Capitalism against Capitalism*, London: Whurr.

Allen, J. (1988) 'Towards a post-industrial economy?', in Allen, J. and Massey, D. (eds) *The Economy in Question*, London: Sage, pp.184–228.

Allen, J. (2000) 'Power/economic knowledge: symbolic and spatial formations', in Bryson, J., Daniels, P., Henry, N. and Pollard, J. (eds) *Knowledge, Space, Economy*, London: Routledge, pp.15–33.

Allen, J. (2003) *Lost Geographies of Power*, Oxford: Blackwell.

Allen, J. and Thompson, G. (1997) 'Think global, then think again: economic globalization in context'. *Area*, 29: 213–27.

Allen, J., Massey, D. and Cochrane, A. (1998) *Rethinking the Region*, London: Routledge.

Amable, B. (2003) *The Diversity of Modern Capitalism*, Oxford: Oxford University Press.

Amin, A. (1999) 'An institutionalist perspective on regional economic development'. *International Journal of Urban and Regional Research*, 23: 365–78.

Amin, A. (2000) 'Industrial districts', in Sheppard, E. and Barnes, T. (eds) *A Companion to Economic Geography*, Oxford: Blackwell, pp.149–68.

Amin, A. (2002) 'Spatialities of globalisation'. *Environment and Planning A*, 34: 385–99.

Amin, A. (2004) Regions abound: towards a new politics of place. *Geografiska Annaler B*, 86: 385–99.

Amin, A. and Thrift, N. (1992) 'Neo-Marshallian nodes in global networks'. *International Journal of Urban and Regional Research*, 16: 571–87.

Amin, A. and Thrift, N. (1994) 'Living in the global', in Amin, A. and Thrift, N. (eds) *Globalization, Institutions and Regional Development in Europe*, Oxford: Oxford University Press, pp.1–22.

Amin, A. and Thrift, N. (2004) 'Introduction', in Amin, A. and Thrift, N. (eds) *The Blackwell Cultural Economy Reader*, Oxford: Blackwell, pp.x–xxx.

Amin, A., Cameron, A. and Hudson, R. (2003) 'The alterity of the social economy', in Leyshon, A., Lee, R. and Williams, C.C. (eds) *Alternative Economic Spaces*, London: Sage, pp.27–54.

Amin, A., Massey, D. and Thrift, N. (2003) *Decentring the Nation: A Radical Approach to Regional Inequality*, London: Catalyst.

Ancien, D. (2005) 'Local and regional development policy in France: of changing conditions and forms, and enduring state centrality'. *Space and Polity*, 9: 217–36.

Arbuthnott, G. (2009) 'Government tells supermarkets it's time to "green up their act"'. *The Herald*, 9 November.

Armstrong, P., Glyn, A. and Harrison, J. (1991) *Capitalism Since 1945*, Oxford: Blackwell.

Arnott, R. and Wrigley, N. (2001) 'Editorial'. *Journal of Economic Geography*, 1, pp.1–4.

Ashcroft, B. (2006) 'Outlook and appraisal'. *Quarterly Economic Commentary*, 30(4): 1–9.

Atkinson, J. (1985) *Flexibility, Uncertainty and Manpower Management*, Brighton: Institute of Manpower Studies, Report No. 89.

Bair, J. (2008) 'Analysing global economic organisation: embedded networks and global chains compared'. *Economy and Society*, 37, 339–64.

Bair, J. and Gereffi, G. (2001) 'Local clusters in global chains: the causes and consequences of export dynamism in Torreon's blue jeans industry'. *World Development*, 29: 1885–903.

Baran, P. and Sweezy, P. (1966) *Monopoly Capital*, New York: Monthly Review Press.

Barnes, T.J. (1997) 'Introduction: theories of accumulation and regulation: bringing life back into economic geography', in Lee and Wills (1997), pp.231–47.

Barnes, T.J. (2000a) 'Inventing Anglo-American economic geography', in Sheppard, E. and Barnes, T. (eds) *A Companion to Economic Geography*, Oxford Blackwell, pp.593–4.

Barnes, T.J. (2000b) 'Political economy', in Johnston, R.J., Gregory, D., Pratt., G. and Watts, M. (eds) *The Dictionary of Human Geography*, 4th edn, Oxford: Blackwell, pp.593–4.

Barnes, T.J. (2001) 'Retheorising economic geography: from the quantitative revolution to the "cultural turn"'. *Annals of the Association of American Geographers*, 91: 546–65.

Barnes, T.J. (2003) 'Introduction: "never mind the economy, here's culture"', in Anderson, K., Domosh, M., Pile, S. and Thrift, N. (eds) *Handbook of Cultural Geography*, London: Sage, pp.89–97.

Barnes, T.J. and Sheppard, E. (2000) 'The art of economic geography', in Sheppard, E. and Barnes, T. (eds) *A Companion to Economic Geography*, Oxford: Blackwell, pp.1–8.

Barrientos, S. and Smith, S. (2007) 'Do workers benefit from ethical trade? Assessing codes of labour practice in global production systems'. *Third World Quarterly*, 28: 713–29.

Bartlett, C. and Ghoshal, S. (1989) *Managing Across Borders*, Boston, MA: Harvard Business School Press.

Bathelt, H. (2006) 'Geographies of production: growth regimes in spatial perspective 3 – towards a relational view of economic action and policy'. *Progress in Human Geography*, 30: 223–36.

Bathelt, H. and Gluckler, J. (2003) 'Toward a relational economic geography'. *Journal of Economic Geography*, 3: 117–44.

Bathelt, H., Malmberg, A. and Maskell, P. (2004) 'Clusters and knowledge: local buzz, global pipelines and the process of knowledge formation'. *Progress in Human Geography*, 28: 31–57.

BBC (2004) 'The battle over trade', BBC News at http://news.bbc.co.uk/1/hi/in_depth/business/2004/world_trade#.

BBC (2005) 'Europe agrees cut in sugar subsidies', BBC News at http://news.bbc.co.uk/1/hi/business/4466388.stm.

Beaverstock, J.V. (2002) 'Transnational elites in global cities: British expatriates in Singapore's financial district'. *Geoforum*, 33(4): 525–38.

Bebbington, A. (2000) 'Reencountering development: livelihood transitions and place transformations in the Andes'. *Annals of the Association of American Geographers*, 90: 495–520.

Bebbington, A. (2009) 'The new extraction: rewriting the political ecology of the Andes'. NACLA Report on the Americas 42, September/October, pp.12–20.

Bebbington, A. and Bebbington, D.H. (2010) 'An Andean Avatar: post-neoliberal and neoliberal strategies for promoting extractive industries'. BWPI Working Paper 117, Brooks World Poverty Institute, University of Manchester.

Bell, D. (1973) *The Coming of Post-industrial Society*, New York: Basic Books.

Bell, M. (1994) 'Images, myths and alternative geographies of the Third World', in Gregory, D., Martin, R. and Smith, G. (eds) *Human Geography: Society, Space and Social Science*, Basingstoke: Macmillan, pp.174–99.

Benwell Community Project (1978) *The Making of a Ruling Class*, Benwell Community Project, Final Series No. 6.

Bergene, A.C., Ednresen, S.B. and Knutsen, H.M. (eds) (2010) *Missing Links in Labour Geography*, Farnham: Ashgate.

Berndt, C. (2000) 'The rescaling of labour regulation in Germany: from national and regional corporatism to intrafirm welfare?' *Environment and Planning A*, 32(9): 1569–92.

Beynon, H. (1984) *Working for Ford*, 2nd edn, Harmondsworth: Pelican.

Beynon, H. (1997) 'The changing practices of work', in Brown, R. (ed.) *The Changing Shape of Work*, Basingstoke: Macmillan, pp.20–54.

Beynon, H., Rubery, J., Grimshaw, D. and Ward, K. (2002) *Managing Employment Change: The New Realities of Work*, Oxford: Oxford University Press.

Birch, K., MacKinnon, D. and Cumbers, A. (2010) 'Old industrial regions in Europe: a comparative assessment of economic performance'. *Regional Studies*, 44: 35–54.

Blackburn, R. (2008) 'The subprime crisis'. *New Left Review*, 50: 63–106.

Blanchflower, D. (2006) 'A cross-country study of union membership', Bonn: IZA (Institute for the Study of Labour) Discussion Paper 2016.

Block, F. (2003) 'Karl Polanyi and the writing of *The Great Transformation*'. *Theory and Society*, 32: 275–306.

Bluestone, B. and Harrison, B. (1982) *The Deindustrialisation of America: Plant Closures, Community Abandonment and the Dismantling of Basic Industry*, New York: Basic Books.

Bowring, F. (1998) 'LETS: an eco-socialist initiative'. *New Left Review*, 232: 91–111.

Bowyer, R. (1990) *The Regulation School: A Critical Introduction*, New York. Columbia University Press.

Bradley, H., Erickson, M., Stephenson, C. and Williams, S. (2000) *Myths at Work*, Cambridge: Polity.

Bristow, G. (2005) 'Everyone's a "winner": problematising the discourse of regional competitiveness'. *Journal of Economic Geography*, 5: 285–304.

Bristow, G. (2010) '*Critical Reflections on Regional Competitiveness*', London: Routledge.

Brohmann, J. (1996) 'Postwar development in the Asian NICs: does the neoliberal model fit reality?' *Economic Geography*, 72: 107–30.

Brown, J.S. and Duguid, P. (2000) *The Social Life of Information*, Boston, MA: Harvard University Press.

Bryson, J.R. (2008) Service economies, spatial divisions of expertise and the second global shift, in Daniels, P.W., Bradshaw, M., Shaw, D. and Sidaway, J. (eds) *Human Geography: Issues for the 21st Century*, 3rd edn, Harlow: Pearson, pp.339–57.

Bryson, J. and Henry, N. (2005) 'The global production system: from Fordism to post-Fordism', in Daniels *et al.* (2005), pp.318–36.

Bryson, J.R. and Rusten, G. (2008) 'Transnational corporations and spatial divisions of "service" expertise as a competitive strategy: the example of 3M and Boeing'. *Service Industries Journal*, 28: 307–23.

Bureau of Labor Statistics (2010) Occupational Employment Statistics (OES) Highlights: An Overview of US Occupational Employment and Wages in 2009. US Bureau of Labor Statistics, Washington, DC, at http://www.bls.gov/oes/highlight_2009.htm.

Burnell, P. (2008) 'Foreign aid in a changing world', in Desai, V. and Potter, R. (eds) *The Companion to Development Studies*, 2nd edn, London: Arnold, pp. 503–8.

Burton, I. (1963) 'The quantitative revolution and theoretical geography'. *Canadian Geographer*, 7: 151–62.

Cable, V. (2009) *The Storm: The World Economic Crisis and What it Means*, London: Atlantic Books.

Carrillo, J. and Zarate, R. (2009) 'The evolution of *Maquiladora* best practices: 1965–2008'. *Journal of Business Ethics*, 88: 335–48.

Cassidy, J. (2008) 'The Minsky moment'. *The New Yorker*, February, at http://www.newyorker.com.

Castells, M. (1989) *The Informational City*, Oxford: Blackwell.

Castells, M. (1996) *The Information Age Volume 1: The Rise of the Network Society*, Oxford: Blackwell.

Castells M. (2000) *The Information Age Volume 3: End of Millennium*, Oxford: Blackwell.

Castells, M. and Hall, P. (1994) *Technopoles of the World*, London: Routledge.

Castree, N. (2000) 'Geographic scale and grassroots internationalism: the Liverpool dock dispute', 1995–8. *Economic Geography*, 76: 272–92.

Castree, N., Coe, N., Ward, K. and Samers, M. (2004) *Spaces of Work: Global Capitalism and Geographies of Labour*, London: Sage.

Chen, H., Gompers, P., Kovner, A. and Lerner, J. (2009) 'Buy local? The geography of successful and unsuccessful venture capital expansion'. Harvard Business School, Working Paper 143.

Clark, G. (1989) *Unions and Communities Under Siege*, Cambridge: Cambridge University Press.

Clark, G.L. (1994) 'Strategy and structure: corporate restructuring and the scope and characteristics of sunk costs'. *Environment and Planning A*, 26: 9–32.

Clark, G.L. (1999) *Pension Fund Capitalism*, Oxford: Oxford University Press.

Clark, G., Gertler, M.S. and Whiteman, J.E.M. (1986) *Regional Dynamics: Studies in Adjustment Theory*, Hemel Hempstead: Allen and Unwin.

Cloke, P., Philo, C. and Saddler, D. (1991) *Approaching Human Geography: An Introduction to Contemporary Theoretical Debates*, London: Paul Chapman.

Coe, N. (2008) 'The geographies of global production networks', in Daniels, P., Bradshaw, M., Shaw, D. and Sidaway, J. (eds) *An Introduction to Human Geography: Issues for the 21st Century*, Harlow: Pearson, pp. 315–38.

Coe, N. and Jordhaus-Lier, D. (2010) 'Constrained agency: re-evaluating the geographies of labour'. *Progress in Human Geography*, online first doi: 10.1177/0309132510366746.

Coe, N. and Kelly, P. (2002) 'Languages of labour: representational strategies in Singapore's labour control regime'. *Political Geography*, 21: 341–71.

Coe, N. and Lee, Y.S. (2006) 'The strategic localisation of transnational retailers; the case of Samsung-Tesco in South Korea'. *Economic Geography*, 82: 61–88.

Coe, N. and Wrigley, N. (2007) 'Host economy impacts of transnational retail: the research agenda'. *Journal of Economic Geography*, 76: 341–71.

Coe, N., Dicken, P. and Hess, M. (2008) Global production networks: realising the potential. *Journal of Economic Geography*, 8: 271–95.

Coe, N., Hess, M., Yeung, H.W., Dicken, P. and Henderson, J. (2004) 'Globalising regional development: a global production networks perspective'. *Transactions of the Institute of British Geographers*, 29: 464–84.

Coe, N., Kelly, P.F. and Yeung, H.W. (2007) *Economic Geography: A Contemporary Introduction*, Oxford: Wiley Blackwell.

Coffey, W, (1996) 'The newer international division of labour', in Daniels, P.W. and Lever, W.F. (eds) *The Global Economy in Transition*, Harlow: Longman, pp.40–61.

Conte-Helm, M. (1999) 'The road from Nissan to Samsung: a historical overview of East Asian investment in a UK region', in Garrahan. P. and Ritchie, J. (eds) *East Asian Direct Investment in Britain*, London: Cass, pp.36–58.

Cook, I. (2002) 'Commodities: the DNA of capitalism, at http://www.exchange-values.org/.

Cook I. *et al.* (2010) 'Geographies of food: "Afters"', *Progress in Human Geography*, online at http://www.phg.sagepub.com/content/early/.

Cook, I. and Crang, P. (1996) 'The world on a plate: culinary culture, displacement and geographical knowledges'. *Journal of Material Culture*, 1: 131–53.

Cooke, P. and Morgan, K. (1998) *The Associational Economy: Firms, Regions, and Innovation*, Oxford: Oxford University Press.

Corbridge, S. (2008) 'Third world debt', in Desai, V. and Potter, R. (eds) *The Companion to Development Studies*. 2nd edn, London: Arnold, pp.508–11.

Corporate Watch (2004) Tesco – A Corporate Profile, at http://www.corporatewatch.org.uk/?lid=252.

Crang, P. (1997) 'Introduction: cultural turns and the (re)constitution of economic geography', in Lee and Wills (1997), pp.3–15.

Crang, P. (2005) 'Consumption and its geographies', in Daniels *et al.* (2005), pp.359–79.

Crang, P. (2008) 'Consumption and its geographies', in Daniels, P., Bradshaw, M., Shaw, D. and Sidaway, J. (eds) *Human Geography: Issues for the 21st Century*, 3rd edn, Harlow: Pearson, pp.376–94.

Cresswell, T. (2004) *Place: A Short Introduction*, Oxford: Blackwell.

Crewe, L. and Gregson, N. (1998) 'Tales of the unexpected: exploring car boot sales as marginal spaces of consumption'. *Transactions, Institute of British Geographers*, NS23: 39–53.

Cronon, W. (1991) *Nature's Metropolis: Chicago and the Great West*, New York: W.W. Norton.

Crotty, J. (2009) 'Structural causes of the global financial crisis: a critical assessment of the "new financial architecture"'. *Cambridge Journal of Economics*, 33: 563–80.

Cumbers, A. (1999) 'The transformation of employment relations in the UK's old industrial regions. A regional comparison of the experience of "Japanization"', in Garrahan, P. and Ritchie, J. (eds) *East Asian Direct Investment in Britain*, London: Cass, pp.183–200.

Cumbers, A. (2004) 'Embedded internationalisms: building transnational solidarity in the British and Norwegian trade union movements'. *Antipode,* 36: 829–50.

Cumbers, A. (2005) 'Genuine renewal or pyrrhic victory? The scale politics of trade union recognition in the UK'. *Antipode,* 37: 116–38.

Cumbers, A. and Martin, S. (2001) 'Changing relationships between multinational companies and their host regions. A case study of Aberdeen and the international oil industry'. *Scottish Geographical Journal,* 117: 31–48.

Cumbers, A., Helms, G. and Keenan, M. (2009) *Beyond Aspiration: Young People and Decent Work in the De-industrialised City*. Discussion paper. Glasgow: Department of Geographical and Earth Sciences and Department of Urban Studies, University of Glasgow.

Cumbers, A., Nativel, C. and Routledge, P. (2008) 'Labour agency and union postionalities in global production networks'. *Journal of Economic Geography*, 8: 369–87.

Currah, A. (2007) 'Hollywood, the internet and the world: a geography of disruptive innovation'. *Industry and Innovation*, 14: 359–84.

Danford, A. (1999) *Japanese Management Techniques and British Workers*, London: Mansell.

Daniels, P., Bradshaw, M., Shaw, D. and Sidaway, J. (eds) (2005) *Human Geography: Issues for the 21st Century*, 2nd edn, Harlow: Pearson.

Dawley, S. (2007) 'Fluctuating rounds of inward investment in peripheral regions: semiconductors in the north east of England'. *Economic Geography*, 83: 51–73.

Dawson, J. (2007) 'Scoping and conceptualising retail internationalisation'. *Journal of Economic Geography*, 7: 373–97.

Dear, M. (2000) 'State', in Johnston, R.J., Gregory, D., Pratt, G. and Watts, M. (eds) *The Dictionary of Human Geography*, 4th edn, Oxford: Blackwell, pp.788–90.

Devine. T.M. (1999) *The Scottish Nation*, London: Penguin.

Dex, S. and McCulloch, J. (1997) 'Unemployment and training histories: findings from the "Family and Working Lives Survey"'. *Labour Market Trends*, 105: 449–54.

Dicken, P. (2003a) *Global Shift: Reshaping the Global Economic Map in the Twenty-first Century*, 4th edn, London: Sage.

Dicken, P. (2003b) 'Placing firms: grounding the debate on the global corporation', in Peck, J. and Yeung, H.W. (eds) *Remaking the Global Economy*, London: Sage, pp.27–44.

Dicken, P. (2007) *Global Shift: Reshaping the Global Economic Map in the 21st Century*, 5th edn, London: Sage.

Dicken, P., Forsgren, M. and Malmberg, A. (1994) 'The local embeddedness of transnational corporations', in Amin, A. and Thrift, N. (eds) *Globalization, Institutions and Regional Development in Europe*, Oxford: Oxford University Press, pp.23–45.

Dicken, P., Kelly, P.F., Olds, K. and Yeung, H.W.-C. (2001) 'Chains and networks, territories and scales: towards a relational framework for analysing the global economy'. *Global Networks*, 1: 89–112.

Dicken, P., Peck, J. and Tickell, A. (1997) 'Unpacking the global', in Lee and Wills (1997), pp.158–66.

Doeringer, P. and Piore, M. (1971) *Internal Labor Markets and Manpower Analysis*, Lexington, MA: Heath Lexington Books.

Dolphin, T. (2009) *The Impact of the Recession on Northern City-regions*, Newcastle upon Tyne: IPPR North.

Domosh, M. (1996) 'The feminised retail landscape: gender, ideology and consumer culture in nineteenth-century New York City', in Wrigley, N. and Lowe, M. (eds) *Retailing, Consumption and Capital: Towards the New Retail Geography*, Harlow: Longman, pp.257–70.

Donaghu, M.T. and Barff, R. (1990) 'Nike just did it: international subcontracting and flexibility in athletic footwear production'. *Regional Studies*, 24: 537–52.

Dorling, D. and Thomas, B. (2004) *People and Places: A 2001 Census Atlas of the UK*, Bristol: Policy Press.

Douglass, M. (1994) 'The "developmental state" and the newly industrialised economies of Asia'. *Environment and Planning A*, 26: 453–66.

Dow, S. (1999) 'The stages of banking development and the spatial evolution of financial systems', in Martin, R. (ed.) *Money and the Space Economy*, Chichester: Wiley, pp.31–48.

Doward, J. (2009) 'UK supermarkets warned over banana price war'. *The Observer*, 11 October.

Dowd, D. (2000) *Capitalism and its Economics: A Critical History*, London: Pluto.

Doyle, M. (2005) 'Two countries' contrasting tales', BBC News, Africa 2005: Time for Change?, at http://newsvote.bbc.co.uk/mpapps/pagetools/print/news.bbc.co.uk/2/hi/africa/4337083.

du Gay, P. (1996) *Consumption and Identity at Work*, London: Sage.

Duncan, S.S. and Goodwin, M. (1988) *The Local State and Uneven Development*, Cambridge: Polity.

Dunford, M. and Perrons, D. (1994) 'Regional inequality, regimes of accumulation and economic development in contemporary Europe'. *Transactions of the Institute of British Geographers*, NS19: 163–82.

Dunning, J.H. (1980) 'Towards an eclectic theory of international production: some empirical tests'. *Journal of International Business Studies*, 11: 9–31.

Dunning, J.H. (1993) *Multinational Enterprises and the Global Economy*, Wokingham: Addison-Wesley.

Elliott, L. (2010) 'IMF has one cure for debt crises – public spending cuts with tax rises'. *The Observer*, 9 May.

Fair Trade Towns http://www.fairtradetowns.org/about/what-is-a-fairtrade-town/.

Fairtrade Foundation (2009) *Unpeeling the Banana Trade*. Briefing Paper. Fairtrade Foundation, London.

Fairtrade Labelling Organizations International (2010) *Growing Stronger Together: Annual Report 2009–10*, Bonn, Germany, at http://www.fairpensions.org.uk/about.

Faludi, A. (2004) 'Territorial cohesion: old (French) wine in new bottles?' *Urban Studies*, 41: 1349–65.

Fields, G. (2006) 'Innovation, time, and territory: space and the business organization of Dell Computer'. *Economic Geography*, 82: 119–46.

Finch, J. (2009) 'Tesco's profits rise to £1.43 billion'. *The Guardian*, 6 October.

Firn, J. (1975) 'External control and regional development: the case of Scotland'. *Environment and Planning A*, 7: 393–414.

Florida, R. (1995) 'Toward the learning region'. *Futures*, 27: 527–36.

Florida, R. (2002) *The Rise of the Creative Classes*, New York: Basic Books.

Florida, R. (2005) *Cities and the Creative Class*, London: Routledge.

Florida, R. and Kenney, M. (1993) *Beyond Mass Production: The Japanese System and its Transfer to the US*, Oxford: Oxford University Press.

French, S., Leyshon, A. and Signoretta, P. (2008) 'All gone now: the material, discursive and political erasure of bank and building society branches in Britain'. *Antipode*, 40: 79–101.

French, S., Leyshon, A. and Thrift, N. (2009) 'A very geographical crisis: the making and breaking of the 2007–8 financial crisis'. *Cambridge Journal of Regions Economy and Society*, 2: 287–302.

Friedman, T. (2005) *The World is Flat: A Brief History of the Twenty-first Century*, New York: Farrar, Straus & Giroux.

Friends of the Earth (2003) Farmers and the supermarket code of practice. Press briefing, 17 March. Friends of the Earth, London.

Froebel, F., Heinrichs, J. and Kreye, O. (1980) *The New International Division of Labour*, Cambridge: Cambridge University Press.

Fukuyama, F. (1992) *The End of History and the Last Man*, London: Penguin.

Fuller, D. and Jonas, A. (2003) 'Alternative financial spaces', in Leyshon, A., Lee, R. and Williams, C.C. (eds) *Alternative Economic Spaces*, London: Sage, pp.55–73.

Fuller, D., Jonas, A. and Lee, R. (eds) (2010) *Interrogating Alterity: Alternative Political and Economic Spaces*, Aldershot: Ashgate.

Galbraith, J.K. (1967) *The New Industrial State*, London: Hamish Hamilton.

Garrahan, P. (1986) 'Nissan in the north east of England'. *Capital and Class,* 27: 5–13.

Garrahan, P. and Stewart, P. (1992) *The Nissan Enigma: Flexibility at Work in a Local Economy*, London: Cassell.

George, S. (1999) *The Lugano Report: On Preserving Capitalism in the Twenty-first Century*, London: Pluto.

Gereffi, G. (1994) 'The organization of buyer-driven commodity chains: how US retailers shape overseas production networks', in Gereffi, G. and Korzeniewicz, M. (eds) *Commodity Chains and Global Capitalism*, Westport, CO: Greenwood Press, pp.95–122.

Gereffi, G., Humphrey, J., Kaplinsky, R. and Sturgeon, T. (2001) 'Introduction: globalisation, value chains and development'. *IDS Bulletin*, 32: 1–8.

Gertler, M. (1995) '"Being there": proximity, organisation and culture in the development and adoption of advanced manufacturing technologies'. *Economic Geography*, 71: 1–26.

Gertler, M. (2003) 'A cultural economic geography of production', in Anderson, K., Domosh, M., Pile, S. and Thrift, N.J. (eds) *The Handbook of Cultural Geography*, London: Sage, pp.131–46.

Gertler, M. (2004) *Manufacturing Culture: The Institutional Geography of Industrial Practice*, Oxford: Oxford University Press.

Gertler, M.S. (2010) 'Rules of the game: the place of institutions in regional economic change'. *Regional Studies*, 41: 1–15.

Ghose, A.K., Majid, N. and Ernst, C. (2008) *The Global Employment Challenge*, Geneva: International Labour Organization.

Gibb, R. (2006) 'The European Union's "Everything but Arms" development initiative and sugar: preferential access or continued protectionism?' *Applied Geography*, 26: 1–17.

Gibson-Graham, J.K. (1996) *The End of Capitalism (As We Knew It): A Feminist Critique of Political Economy*, Oxford: Blackwell.

Gibson-Graham, J.K. (2006) *Postcapitalist Politics*, Minneapolis: University of Minnesota Press.

Gibson-Graham, J.K. (2008) 'Diverse economies: performative practices for "other worlds"'. *Progress in Human Geography*, 32, 613–32.

Giddens, A. (1985) *The Nation State and Violence*, Cambridge: Polity.

Giddens, A. (1998) *The Third Wave: The Renewal of Social Democracy*, Cambridge: Polity.

Gilbert, E. (2005) 'Common cents: mapping money in time and space'. *Economy and Society*, 34: 356–87.

Gomez, G. and Helmsing, A.H.J. (2008) 'Selective spatial closure and local economic development: what do we learn from the Argentine local currency systems?' *World Development*, 36(11): 2489–511.

Goodin, R.E. and Le Grand, J. (1987) *Not Only the Poor: The Middle Classes and the Welfare State*, London: Allen & Unwin.

Goodman, M., Maye, D. and Holloway, L. (2010) 'Ethical Foodscapes?: Premises, Promises and Possibilities', King's College London, Working Paper 29.

Gordon, D. (1996) *Fat and Mean: The Corporate Squeeze of Working Americans and the Myth of Managerial Downsizing*, New York: Free Press.

Gordon, I., Haslam, C., McCann, P. and Scott-Quinn, B. (2005) *Offshoring and the City of London*, London: The Corporation of London.

Gorz, A. (1982) *Farewell to the Working Class*, London: Pluto.

Goss, J. (1993) 'The "magic of the mall": an analysis of form, function and meaning in the contemporary retail built environment'. *Annals of the Association of American Geographers*, 83: 18–47.

Goss, J. (1999) 'Once upon a time in the commodity world: an unofficial guide to the Mall of America'. *Annals of the Association of American Geographers,* 89: 45–75.

Goss, J. (2005) 'Consumption geographies', in Cloke, P., Crang, P. and Goodwin, M. (eds) *Introducing Human Geographies*, 2nd edn, London: Arnold.

Grabher, G. (1993) 'The weakness of strong ties: the lock-in of regional development in the Ruhr area', in Grabher, G. (ed.) *The Embedded Firm: On the Socio-economics of Industrial Networks*, London: Routledge, pp. 255–77.

Granovetter, M. (1985) 'Economic action and social structure: the problem of embeddedness'. *American Journal of Sociology,* 91: 481–510.

Gray, J. (1999) *False Dawn: The Delusions of Global Capitalism*, London: Grant.

Green, F. (2001) 'It's been a hard day's night: the concentration and intensification of work in late twentieth-century Britain'. *British Journal of Industrial Relations,* 39: 53–80.

Green, F. (2004) 'Why has work effort become more intense?' *Industrial Relations,* 43: 709–41.

Gregory, D. (1996) 'A real differentiation and post-modern human geography', in Agnew, J., Livingstone, D.N. and Rogers, A. (eds) *Human Geography: An Essential Anthology*, London: Blackwell, pp.211–32.

Gregory, D. (2000) 'Positivism', in Johnston, R.J., Gregory, D., Pratt, G. and Watts, M. (eds) *The Dictionary of Human Geography*, 4th edn, Oxford: Blackwell, pp.606–8.

Gregson, N. (2000) 'Family, work and consumption: mapping the borderlands of economic geography', in Sheppard, E. and Barnes, T. (eds) *A Companion to Economic Geography*, Oxford: Blackwell, pp.311–24.

Gregson, N. and Lowe, M. (1994) *Servicing the Middle Classes: Class, Gender and Waged Domestic Labour in Contemporary Britain*, London: Routledge.

Gremple, G. (undated) 'The Industrial Revolution', at http://mars.acnet.wnec.edu/~grempel/courses/wc2/lectures/industrialrev.html.

Grimshaw, D., Ward, K.G., Rubery, J. and Beynon, H. (2001) 'Organisations and the transformation of the internal labour market'. *Work Employment and Society,* 15: 25–54.

Gwynne, R.N. (1990) *New Horizons?: Third World Industrialisation in an International Framework*, Harlow: Longman.

Gwynne, R. (2006) 'Governance and the wine commodity chain: upstream and downstream strategies in New Zealand and Chilean wine firms'. *Asia Pacific Viewpoint*, 47: 381–95.

Gwynne, R. (2008) 'UK retail concentration, Chilean wine producers and value chains'. *Geographical Journal*, 174: 970–1108.

Habermas, J. (1984) *The Theory of Communicative Action*, Vol. 1, Boston, MA: Beacon Press.

Hall, P. (1985) 'The geography of the fifth Kondratieff', in Hall, P. and Markusen, A. (eds) *Silicon Landscapes*, Boston, MA: Allen & Unwin, pp.1–19.

Hall, P. and Soskice, D. (eds) (2001) *Varieties of Capitalism: The Institutional Foundations of Comparative Advantage*, Oxford: Oxford University Press.

Hanlon, P., Walsh, D. and Whyte, B. (2006) 'Let Glasgow flourish: a comprehensive report on health and its determinants in Glasgow and west-central Scotland.' Glasgow: Glasgow Centre for Population Health.

Hardagon, A. (2005) 'Technology brokering and innovation: linking strategy, practice and people'. *Strategy and Leadership*, 33: 32–6.

Harrison, B. (1994) *Lean and Mean: The Changing Landscape of Corporate Power in the Age of Flexibility*, New York: Basic Books.

Hartshorne, R. (1939) *The Nature of Geography: A Critical Survey of the Present in the Light of the Past*, Lancaster, PA: The Association of American Geographers.

Harvey, D. (1973) *Social Justice and the City*, London: Arnold.

Harvey, D. (1982) *The Limits to Capital*, Oxford: Blackwell.

Harvey, D. (1989a) *The Condition of Postmodernity*, Oxford: Blackwell.

Harvey, D. (1989b) 'The transition from managerialism to entrepreneurialism: the transformation of urban governance in late capitalism'. *Geografiska Annaler,* 71B: 3–17.

Harvey, D. (1989c) 'Editorial: a breakfast vision'. *Geography Review*, 3: 1.

Harvey, D. (1996) *Justice, Nature and the Geography of Difference*, Oxford: Blackwell.

Harvey, D. (2003) *The New Imperialism*, Oxford: Oxford University Press.

Harvey, D. (2005) *A Brief History of Neoliberalism*, Oxford: Oxford University Press.

Hayter, R. (1997) *The Dynamics of Industrial Location: The Factory, the Firm and the Production System*, London: Wiley.

Healey, M. and Clark, D. (1986) 'Industrial decline and government response in the West Midlands: the case of Coventry'. *Regional Studies*, 18(4): 303–18.

Henderson, J., Dicken, P., Hess, M., Coe, N. and Yeung, H.W. (2002) 'Global production networks and economic development'. *Review of International Political Economy*, 9: 436–64.

Hendriskse, R.P. and Sidaway, J.D. (2010) 'Neoliberalism 3.0.' *Environment and Planning A*, 42, 2037–42.

Henwood, D. (1998) 'Talking about work', in Meiskins, E., Wood, P. and Yates, M. (eds) *Rising from the Ashes? Labor in the Age of Global Capitalism*, New York: Monthly Review Press.

Herod, A. (1997) 'From a geography of labor to a labor geography: labor's spatial fix and the geography of capitalism'. *Antipode*, 29: 1–31.

Herod, A. (2001) *Labor Geographies: Workers and the Landscapes of Capitalism*, New York: Guilford.

Herod, A., Peck, J. and Wills, J. (2003) 'Geography and industrial relations', in Ackers, P. and Wilkinson, A. (eds) *Understanding Work and Employment: Industrial Relations in Transition*, Oxford: Oxford University Press, pp.176–94.

Hess, M. (2004) '"Spatial" relationships? Towards a reconceptualisation of embeddedness'. *Progress in Human Geography*, 28: 165–86.

Hess, M. and Yeung, H.W. (2006) 'Whither global production networks in economic geography? Past, present and future'. *Environment and Planning A*, 38: 1193–204.

Hines, C. (2000) *Localization: A Global Manifesto*, London: Earthscan.

Hobsbawm, E.J. (1962) *The Age of Revolution, 1789–1848*, London: Weidenfeld & Nicolson.

Hobsbawm, E.J. (1987) *The Age of Empire, 1875–1914*, London: Weidenfeld & Nicolson.

Hobsbawm, E.J. (1999) *Industry and Empire*, new edn, Harmondsworth: Penguin.

Hodgson, G.M. (1993) *Economics and Evolution: Bringing Life Back into Economics*, Cambridge: Polity.

Hodgson, G. (1999) *Economics and Utopia: Why the Learning Economy is not the End of History*, London: Routledge.

Holmes, J. (2004) 'Rescaling collective bargaining: union responses to restructuring in the North American auto industry'. *Geoforum*, 35: 9–21.

Hopkins, T. and Wallerstein, I. (1986) 'Commodity chains in the world economy prior to 1800'. *Review*, 10: 157–70.

Howells, J.R.L. (2002) 'Tacit knowledge, innovation and economic geography'. *Urban Studies*, 39: 871–84, at http://www.tnsglobal.com/.

Hudson, R. (1989) 'Labor-market changes and new forms of work in old industrial regions: maybe flexibility for some but not flexible accumulation'. *Environment and Planning D: Society and Space*, 7: 5–30.

Hudson, R. (1997) 'The end of mass production and of the mass collective worker? Experimenting with production, employment and their geographies', in Lee and Wills (1997), pp.302–10.

Hudson, R. (2001) *Producing Places*, London: Guilford.

Hudson, R. (2003) 'Geographers and the regional problem', in Johnston, R.J. and Williams, M. (eds) *A Century of British Geography*, Oxford: Oxford University Press, pp.583–602.

Hudson, R. (2005) *Economic Geographies: Circuits, Flows and Spaces*, London: Sage.

Hudson, R. (2006) 'On what's right and keeping left: or why geography still needs Marxian political economy'. *Antipode*, 38: 374–95.

Hudson, R. (2008) 'Cultural political economy meets global production networks: a productive meeting?' *Journal of Economic Geography*, 8: 421–40.

Hudson, R. (2009) 'Life on the edge: navigating the competitive tensions between the "social" and the "economic" in the social economy and in its relations to the mainstream'. *Journal of Economic Geography*, 9: 493–510.

Hudson, R. and Sadler, D. (1986) 'Contesting works closures in western Europe's old industrial regions: defending place or betraying class?', in Scott, A. and Storper, M. (eds) *Production, Work, Territory: The Geographical Anatomy of Industrial Capitalism*, Boston, MA: Allen & Unwin, pp.172–93.

Huff, W.G. (1995) 'The developmental state, government and Singapore's economic development since 1960'. *World Development*, 23: 1421–38.

Hughes, A. (2000) 'Retailers, knowledge and changing commodity networks: the case of the cut flower trade'. *Geoforum*, 31: 175–90.

Hughes, A., Wrigley, N. and Buttle, M. (2008) 'Global production networks, ethical campaigning and the embeddedness of responsible governance'. *Journal of Economic Geography*, 8: 345–67.

Hulme, M. and Scott, J. (2010) The political economy of the MDGs: retrospect and prospect for the world's biggest promise. BWPI Working Paper 111, Brooks World Poverty Institute, University of Manchester.

Humphrey, J. and Schmitz, H. (2001) 'Governance in global value chains'. *IDS Bulletin*, 32: 19–29.

Humphrey, J. and Schmitz, H. (2002) 'How does insertion in global value chains affect upgrading in industrial clusters?' *Regional Studies*, 36: 1017–27.

Hutton, W. (1995) *The State We're In*, London: Vintage.

Hymer, S. (1972) 'The multinational corporation and the law of uneven development', in Bhagwati, J.H. (ed.) *Economics and World Order*, London: Macmillan, pp.113–40.

ILO (2000) *Labour Practices in the Footwear, Leather, Textiles and Clothing Industries*, Geneva: International Labour Organization.

ILO (2005a) *A Global Alliance against Forced Labour*, Geneva: International Labour Organization.

ILO (2005b) *World Employment Report*, Geneva: International Labour Organization.

ILO (2010) *World of Work Report 2010*, Geneva: International Labour Organization.

IMF (2009) *World Economic Outlook: Sustaining the Recovery*, Washington, DC: International Monetary Fund.

Imrie, R. and Thomas, H. (eds) (1999) *British Urban Policy and the Urban Development Corporations*, 2nd edn, London: Sage.

Industrial Communities Alliance (2009) *The Impact of the Recession on Unemployment in Industrial Britain*, Barnsley: Industrial Communities Alliance.

Ingham, G. (1994) 'States and markets in the production of world money', in Corbridge, S., Thrift, N. and Martin, R. (eds) *Money, Power and Space*, Oxford: Blackwell, pp.29–49.

Ingham, G. (1996) 'Some recent changes in the relationship between economics and sociology'. *Cambridge Journal of Economics*, 20: 243–75.

Internet World Statistics (2009) Internet Usage Statistics, at http://www.internetworldstats.com/stats.htm.

Ironmonger, D. (1996) 'Counting outputs, capital inputs and caring labor: estimating gross household product'. *Feminist Economics*, 2: 37–64.

Ismi, A. (2004) *Impoverishing a Continent: The World Bank and IMF in Africa*, British Columbia: Canadian Centre for Policy Alternatives Report.

Jackson, P. (1989) *Maps of Meaning: An Introduction to Cultural Geography*, London: Unwin Hyman.

Jacobskind, M.A. (2010) A republican and democratic reform is necessary in Brazil – Pagina 64 interview with Joao Pedro Stedile, at http://www.mstbrazil.org/?q=node/649.

Jacoby, N. (1974) *Multinational Oil: A Study in Industrial Dynamics*, London: Macmillan.

Jeffreys, S. (2001) 'Western European trade unionism at 2000'. *Socialist Register*, 37: 143–70.

Jenkins, P. (1987) *Mrs Thatcher's Revolution: The End of the Socialist Era*, London: Pan Books.

Jessop, B. (1994) 'Post-Fordism and the state', in Amin, A. (ed.) *Post-Fordism: A Reader*, Oxford: Blackwell, pp.251–79.

Jessop, B. (2002) *The Future of the Capitalist State*, Cambridge: Polity.

Johnston, R.J. (1984) 'The world is our oyster'. *Transactions of the Institute of British Geographers*, NS9: 443–59.

Johnston, R.J. (1986) 'The state, the region and the division of labour', in Scott, A.J. and Storper, M. (eds) *Production, Work, Territory: The Geographical Anatomy of Industrial Capitalism*, Boston, MA: Allen & Unwin, pp.265–80.

Johnston, R.J. (2000) 'Central place theory', in Johnston, R.J., Gregory, D., Pratt, G. and Watts, M. (eds) *The Dictionary of Human Geography*, 4th edn, Oxford: Blackwell, pp.72–3.

Johnston, R.J. and Sidaway, J.D. (2004) *Geography and Geographers: Anglo-American Human Geography Since 1945*, 6th edn, London: Arnold.

Jones, M., Jones, R. and Woods, R. (2004) *An Introduction to Political Geography: Space, Place and Politics*, London: Routledge.

Keen, S. (2009a) 'Mad, bad and dangerous to know'. *Real-World Economics Review*, 49: 2–7.

Keen, S. (2009b) The Roving Cavaliers of Credit. DebtWatch Blog, at http://www.debtdeflation.com/blogs/.

Kelly, P. (2001) 'The political economy of local labor control in the Philippines'. *Economic Geography*, 77: 1–22.

Kempson, E., Whyley, C., Caskey, J. and Collard, S. (2000) *In or Out? Financial Exclusion: A Literature and Research Review*, London: Financial Services Authority, Consumer Research 3.

Khan, M.H. (2002) *When is Economic Growth Pro-poor? Experiences in Malaysia and Pakistan*, Working Paper WP/02/85, Washington, DC: International Monetary Fund (IMF).

Kitson, M., Martin, R. and Tyler, P. (2004) 'Regional competitiveness: an elusive yet critical concept?' *Regional Studies,* 38: 991–9.

Klagge, B. and Martin, R. (2005) 'Decentralized versus centralized financial systems: is there a case for local capital markets?' *Journal of Economic Geography*, 5: 387–421.

Klare, M.T. (2002) *Resource Wars: The New Landscape of Global Conflict*, New York: Owl Books.

Knox, P. and Agnew, J. (1994) *The Geography of the World Economy*, 2nd edn, London: Arnold.

Knox, P., Agnew, J. and McCarthy, L. (2003) *The Geography of the World Economy*, 4th edn, London: Arnold.

Konadu-Agyemang, K. (2000) 'The best of times and the worst of times: structural adjustment programmes and uneven development in Africa: the case of Ghana'. *Professional Geographer,* 52: 469–83.

Krugman, P. (1996) *Pop Internationalism*, Cambridge, MA: MIT Press.

Krugman, P. (1998) 'What's new about the new economic geography?' *Oxford Review of Economic Policy,* 14: 7–17.

Krugman, P. (2000) 'Where in the world is the "new economic geography"?', in Clark, G., Feldmann, M. and Gertler, M. (eds) *The Oxford Handbook of Economic Geography*, Oxford: Oxford University Press, pp.49–60.

Krugman, P. (2010) The new economic geography, now middle aged. The Regional Studies Lecture, presented at the 2010 Annual Meeting of the Association of American Geographers, Washington, DC, 14–18 April.

Kujis, L. (2010) *China Through 2020 – A Macroeconomic Scenario*. Research Working Paper No. 9, The World Bank China Office. The World Bank Group, Washington, DC.

Lapavitsas, C. (2010) 'Trapped in the eurozone'. *The Guardian*, 1 October, at http://www.guardian.co.uk.

Lash, S. and Urry, J. (1994) *Economies of Signs and Spaces*, London: Sage.

Lawrence, F. (2009) 'The banana war's collateral damage is many miles away'. *The Guardian*, 13 October.

Leadbetter, C. (1999) *Living on Thin Air: The New Economy*, London: Viking.

Lee, C. (1986) 'Regional structure and change', in Langton, J. and Morris, R.J. (eds) *Atlas of Industrialising Britain 1780–1914*, London and New York: Methuen, pp.30–3.

Lee, R. (1999) 'Local money: geographies of autonomy and resistance?, in Martin, R. (ed.) *Money and the Space Economy*, Chichester: Wiley, pp.207–24.

Lee, R. (2000) 'Economic geography', in Johnston, R.J., Gregory, D., Pratt, G. and Watts, M. (eds) *The Dictionary of Human Geography*, 4th edn, Oxford: Blackwell, pp.195–9.

Lee, R. (2002) 'Nice maps, shame about the theory? Thinking geographically about the economic'. *Progress in Human Geography,* 26: 333–55.

Lee, R. and Wills, J. (eds) (1997) *Geographies of Economies*, London: Arnold.

Lees, I., Leyshon, A., Graham, J. and Gibson, K. (2010) *The End of Capitalism (as we knew it): A Feminist Critique of Political Economy, Progress in Human Geography*, Oxford: Blackwell, 34: 1201–23.

Ley, D. (1994) 'Postmodernism', in Johnston, R.J., Gregory, D. and Smith, D.M. (eds) *The Dictionary of Human Geography*, 3rd edn, Oxford: Blackwell, pp.66–8.

Leyshon, A. (1995) 'Annihilating space? The speed-up of communications', in Allen, J. and Hamnett, C. (eds) *A Shrinking World? Global Unevenness and Inequality*, Oxford: Oxford University Press, pp.11–54.

Leyshon, A. (2000) 'Money and finance', in Sheppard, B. and Barnes, T. (eds) *A Companion to Economic Geography*, Oxford: Blackwell, pp.432–49.

Leyshon, A. and Thrift, N. (1995) 'Geographies of financial exclusion: financial abandonment in Britain and the United States'. *Transactions of the Institute of British Geographers,* NS 20: 312–41.

Leyshon, A., Lee, R. and Williams, C.C. (eds) (2003) *Alternative Economic Spaces*, London: Sage.

Liu, W. and Dicken, P. (2006) 'Transnational corporations and "obligated embeddedness": foreign direct investment in China's automobile industry'. *Environment and Planning A*, 38: 1229–47.

Livingstone, D.N. (1992) *The Geographical Tradition*, Oxford: Blackwell.

Lloyd, P. and Shutt, J. (1985) 'Recession and restructuring in the north west region 1974–82: the implications of recent events', in Massey, D. and Meegan, R. (eds) *Politics and Method*, London: Methuen, pp.16–60.

Lowe, M. and Wrigley, N. (1996) 'Towards the new retail geography', in Wrigley, N. and Lowe, M. (eds) *Retailing, Consumption and Capital: Towards the New Retail Geography*, Harlow: Longman, pp.1–30.

Lundvall, B.-A. (1992) *National Systems of Innovation*, London: Pinter.

Lundvall, B.-A. (1994) 'The learning economy: challenges to economic theory and policy', paper presented at the European Association for Evolutionary Political Economy Conference, Copenhagen, October.

Lundvall, B.-A. and Johnson, B. (1994) 'The learning economy'. *Journal of Industry Studies,* 1: 23–43.

MacKinnon, D. and Phelps, N.A. (2001) 'Devolution and the territorial politics of foreign direct investment'. *Political Geography*, 20: 353–79.

MacLean, K. (1988) 'George Goudie Chisholm 1850–1930', in Freeman, T.W. (ed.) *Geographers Biobibliographic Studies*, vol. 12, London: Mansell Information Publishing, pp.21–33.

MacLeod, G. (1999) 'Place, politics and "scale dependence": exploring the structuration of Euro-regionalism'. *European Urban and Regional Studies,* 6: 231–53.

Malanga, S. (2004) 'The curse of the creative class'. *City Journal,* Winter: 1–9.

Malcolmson, R.W. (1988) 'Ways of getting a living in eighteenth-century England', in Pahl, R. (ed.) *On Work: Historical, Comparative and Theoretical Approaches*, Oxford: Blackwell, pp.48–60.

Malmberg, A. and Maskell, P. (2002) 'The elusive concept of localisation economies: towards a knowledge-based theory of spatial clustering'. *Environment and Planning A*, 34: 429–49.

Mann, M. (1984) 'The autonomous power of the state'. *European Journal of Sociology*, 25: 185–213.

Mansvelt, J. (2005) *Geographies of Consumption*, London: Sage.

Markusen, A. (2006) 'Urban development and the politics of a creative class – evidence from a study of artists'. *Environment and Planning A*, 38, 1921–40.

Marshall, J.N., Pike, A., Pollard, J., Tomaney, J., Dawley, S., Gray, J. and De-Cecco, F. (2010) Placing the run on Northern Rock. Mimeograph, Centre for Urban and Regional Development and School of Law, University of Newcastle.

Martin, R. (1988) 'The political economy of Britain's north–south divide'. *Transactions of the Institute of British Geographers*, NS13: 389–418.

Martin, R. (1989) 'The new regional economics and the politics of regional restructuring', in Albrechts, L., Moulaert, F., Roberts, P. and Swyngedouw, E. (eds) *Regional Policy at the Crossroads: European Perspectives*, London: Jessica Kingsley, pp.27–51.

Martin, R. (1999a) The economic geography of money, in Martin, R.L. (ed.) *Money and the Space Economy*, London: Wiley, pp.3–28.

Martin, R. (1999b) 'The new "geographical turn" in economics: some critical reflections'. *Cambridge Journal of Economics,* 23: 65–91.

Martin, R. (1999c) 'The "new economic geography": challenge or irrelevance?', *Transactions of the Institute of British Geographers*, NS24: 387–91.

Martin, R. (2000) 'Institutionalist approaches to economic geography', in Sheppard, E. and Barnes, T. (eds) *A Companion to Economic Geography*, Oxford: Blackwell, pp.77–97.

Martin, R. and Sunley, P. (1997) 'The post-Keynesian state and the space economy', in Lee and Wills (1997), pp.278–89.

Martin, R. and Sunley, P. (2003) 'Deconstructing clusters: chaotic concept or policy panacea?' *Journal of Economic Geography*, 3: 5–35.

Marx, K. [1867] (1976) *Das Capital*, Vol. 1, New York: International Publishers.

Marx, K. and Engels, F. [1848] (1967) *The Communist Manifesto*, Harmondsworth: Penguin.

Maskell, P. (2001) 'The firm in economic geography'. *Economic Geography*, 77: 329–43.

Maskell, P. and Malmberg, A. (1999) 'The competitiveness of firms and regions: "ubiquitification" and the importance of localised learning'. *European Urban and Regional Studies*, 5(2): 119–37.

Mason, C. (2007) 'Venture capital: a geographical perspective', in Landstrom, E. (ed.) *Handbook of Research on Venture Capital*, Cheltenham: Edward Elgar, pp.86–112.

Mason, C.M. and Harrison, R.T. (1999) 'Financing entrepreneurship: venture capital and regional development', in Martin, R. (ed.) *Money and the Space Economy*, Chichester: Wiley, pp.157–84.

Mason, C.M. and Harrison, R.T. (2002) 'The geography of venture capital investments in the UK'. *Transactions of the Institute of British Geographers*, NS27: 427–51.

Mason, M. (1994) 'Historical perspectives on Japanese direct investment in Europe', in Mason, M. and Encarnation, D. (eds) *Does Ownership Matter? Japanese Multinationals in Europe*, Oxford: Clarendon, Chapter 1.

Massey, D. (1984) *Spatial Divisions of Labour: Social Structures and the Geography of Production*, London: Macmillan.

Massey, D. (1988) 'Uneven development: social change and spatial diversions of labour', in Allen, J. and Massey, D. (eds) *Uneven Redevelopment: Cities and Regions in Transition*, London: Hodder & Stoughton, pp.250–76.

Massey, D. (1994) 'A global sense of place', in Massey, D. (ed.) *Place, Space and Gender*, Cambridge: Polity, pp.146–56.

Massey, D. (2001) 'Geography on the agenda', *Progress in Human Geography*, 25: 5–17.

Massey, D. (2004) 'Geographies of responsibility'. *Geografisker Annaler B*, 86: 5–18.

Massey, D. (2005) *For Space*, London: Sage.

Massey, D. (2007) *World City*, Cambridge: Polity.

Mawdsley, E. and Rigg, J. (2003) 'The World Development Report II: continuity and change in development orthodoxies'. *Progress in Development Studies*, 3: 271–86.

McCann, E. and Ward, K. (2010) 'Relationality/territoriality: towards a conceptualisation of cities in the world'. *Geoforum*, 41: 175–84.

McChesney, R.W. (1999) 'Noam Chomsky and the struggle against neoliberalism'. *Monthly Review*, 50: 40–7.

McDonald, B. (2009) '20,000 demand Diageo rethink on plant closure'. *The Herald*, 27 July.

McDowell, L. (1997) *Capital Culture: Gender at Work in the City*, Oxford: Blackwell.

McDowell, L. (2003) *Redundant Masculinities: Employment Change and White Working Class Youth*, Oxford: Blackwell.

McFarlane, C. (2004) 'Geographical imaginations and spaces of political engagement: examples from the Indian Alliance'. *Antipode*, 36: 890–916.

Meegan, R. (1988) 'A crisis of mass production?', in Allen, J. and Massey, D. (eds) *The Economy in Question*, London: Sage, pp.136–83.

Miller, D. (1995) 'Consumption as the vanguard of history: a polemic by way of introduction', in Miller, D. (ed.) *Acknowledging Consumption: A Review of New Studies*, London: Routledge, pp.1–57.

Miller, D. (1998a) 'Coca-Cola: a black sweet drink from Trinidad', in Miller, D. (ed.) *Material Culture: Why Some Things Matter*, Chicago, IL: University of Chicago Press, pp.169–88.

Miller, D. (1998b) *A Theory of Shopping*, Cambridge: Polity.

Minsky, H. (1975) *John Maynard Keynes*, New York: Columbia University Press.

Minsky, H. (1986) *Stabilizing an Unstable Economy*, New Haven, CT: Yale University Press.

Minsky, H. (1992) The Financial Instability Hypothesis. Working Paper No. 74, Jerome Levy Economics Institute, Bard College.

Mohan, J. (1999) *A United Kingdom? Economic, Social and Political Geographies*, London: Arnold.

Molina, G.C. and Purser, M. (2010) Human development trends since 1970: a social convergence story. UNDP, Human Development Reports, Research Paper 2010/02. UNDP, New York, at http://hdr.undp.org/reports.

Monbiot, G. (2003) *The Age of Consent: A Manifesto for a New World Order*, London: Flamingo.

Moneyweek (2004) 'Who makes money out of Christmas?' *Moneyweek: The Best of the International Financial Media*, 2 January.

Moody, K. (1997) *Workers in a Lean World*, London: Verso.

Moore, G. (2004) 'The fair trade movement: parameters, issues and future research'. *Journal of Business Ethics*, 53: 73–86.

Morgan, K. (1997) 'The learning region: institutions, innovation and regional renewal'. *Regional Studies*, 31: 491–504.

Morgan, K. (2004) 'The exaggerated death of geography: learning, proximity and territorial innovation systems'. *Journal of Economic Geography*, 4: 3–21.

Morgan, K. (2006) 'Devolution and development: territorial justice and the north–south divide'. *Publius: The Journal of Federalism*, 36: 189–206.

Morgan, K., Marsden, T. and Murdoch, J. (2006) *Worlds of Food: Place, Power and Provenance in the Food Chain*, Oxford: Oxford University Press.

Munck, R. (1999) 'Labour dilemmas and labour futures', in Munck, R. and Waterman, P. (eds) *Labour Worldwide in the Era of Globalization*, Basingstoke: Macmillan, pp.3–26.

Munday, M., Morris, J. and Wilkinson, B. (1995) 'Factories or warehouses? A Welsh perspective on Japanese transplant manufacturing'. *Regional Studies*, 29: 1–17.

Myrdal, G. (1957) *Economic Theory and the Under-developed Regions*, London: Duckworth.

Nambisan, S. (2005) 'How to prepare tomorrow's technologists for global networks of innovation'. *Communications of the Association for Computing Machinery*, 48(5): 29–31.

Nathan, M. (2005) The wrong stuff: creative class theory, diversity and city performance. Centre for Cites, Discussion Paper No. 1, September.

Nelson, R.R. and Winter, S.G. (1982) *An Evolutionary Theory of Economic Change*, Cambridge, MA: Harvard University Press.

Nonaka, I. and Takeuchi, H. (1995) *The Knowledge-creating Company: How Japanese Companies Create the Dynamics of Innovation*, Oxford: Oxford University Press.

Nonaka, I., Toyama, R. and Konno, N. (2001) 'SECI, *ba* and leadership: a unified model of dynamic knowledge creation', in Nonaka, I. and Teece, D. (eds) *Managing Industrial Knowledge: Creation, Transfer and Utilisation*, London: Sage, pp.13–43.

North, P. (1998) 'Exploring the politics of social movements through "sociological intervention": a case study of local exchange trading schemes'. *Sociological Review*, 46: 564–82.

North, P. (2005) 'Scaling alternative economic practices? Some lessons from alternative currencies'. *Transactions of the Institute of British Geographers*, 30(2): 233–5.

Noyes, K. (2010) 'Reforms put Wall Street in its place'. *The Guardian*, 21 May.

O'Brien, R. (1992) *Global Financial Integration: The End of Geography*, London: Pinter.

O'Connor, J. (1973) *The Fiscal Crisis of the State*, New York: St Martin's Press.

O'Neill, P. (1997) 'Bringing the qualitative state into economic geography', in Lee and Wills (1997), pp.290–301.

O'Neill, P. and Gibson-Graham, J.K. (1999) 'Enterprise discourse and executive talk: stories that destabilize the company'. *Transactions of the Institute of British Geographers*, 24: 11–22.

O'Sullivan, A. and Sheffrin, S.M. (2003) *Economics: Principles in Action*, Upper Saddle River, NJ: Pearson Prentice Hall.

OECD (2005a) *Measuring Globalization*, Paris: Organization for Economic Cooperation and Development.

OECD (2005b) *Labour Force Statistics 2005*, Paris: Organization for Economic Cooperation and Development

OECD (2006) 'Ghana', in *Africa Economic Outlook 2005-6*, Paris: Organization for Economic Cooperation and Development.

OECD (2009) *Labour Force Statistics 2009*, Paris: Organization for Economic Cooperation and Development.

OECD (2010) OECD statistics, at http://www.oecd.org.

Offe, C. and Heinze, R. (1992) *Beyond Employment: Time, Work, and the Informal Economy*, Cambridge: Polity Press.

Office for National Statistics (2009) 'Consumer Trends', at http://www.statistics.gov.uk/STATBASE/Product.asp?vlnk=242/.

Ohmae, K. (1990) *The Borderless World*, London: Collins.

Olesen, A. (2006) 'China's upwards spiral sparks fears of overheating'. *The Herald*, 17 April.

Oliver, N. and Wilkinson, B. (1992) *The Japanisation of British Industry: New Developments in the 1990s*, Oxford: Blackwell.

ONS (2005a) *Labour Market Statistics*, March, London: Office for National Statistics.

ONS (2005b) *Travel Trends 2004*, London: Office for National Statistics.

ONS (2010) Regional Profiles: London Economy, at http://www.statistics.gov.uk/cci/nugget.asp?id=2286.

Oxfam (2005) *What Happened in Hong Kong? Initial Analysis of the WTO Ministerial, December 2005*, Oxfam Briefing Paper 85, Oxfam International.

Oxfam (2007) Strategic Plan 2007–2010, Oxford: Oxfam.

Oxfam (2009) Annual Report and Accounts 2008–2009, Oxford: Oxfam.

Pahl, R. (1988) 'Historical aspects of work, employment, unemployment and the sexual division of labour', in Pahl, R. (ed.) *On Work: Historical, Comparative and Theoretical Approaches*, Oxford: Blackwell, pp.1–7.

Painter, J. (2000) 'State and governance', in Sheppard, E. and Barnes, T. (eds) *A Comparison to Economic Geography*, Oxford: Blackwell, pp.359–76.

Painter, J. (2002) 'The rise of the workfare state', in Johnston, R.J., Taylor, P. and Watts, M. (eds) *Geographies of Global Change: Remapping The World*, 2nd edn, Oxford: Blackwell, pp.158–73.

Painter, J. (2006) 'Prosaic geographies of stateness'. *Political Geography*, 25: 752–74.

Palma, J.G. (2009) 'The revenge of the market on the rentiers. Why neo-liberal reports of the end of history turned out to be premature'. *Cambridge Journal of Economics*, 33: 829–69.

Pavlinek, P. (2004) 'Regional development implications of foreign direct investment in Central Europe'. *European Urban and Regional Studies*, 11: 47–70.

Peck, J. (1996) *Work-Place: The Social Regulation of Labour Markets*, New York: Guilford.

Peck, J. (1999) 'Grey geography?' *Transactions of the Institute of British Geographers*, NS24: 131–5.

Peck, J. (2000) 'Places of work', in Sheppard, E. and Barnes, T. (eds) *A Companion to Economic Geography*, Oxford: Blackwell, pp.133–48.

Peck, J. (2001) 'Neoliberalising states: thin policies/hard outcomes'. *Progress in Human Geography*, 25: 445–55.

Peck, J. (2005a) 'Economic sociologies in space'. *Economic Geography*, 81: 129–78.

Peck, J. (2005b) 'Struggling with the creative classes'. *International Journal of Urban and Regional Research*, 29: 740–70.

Peck, J. and Theodore, N. (1998) 'The business of contingent work: growth and restructuring in Chicago's temporary employment industry'. *Work, Employment and Society*, 12: 655–74.

Peck, J. and Theodore, N. (2001) 'Contingent Chicago: restructuring the spaces of temporary labor'. *International Journal of Urban and Regional Research*, 25(3): 471–96.

Peck, J. and Theodore, N. (2007) 'Variegated capitalism'. *Progress in Human Geography*, 31: 731–72.

Peck, J. and Theodore, N. (2010) 'Recombitant workfare, across the Americas: transnationalising 'fast' social policy'. *Geoforum*, 41: 195–208.

Peck, J., Theodore, N. and Brenner, N. (2010) 'Postneoliberalism and its malcontents'. *Antipode*, 42: 94–116.

Peck, J. and Tickell, A. (1994) 'Searching for a new institutional fix: the after-Fordist crisis and the global-local disorder', in Amin, A. (ed.) *Post-Fordism: A Reader*, Oxford: Blackwell, pp.280–315.

Peck, J. and Tickell, A. (2000) 'Labour markets', in Gardiner, V. and Matthews, H. (eds) *The Changing Geography of the United Kingdom*, 3rd edn, London: Routledge, pp.150–68.

Peck, J.A. and Tickell, A. (2002) 'Neoliberalising space'. *Antipode*, 34: 380–404.

Peet, R. and Hardwick, E. (1999) *Theories of Development*, New York: Guilford.

Penrose, E.T. (1959) *The Theory of the Growth of Firms*, New York: Oxford University Press.

Phelps, N.A. (1993) 'Branch plants and the evolving spatial division of labour: a study of material linkage change in the north east of England'. *Regional Studies*, 27: 87–101.

Phelps, N.A. and Fuller, C. (2000) 'Multinationals, intracorporate competition and regional development'. *Economic Geography*, 76: 224–43.

Phelps, N.A., Lovering, J. and Morgan, K. (1998) 'Tying the firm to the region or tying the region to the firm? Early observations on the case of LG in South Wales'. *European Urban and Regional Studies*, 5: 119–37.

Phelps, N.A., MacKinnon, D., Stone, I. and Braidford, P. (2003) 'Embedding the multinationals? Institutions and the development of overseas manufacturing affiliates in Wales and north-east England'. *Regional Studies*, 37: 27–40.

Pike, A. (2006) 'Shareholder value versus the regions: the closure of the Vaux Brewery in Sunderland'. *Journal of Economic Geography*, 6: 201–22.

Pike, A., Rodríguez-Pose, A. and Tomaney, J. (2006) *Local and Regional Development*, London: Routledge.

Polanyi, K. (1944) *The Great Transformation: The Political and Economic Origins of Our Time*, Boston, MA: Beacon Press.

Polanyi, K. [1959] (1982) 'The economy as an instituted process', in Granovetter, M. and Swelberg, R. (eds) *The Sociology of Economic Life*, Boulder, CO: Westview Press, pp.29–51.

Pollard, J. (1996) 'Banking at the margins: a geography of financial exclusion in Los Angeles'. *Environment and Planning A*, 28: 1209–32.

Pollard, J. (2005) 'The global financial system: worlds of monies', in Daniels *et al.* (2005), pp.337–58.

Pollard, J. (2008) 'The global financial system: world of monies', in Daniels, P., Bradshaw, M., Shaw, D. and Sidaway, J. (eds) *An Introduction to Human Geography*, 3rd edn, London: Prentice Hall/Pearson, pp. 358–75.

Pollard, S. (1971) 'Introduction', in Pollard, S. and Salt, J. (eds) *Robert Owen, Prophet of the Poor*, London: Redfern Percy.

Pollard, S. (1981) *Peaceful Conquest: The Industrialisation of Europe 1760–1870*, Oxford: Oxford University Press.

Porter, M.E. (1998) 'Clusters and the new economics of competition'. *Harvard Business Review*, December, 77–90.

Potter, R.B. (1983) 'Tourism and development: the case of Barbados, West Indies'. *Geography*, 68: 46–50.

Potter, R.B., Binns, T., Elliott, J.A. and Smith, D. (2004) *Geographies of Development*, 2nd edn, Harlow: Pearson.

Potter, R.B., Binns, T., Elliott, J.A. and Smith, D. (2008) *Geographies of Development*, 3rd edn, Harlow: Pearson.

Power, M. (2008) 'Worlds apart: global difference and inequality', in Daniels, P., Bradshaw, M., Shaw, D. and Sidaway, J. (eds) *Human Geography: Issues for the 21st Century*, 3rd edn, Harlow: Pearson, pp.180–202.

Pratt, G. (1999) 'From registered nurse to registered nanny: discursive geographies of Filipina domestic workers in Vancouver, BC'. *Economic Geography*, 75: 215–36.

Prynn, J. (2006) 'Experts warn of Tesco expansion'. *Evening Standard*, 21 April.

Ramalho, J.R. (1999) 'Restructuring of labour and trade union responses in Brazil', in Munck, R. and Waterman, P. (eds) *Labour Worldwide in the Era of Globalization*, Basingstoke: Macmillan, pp.158–74.

Rees, J. (1998) 'The return of Marx?' *International Socialism*, 79, at http://pubs.socialistreviewindex.org.uk/isj79.rees.htm.

Reich, R. (1991) *The Work of Nations*, New York: Knopf.

Rigby, E. (2008) Christmas retail slump likely. *Financial Times*, 23 October.

Ritzer, G. (2004) *The Globalization of Nothing*, Thousand Oaks, CA, and London: Pine Forge Press.

Robinson, P.K. (2009) 'Responsible retailing: regulating fair and ethical trade'. *Journal of International Development*, 21: 1015–26.

Rodríguez-Pose, A. and Gill, N. (2004) 'Is there a global link between regional disparities and devolution?' *Environment and Planning A,* 36: 2097–117.

Routledge, P. (2005) 'Survival and resistance', in Cloke, P., Crang, P. and Goodwin, M. (eds) *Introducing Human Geographies*, 2nd edn, London: Arnold, pp.211–24.

Routledge, P. and Cumbers, A. (2009) *Global Justice Networks*, Manchester: Manchester University Press.

Russell, B. and Thite, M. (2008) 'The next division of labour: work skills in Australian and Indian call centres'. *Work, Employment and Society*, 22: 615–634.

Rutherford, T. (2010) 'De-centring work and class? A review and critique of labour geography'. *Geography Compass*, 4: 768–77.

Rutherford, T. and Gertler, M. (2002) 'Labour in "lean" times: geography, scale and national trajectories of workplace change'. *Transactions of the Institute of British Geographers,* NS27: 195–212.

Ryland, R. (2010) 'Exploring the grassroots perspective on labour internationalism', in Bergene, A.C., Endresen, S.B. and Knutsen, H.G. (eds) *Missing Links in Labour Geography*, Aldershot: Ashgate, pp.57–70.

Sader, E. (2009) 'Postneoliberalism from and as a counter-hegemonic perspective'. *Development Dialogue*, January: 59–71.

Sawyer, M.C. and O'Donnell, K. (1999) *A Future for Public Ownership*, London: Lawrence & Wishart.

Saxenian, A.L. (1994) *Regional Advantage: Culture and Competition in Silicon Valley and Route 128*, Cambridge, MA: Harvard University Press.

Saxenian, A.L. (2006) The *New Argonauts: Regional Advantage in a Global Economy*, Cambridge, MA: Harvard University Press.

Sayer, A. (1985) 'Industry and space: a sympathetic critique of radical research', *Environment and Planning D: Society and Space*, 3: 3–39.

Sayer, A. (1995) *Radical Political Economy: A Critique*, Oxford: Blackwell.

Sayer, A. (2000) 'Moral economy and political economy'. *Studies in Political Economy*, 61: 79–103.

Sayer, A. and Walker, R. (1992) *The New Social Economy: Reworking the Division of Labour*, Oxford: Blackwell.

Schoenberger, E. (1994) 'The firm in the region and the region in the firm', paper presented at Regions, Institutions, and Technology Conference, Toronto, September.

Schoenberger, E. (1997) *The Cultural Crisis of the Firm*, Oxford: Blackwell.

Schumpeter, J.A. (1943) *Capitalism, Socialism and Democracy*, London: Allen & Unwin.

Scott, A.J. (2000) 'Economic geography: the great half century', in Clark, G., Feldmann, M. and Gertler, M. (eds) *The Oxford Handbook of Economic Geography*, Oxford: Oxford University Press, pp.18–44.

Scott, A.J. (2002) 'A new map of Hollywood: the production and distribution of American motion pictures'. *Regional Studies*, 36: 957–75.

Scott, A.J. (2006) 'A perspective of economic geography', in Bagchi-Sen, S. and Lawton-Smith, H. (eds) *Economic Geography: Past, Present and Future*, Abingdon: Routledge, pp.56–79.

Scottish Government (2003) *Scottish Economic Statistics 2003*, Edinburgh: The Scottish Government.

Scottish Government (2004) *Scottish Economic Report, March 2004*, Edinburgh: The Scottish Government.

Scottish Government (2006) *Progress Report on the Fresh Talent Initiative*, Edinburgh: The Scottish Government.

Scottish Government (2008) *Fresh Talent: Working in Scotland Scheme: An Evidence Review*, Edinburgh: The Scottish Government.

Seddon, M. (2005) 'Kapital gain: Karl Marx is the Home Counties favourite'. *The Guardian*, 14 July.

Seyfang, G. (2006) 'Sustainable consumption, the new economics and community currencies: developing new institutions for environmental governance. *Regional Studies*, 40: 781–91.

Shaw, M., Davey Smith, G. and Dorling, D. (2005) 'Health inequalities and new Labour: how the promises compare with real progress'. *British Medical Journal*, 330: 1016–21.

Sheppard, E. and Leitner, H. (2010) 'Quo vadis neoliberalism? The remaking of global capitalist governance after the Washington Consensus'. *Geoforum*, 45: 185–94.

Sheppard, E., Barnes, T.J., Peck, J. and Tickell, A. (2004) 'Introduction: reading economic geography', in Barnes T.J., Peck, J., Sheppard, E. and Tickell, A. (eds) *Reading Economic Geography*, Oxford: Blackwell, pp.1–11.

Simon, D. (2008) 'Neoliberalism, structural adjustment and poverty reduction strategies', in Desai, V. and Potter, R. (eds) *The Companion to Development Studies*, 2nd edn, London: Arnold, pp.86–92.

Skidelsky, R. (2003) *John Maynard Keynes: Economist, Philosopher, Statesman*, London: Pan Books.

Slater, D. (2003) 'Cultures of consumption', in Anderson, K., Domosh. M., Pile, S. and Thrift, N. (eds) *Handbook of Cultural Geography*, London: Sage, pp.146–63.

Slaven, A. (1986) 'Shipbuilding', in Langton, J. and Morris, R.J. (eds) *Atlas of Industrialising Britain 1780–1914*, London and New York: Methuen, pp.132–5.

Smith, A. [1776] (1991) *The Wealth of Nations. Inquiry into the Nature and Causes of the Wealth of Nations*, Buffalo, CO: Prometheus Books.

Smith, A., Rainnie, A., Dunford, M., Hardy, J., Hudson, R. and Sadler, D. (2002) 'Networks of value, commodities and regions: reworking divisions of labour in macro-regional economies'. *Progress in Human Geography*, 26: 41–63.

Smith, D.M. (2000) 'Export platform', in Johnston, R.J., Gregory, D., Pratt, G. and Watts, M. (eds) *The Dictionary of Human Geography*, 4th edn, Oxford: Blackwell, pp.249–50.

Smith, N. (1984) *Uneven Development: Nature, Capital and the Production of Space*, Oxford: Blackwell.

Smith, N. (1996) *The New Urban Frontier: Gentrification and the Revanchist City*, London: Routledge.

Smith, N. (2001) 'Marxism and geography in the Anglophone world'. *Geographische Revue*, 3: 5–22, at http://www.geographische-revue.de/gr2-01.htm.

Smith, S.J., Munro, M. and Christie, H. (2006) 'Performing (housing) markets'. *Urban Studies*, 43: 81–98.

Sorrentino, C. and Moy, J. (2002) 'US labor market performance in international perspective'. *Monthly Labor Review*, June, pp.15–35.

Sparke, M., Sidaway, J.D., Bunnell, T. and Grundy-Warr, C.V. (2004) 'Triangulating the borderless world: geographies of power in the Indonesia–Malaysia– Singapore growth triangle'. *Transactions of the Institute of British Geographers*, NS29: 485–98.

Stiglitz, J. (2002) *Globalization and its Discontents*, London: Penguin.

Stiglitz, J. (2003) 'Trade imbalances'. *The Guardian*, 15 August.

Stohr, W. and Todling, F. (1978) 'Spatial equity – some anti-theses to current regional development doctrine'. *Papers of the Regional Science Association*, 38: 33–54.

Stoker, G. (ed.) (1999) *The New Politics of British Local Governance*, Basingstoke: Macmillan.

Stone, I. (1998) 'East Asian FDI and the UK periphery working miracles? Regional renewal and East Asian inter-linkages', in Garrahan, P. and Ritchie, J. (eds) *East Asian Direct Investment in Britain*, London: Cass, pp.59–94.

Storper, M. (1997) *The Regional World: Territorial Development in a Global Economy*, London: Guilford.

Storper, M. and Christopherson, S. (1987) 'Flexible specialisation and regional industrial agglomeration: the case of the US motion picture industry'. *Annals of the Association of American Geographers*, 77: 104–17.

Storper, M. and Walker, R. (1989) *The Capitalist Imperative: Territory, Technology and Industrial Growth*, Oxford: Blackwell.

Sturgeon, T. (2001) 'How do we define value chains and production networks?' *IDS Bulletin*, 32: 9–18.

Sutherland, E. (1995) 'Silicon Glen – fifty years of the electronics industry', in Green, M.B. and McNaughton, R.B. (eds) *The Location of Foreign Direct Investment*, Avebury: Aldershot, pp.59–84.

Swyngedouw, E. (2000) 'The Marxian alternative: historical–geographical materialism and the political economy of capitalism', in Sheppard, E. and Barnes, T. (eds) *A Companion to Economic Geography*, Oxford: Blackwell, pp.40–59.

Taylor, M. and Asheim, B. (2001) 'The concept of the firm in economic geography'. *Economic Geography*, 77: 315–28.

Taylor, M. and Thrift, N. (eds) (1982) *The Geography of Multinationals*, London: Croom Helm.

Taylor. P.J. (2002) 'Geography/globalisation', in Taylor, P.J., Watts, M. and Johnston, R.J. (eds) *Geographies of Global Change, Remapping the World* 2nd edn, Oxford: Blackwell, pp.1–17.

Taylor, P.J. and Bain, P. (1999) 'An assembly-line in the head: work and employee relations in the call centre'. *Industrial Relations Journal*, 30: 101–17.

Taylor P.J. and Flint, C. (2000) *Political Geography: World Economy, Nation-State and Locality*, Harlow: Pearson.

Tesco (2010) *A Business For a New Decade. Annual Report and Financial Statements 2010*. Tesco plc, Cheshunt, Hertfordshire, at http://ar2010.tescoplc.com/.

The Economist (2003) 'The WTO under Fire', 18 September.

The Economist (2004) 'Exit strategy: can Silicon Valley's magic last?' 1 May, p.9.

The Economist (2005a) 'Things go from bad to worse in retailing'. 14 May, p.34.

The Economist (2005b) 'Consumers have stopped spending: for how long?' 12 November, p.51.

The Economist (2006) 'Podtastic, Apple'. 14 January, p.68.

Thompson, E.P. (1963) *The Making of the English Working Class*, Harmondsworth: Penguin.

Thompson, P. (2003) 'Disconnected capitalism: or why employers can't keep their side of the bargain'. *Work Employment and Society,* 17(2): 359–78

Thompson, P. (2004) *Skating on Thin Ice: The Knowledge Economy Myth*, Glasgow: Big Thinking.

Thrift, N. (1994) 'On the social and cultural determinants of international financial centres: the case of the City of London', in Corbridge, S., Thrift, N. and Martin, R. (eds) *Money, Power and Space*, Oxford: Blackwell, pp.327–55.

Thrift, N. (2000) 'Pandora's box?: cultural geographies of economies', in Clark, G., Feldmann, M. and Gertler, M. (eds) *The Oxford Handbook of Economic Geography*, Oxford: Oxford University Press, pp.689–704.

Thrift, N. (2002) 'A hyperactive world', in Johnston, R.J., Taylor, P.J. and Watts, M.J. (eds) *Geographies of Global Change: Remapping the World*, Oxford: Blackwell, pp.29–42.

Thrift, N. and Olds, K. (1996) 'Refiguring the economic in economic geography'. *Progress in Human Geography,* 20: 311–37.

Tian, L., Wang, H.H. and Chen, Y. (2010) 'Spatial externalities in China regional economic growth'. *China Economic Review*, 21, Supplement 1: S20–31.

Tickell, A. (2000) 'Dangerous derivatives: controlling and creating risks in international money'. *Geoforum,* 31: 87–99.

Tickell, A. (2003) 'Cultures of money', in Anderson, K., Domosh, M., Pile, S. and Thrift, N. (eds) *Handbook of Cultural Geography*, London: Sage, pp.116–30.

Tickell, A. (2005) 'Money and finance', in Cloke, P., Crang, P. and Goodwin, M. (eds) *Introducing Human Geographies*, 2nd edn, London: Arnold, pp.244–52.

TNS Worldpanel (undated) Market Share data for UK grocery retailers.

Tomaney, J., Pike, A. and Rodríguez-Pose, A. (2010) 'Local and regional development in times of crisis'. *Environment and Planning A*, 42, 771–9.

Tormey, S. (2004) *Anti-capitalism: A Beginner's Guide*, Oxford: Oneworld.

Townsend, J.G., Porter, G. and Mawdsley, E. (2004) 'Creating spaces of resistance: development NGOs and their clients in Ghana, India and Mexico'. *Antipode,* 36: 781–889.

Tufts, S. (2007) 'Emerging labour strategies in Toronto's Hotel Sector: toward a spatial circuit of union renewal'. *Environment and Planning A*, 39(10): 2383–404.

Tuppen, J.N. and Thompson, I.B. (1994) 'Industrial restructuring in contemporary France: spatial priorities and policies'. *Progress in Planning*, 44: 99–172.

Turner, W. (undated) 'The decline of the handloom weaver', at http://www.cottontown.org/page.cfm?pageid=456&language=eng.

Turok, I. and Bailey, N. (2004) 'Glasgow's recent trajectory: partial recovery and its consequences', in Newlands, D., Danson, M. and McCarthy, J. (eds) *Divided Scotland? The Nature, Causes and Consequences of Economic Disparities within Scotland*, Aldershot: Ashgate, pp.35–59.

UNCTAD (2004) *World Investment Report 2004: The Shift Towards Services*, New York and Geneva: United Nations Conference on Trade and Development.

UNCTAD (2009) *World Investment Report 2009: Transnational Corporations, Agricultural Production and Development*. United Nations Conference on Trade and Development, New York and Geneva.

UNDP (2001) *Making New Technologies Work For Human Development*, Human Development Report, New York: UNDP, at http://hdr.undp.org/reports.

UNDP (2004) *Cultural Liberty in Today's Diverse World*, Human Development Report, New York: UNDP, at http://hdr.undp.org/reports.

UNDP (2005) *International Cooperation at a Crossroads: Aid, Trade and Security in an Unequal World*, Human Development Report, New York: UNDP, at http://hdr.undp.org/reports.

UNDP (2008) *Access for All: Basic Public Services for 1.3 Billion People*. Human Development Report, China 2007/2008. UNDP, New York.

UNDP (2009) *Overcoming Barriers: Human Mobility and Development*. Human Development Report 2009. UNDP, New York.

UNICEF (2006) *Excluded and Invisible: The State of the World's Children*, New York: United Nations Children's Fund.

United Nations (2010) *The Millennium Development Goals Report 2010*. United Nations, New York and Geneva.

Urry, J. (2000) *Sociology Beyond Societies: Mobilities for the Twenty-first Century*, London: Routledge.

Vernon, R. (1971) *Sovereignty at Bay: The Multinational Spread of US Enterprises*, New York: Basic Books.

Vidal, J. (1999) 'Secret life of a banana'. *The Guardian*, 10 November.

Wade, R. (2010) 'Is the globalisation consensus dead?' *Antipode*, 41: 142–65.

Wainright, J. (2007) 'Spaces of resistance in Seattle and Cancun', in Leitner, H., Peck, J. and Sheppard, E.S. (eds) *Contesting Neoliberalism: Urban Frontiers*, New York: Guilford, pp.179–203.

Walker, R. (2000) 'The geography of production', in Sheppard, E. and Barnes, T. (eds) *A Companion to Economic Geography*, Oxford: Blackwell, pp.111–32.

Wallop, H. (2007) 'Disposable income at lowest level for 10 years'. *The Telegraph*, 8 October.

Walsh, J. (2000) 'Organising the scale of labour regulation in the United States: service sector activism in the city'. *Environment and Planning A*, 32: 1593–610.

Ward, K. (2004) 'Going global? Internationalization and diversification in the temporary staffing industry'. *Journal of Economic Geography*, 4(3): 251–73.

Warf, B. (1995) 'Telecommunications and the changing geographies of knowledge transmission in the late twentieth century'. *Urban Studies*, 32: 361–78.

Waterman, P. (2000) *Globalization, Social Movements and the New Internationalisms*, New York: Continuum.

Watts, M. (2005) 'Commodities', in Cloke, P., Crang, P. and Goodwin, M. (eds) *Introducing Human Geographies*, 2nd edn, London: Arnold, pp.527–46.

Wei, Y.D. (1999) 'Regional inequality in China'. *Progress in Human Geography*, 23: 49–59.

Wei, Y.H.D. and Ye, X. (2009) 'Beyond convergence: space, scale and regional inequality in China'. *Tijdschrift voor Economische en Sociale Geografie*, 100: 59–80.

Weiss, L. (1997) 'Globalization and the myth of the powerless state'. *New Left Review*, 225: 3–27.

Weiss, L. (2000) 'Developmental states in transition: adapting, dismantling, innovating, not "normalising"'. *The Pacific Review*, 13: 21–55.

Weller, S. (2006) 'The embeddedness of global production networks: the impact of crisis in Fiji's garment export sector'. *Environment and Planning A*, 38: 1249–67.

Whitley, R. (1999) *Divergent Capitalisms: The Social Structuring and Change of Business Systems*, Oxford: Oxford University Press.

Wilkinson, R. (2009) Concluding the Doha Round. development@manchester policy briefs, issue 2, November.

Williams, A.M. (1995) *The European Community: The Contradictions of Integration*, 2nd edn, Oxford: Blackwell.

Williams, C.C., Aldridge, T., Lee, R., Leyshon, A., Thrift, N. and Tooke, J. (2001) *Bridges into Work? An Evaluation of Local Exchange and Trading Schemes (LETS)*, Bristol: Policy Press.

Williams, K., Haslam, C., Williams, J., Cutler, A. and Johal, S. (1992) 'Against lean production'. *Economy and Society*, 21: 321–54.

Willis, K. (2005) 'Theories of development', in Cloke, P., Crang, P. and Goodwin, M. (eds) *Introducing Human Geographies*, 2nd edn, London: Arnold, pp.187–99.

Wills, J. (2002) 'Bargaining for the space to organize in the global economy: a review of the Accor–IUF trade union rights agreement'. *Review of International Political Economy*, 9: 675–700.

Wills, J. (2003) 'Community unionism and trade union renewal in the UK: moving beyond the fragments at last?' *Transactions of the Institute of British Geographers*, 26: 465–83.

Wills, J. (2005) 'The geography of union organising in low-paid service industries in the UK: lessons from the T&G's campaign to unionise the Dorchester Hotel, London'. *Antipode*, 37: 139–59.

Wills, J. and Lee, R. (1997) 'Introduction', in Lee and Wills (1997), pp.302–10.

Wolf, M. (2005) 'On the move: different routes in pursuit of economic greatness', in *The Financial Times Asia's Emerging Giants. India and China: Prospects for Growth, Cooperation and Competition*, London: *Financial Times*, p.4.

Wolfe, D.A. (2010) 'The strategic management of core cities: path dependence and economic adjustment in resilient regions. *Cambridge Journal of Regions, Economy and Society*, 3: 139–52.

Wolfe, D.A. and Gertler, M.S. (2001) 'Globalization and economic restructuring in Ontario: from industrial heartland to learning region?' *European Planning Studies*, 9: 575–92.

Wolfson, M. (1986) *Financial Crises*, New York: M.E. Sharpe.

Womack, J.P., Jones, D.T. and Roos, D. (1990) *The Machine that Changed the World*, New York: Macmillan.

Woolfson, C., Foster, J. and Beck, M. (1997) *Paying for the Piper: Capital and Labour in Britain's Offshore Oil Industry*, London: Mansell.

World Bank (2006) *World Development Indicators*, Washington, DC: World Bank.

World Bank (2009) *World Development Report 2010: Development and Climate Change*, Washington, DC: World Bank.

Wright, M. (2001) 'A manifesto against femicide'. *Antipode*, 33: 550–66.

Wright, R. (2002) 'Transnational corporations and global divisions of labor', in Johnston, R.J., Taylor, P. and Watts, M. (eds) *Geographies of Global Change: Remapping the World*, Oxford: Blackwell, pp.68–77.

Yeoh, B. and Chang, T.C. (2001) 'Globalising Singapore: debating transnational flows in the city'. *Urban Studies*, 38: 2025–44.

Yeung, H.W. (2009) 'Transnational corporations, global production networks and urban and regional development: a geographer's perspective on multinational enterprises and the global economy'. *Growth and Change*, 40: 197–226.

Yeung, H.W.C. (2005) 'Rethinking relational economic geography'. *Transactions of the Institute of British Geographers*, 30: 37–51.

Zeller, C. (2000) 'Rescaling power relations between trade unions and corporate management in a globalising pharmaceutical industry: the case of the acquisition of Boehringer Mannheim by Hoffman–La Roche'. *Environment and Planning A*, 32: 1545–67.

Index

CPSIA information can be obtained
at www.ICGtesting.com
Printed in the USA
BVOW10s1504090616
451232BV00003B/12/P